Model-Based Reasoning in Scientific Discovery

Model-Based Reasoning in Scientific Discovery

Edited by

Lorenzo Magnani
University of Pavia
Pavia, Italy

Nancy J. Nersessian
Georgia Institute of Technology
Atlanta, Georgia

and

Paul Thagard
University of Waterloo
Waterloo, Ontario, Canada

Kluwer Academic / Plenum Publishers
New York, Boston, Dordrecht, London, Moscow

Library of Congress Cataloging-in-Publication Data

Model-based reasoning in scientific discovery/edited by Lorenzo Magnani, Nancy J. Nersessian and Paul Thagard.
p. cm.
Proceedings of an International Conference on Model-Based Reasoning in Scientific Discovery, held December 17–19, 1998, in Pavia, Italy—T.p. verso.
Includes bibliographical references and index.
ISBN 0-306-46292-3
1. Science—Methodology Congresses. 2. Discoveries in science Congresses. 3. Qualitative reasoning Congresses. I. Magnani, Lorenzo. II. Nersessian, Nancy J. III. Thagard, Paul.
IV. International Conference on Model-Based Reasoning in Scientific Discovery (1998: Pavia, Italy)
Q175.32.R45M65 1999
501'.1—dc21 99-44860
CIP

Proceedings of the International Conference on Model-Based Reasoning in Scientific Discovery, held December 17–19, 1998, in Pavia, Italy

ISBN: 0-306-46292-3

233 Spring Street, New York, N.Y. 10013

10 9 8 7 6 5 4 3 2 1

A C.I.P. record for this book is available from the Library of Congress.

Printed in the United States of America

Preface

The volume is based on the papers that were presented at the International Conference *Model-Based Reasoning in Scientific Discovery* (MBR'98), held at the Collegio Ghislieri, University of Pavia, Pavia, Italy, in December 1998.

The papers explore how scientific thinking uses models and explanatory reasoning to produce creative changes in theories and concepts.

The study of diagnostic, visual, spatial, analogical, and temporal reasoning has demonstrated that there are many ways of performing intelligent and creative reasoning that cannot be described with the help only of traditional notions of reasoning such as classical logic. Traditional accounts of scientific reasoning have restricted the notion of reasoning primarily to deductive and inductive arguments. Understanding the contribution of modeling practices to discovery and conceptual change in science requires expanding scientific reasoning to include complex forms of creative reasoning that are not always successful and can lead to incorrect solutions. The study of these heuristic ways of reasoning is situated at the crossroads of philosophy, artificial intelligence, cognitive psychology, and logic; that is, at the heart of cognitive science.

There are several key ingredients common to the various forms of model-based reasoning to be considered in this book. The models are intended as interpretations of target physical systems, processes, phenomena, or situations. The models are retrieved or constructed on the basis of potentially satisfying salient constraints of the target domain. In the modeling process, various forms of abstraction are utilized. Evaluation and adaptation take place in light of structural, causal, and/or functional constraints. Simulation can be used to produce new states and enable evaluation of behaviors and other factors.

The book also addresses some of the main aspects of the nature of abduction, connecting it to the central epistemological question of hypothesis withdrawal in science and model-based reasoning, where abductive inferences exhibit their most appealing cognitive virtues.

The various contributions of the book are written by interdisciplinary researchers in philosophy, artificial intelligence, and cognitive science who are active in the area of creative reasoning in science: the most recent results and achievements about the topics above are illustrated in detail in their papers.

The book is divided in three parts. The first part *Models, mental models, and representations* contains the contributions of N.J. Nersessian; D. Bailer-Jones; R. Giere; K. Knoespel; M. Suarez; and K. Dunbar. The second part *Discovery processes and mechanism* is composed of the papers of D. Gooding and T.R. Addis; P. Thagard and D. Croft; T. Harris; V. Raisis; S. Krauss, L. Martignon, and U. Hoffrage; and F.T. Arecchi. Finally, the last part, *Creative inferences and abduction*, include the contributions of J. Meheus; L. Magnani; I. Niiniluoto; E. Winsberg; F. Hendricks and J. Faye; C. Pizzi; J. Zytkow.

The conference, and thus indirectly this book, was made possible through the generous financial support of the University of Pavia and of CARIPLO (Cassa di Risparmio delle Provincie Lombarde). Their support is gratefully acknowledged.

The editors, and co-chairs of the conference, would like to express their appreciation to the other members of the Scientific Committee for their suggestions and assistance: R. Giere, Department of Philosophy, University of Minnesota, Minneapolis, USA; David Gooding, Department of Psychology, University of Bath, Bath, United Kingdom; Joke Meheus, University of Ghent, Ghent, Belgium; Claudio Pizzi, Department of Philosophy and Social Sciences, University of Siena, Siena, Italy; Mario Stefanelli, Department of Computer Science, University of Pavia, Pavia, Italy; Raul Valdés-Pérez, Department of Computer Science, Carnegie Mellon University, Pittsburgh, USA.

Special thanks to the members of the Local Organizing Committee Maria Teresa Oldani, Stefania Pernice, Giulio Poletti, Stefano Rini, for their contribution in organizing the conference, to Massimo Manganaro, Stefania Pernice, Anna Maria Marchini for their contribution in the preparation of this volume, and to the copy-editor Linda d'Arrigo. The preparation of the volume would not have been possible without the contribution of resources and facilities of the Computational Philosophy Laboratory and of the Department of Philosophy, University of Pavia.

The remaining papers deriving from the presentations given at the Conference will be published in three Special Issues of Journals: in *Foundations of Science*, Model-based reasoning in science: learning and discovery; in *Philosophica*, Abduction and scientific discovery; in *Philosophica*, Analogy and mental modeling in scientific discovery.

Lorenzo Magnani, Pavia, Italy
Nancy J. Nersessian, Atlanta, GA, USA
Paul Thagard, Waterloo, Canada
May 1999

Contents

Model-Based Reasoning in Scientific Discovery

This schematism of our understanding in regard to phenomena and their mere form, is an art, hidden in the depths of the human soul, whose true modes of action we shall only with difficulty discover and unveil

Immanuel Kant, *The Critique of Pure Reason*

Models, Mental models, and Representations

Model-Based Reasoning in Conceptual Change

Nancy J. Nersessian
School of Public Policy and College of Computing, Georgia Institute of Technology, Atlanta, GA, USA, nancyn@cc.gatech.edu

Abstract: This paper addresses how specific modeling practices employed by scientists are productive methods of conceptual change in science. Within philosophy, where the identification of reasoning with argument and logic is deeply ingrained, these practices have not traditionally been considered significant forms of scientific reasoning. Embracing these modeling practices as "methods" of conceptual change in science requires expanding philosophical notions of scientific reasoning to encompass forms of creative reasoning. I focus on three forms of model-based reasoning demonstrated in my previous work as generative of conceptual change in science: analogical modeling, visual modeling, and thought experimenting. The models are intended as interpretations of target physical systems, processes, phenomena, or situations. The models are retrieved or constructed on the basis of potentially satisfying salient constraints of the target domain. In the modeling process, various forms of abstraction, such as limiting case, idealization, generalization, generic modeling, are utilized. Evaluation and adaptation take place in light of structural, causal, and/or functional constraint satisfaction. Simulation can be used to produce new states and enable evaluation of behaviors, constraint satisfaction, and other factors.

1. INTRODUCTION

The problem of how new concepts, in general, are formed and how they related existing concepts has vexed philosophers for centuries. The problem, with respect to science, came to the forefront of concerns in late 19th century, with the development of new concepts like "energy" and "field" and became critical with the radical conceptual changes in physics early in this century. On what we have come to know as "the standard view" – that associated with logical positivism – conceptual structures are treated as languages and the problem of conceptual change is that of explaining the nature of the logical and interpretive relations between the old and new conceptual structures and between concepts and the world. Possible "methods" of concept formation, as discussed, e.g. in Carl Hempel's classic monograph (1952), included setting up correlations with empirical phenomena via definition and forms of operationalization. Arguably, the spirit of this project had continued through more recent and innovative work, such as Thomas Kuhn's notion of incom-

Model-Based Reasoning in Scientific Discovery, edited by L. Magnani, N.J. Nersessian, and P. Thagard, Kluwer Academic/Plenum Publishers, New York, 1999

mensurability as "untranslatability" and his later approach to the problem through analysis of the lexicon of a scientific theory (1991).

In contrast to the standard approach, my approach in constructing a theory of concept formation and change in science has been to shift focus of analysis from the products of conceptual change to the processes whereby such change comes about. This shift in focus highlights the nature of the scientific practices, specifically, of the *methods* or kinds of *reasoning* through which concepts are constructed. We will focus on three such practices that have been, as claimed in my previous work (1984, 1988, 1992, 1993), generative of conceptual change in science: creating analogies, employing visual representations, and thought experimenting. On my account, what these practices have in common is that they are all forms of "model-based reasoning". What this means will be developed in the course of the paper.

Examining the historical records of major conceptual changes establishes these practices to be employed across the sciences. Since my analysis draws from physics, I will not here claim to be providing an account of how they function in other sciences. However, the hypothesis that there are significant commonalities among these practices as employed across the sciences garners support from the *cognitive* foundations indicated for them here. Specifically how they might function in other domains, though, needs to be developed by researchers with expertise in those domains. Further, although these practices are ubiquitous they are not exhaustive. Clearly, there are other modes of reasoning that generate conceptual change.

Our concern here is to determine how these specific forms of what I have called "model-based reasoning" are productive *methods* of conceptual change. From the perspective of traditional philosophy of science, the modeling practices discussed here have not been considered significant forms of scientific reasoning – in some cases, not as "reasoning" at all. Philosophical accounts of scientific reasoning have restricted the notion of reasoning primarily to deductive and inductive arguments. The practices investigated here have been understood as performing an ancillary and inessential role, with the productive reasoning being carried out by some other means. The position taken in this paper is that by developing a cognitive foundation for modeling practices as productive forms of reasoning, more general and widely applicable than the specific contexts in which they are found to be employed in science, one can mount a case for how they are extremely productive forms of reasoning in conceptual change in science.

2. THE TRADITIONAL VIEW OF REASONING

Embracing these modeling practices as the reasoning through which conceptual change can take place requires expanding philosophical notions of scientific reasoning to encompass forms of creative reasoning, most of which cannot be reduced to an algorithm in application, are not always productive of solutions, and can lead to incorrect solutions. To do this requires challenging one of the most sacrosanct notions in philosophy: "reasoning". In the traditional view, the identification of reasoning with argument, and thus with logic, is deeply ingrained. So, before we can develop a notion of "model-based reasoning" in science, we need first to address the question "What is reasoning?" and, specifically, "What is scientific reasoning?"

In standard philosophical accounts reasoning is employing deductive or inductive algorithms to sets of propositions. The understanding of deductive reasoning provided by classical logic stands as the model. Here the essential notion is soundness: true premises plus good reasoning yields true conclusions. A major objective of logical positivism in this century was to develop a notion of soundness for induction similar to that for deduction, to be applied in inductive justification of rationally reconstructed scientific reasoning. Loosely construed, starting from maximally probable premises and using correct inductive logic one should arrive at maximally probable conclusions.

Note that I have switched to talking about "justification" and "rationally reconstructed scientific reasoning" in discussing induction. On extending the traditional notion of reasoning to what goes on in the domain of scientific discovery, in which lies conceptual innovation, a problem arises immediately. Good reasoning, with *T* premises can lead to incorrect solutions or to no solution at all. For example, Newton's path to the concept of universal gravitational force was largely through analogy. Analogical argument is a notoriously weak form of argument and one could hold that concerns about it have been borne out in this case. According to the general theory of relativity, that conclusion is wrong. There is no gravitational force, falling bodies are just following their natural trajectory in a curved space-time. But, of course, we know this could prove to be wrong as well. The problem of unsoundness has been a factor in the contention of philosophers of various persuasions that there is no "logic of discovery". Some nontraditional philosophical accounts have allowed for the possibility of "abductive" inference, but these accounts leave mysterious the nature of the reasoning *processes* underlying abductive inference and hypothesis generation. Analyzing modeling practices provides a way of specifying the nature of some abductive reasoning processes.

A quick review establishes that, in traditional philosophical accounts, what I am calling model-based reasoning practices are considered ancillary, inessential aids to thinking. At most they have constituted – and continue to constitute – fringe topics in the literature in philosophy of science. Analogy has received the most attention and there have been some attempts to countenance it as a form of inductive reasoning. To take a classic example, in Rudolph Carnap's *Logical Foundations of Probability* (1950) analogy occupies an entire page in the Appendix. He concludes that "reasoning by analogy, although admissible, can usually yield only weak results." Even those who have taken analogy in science as a serious topic of discussion have not given made a strong case for its generative role in conceptual change. For Norman Campbell (1920) and Mary Hesse (1966) analogy is constitutive of theories in that analogies provide meaning for new theoretical terms. Thus analogy serves an explanatory function and provides some basis for hypothesis generation. Hesse's analysis of analogies treats them as propositional and focuses on the nature of the arguments enabled by specific mappings of properties and relations. Visual representation has not fared even this well, since most of the discussion from the late 19th century until quite recently has concurred that such representations are fundamentally misleading in reasoning processes and have such a detrimental effect that it is better to eliminate them entirely, as, e.g., David Hilbert did in axiomatizing geometry.

Finally, traditional empiricist accounts of thought experiment have been weighted in favor of their being eliminable from the reasoning process. Pierre Duhem (1902), for example, argued that they can either be transformed into real experiments or are simply bogus. A modern version of this view has been developed by John Norton (1991) who argues that a thought experiment can be reformulated as and replaced by a deductive or inductive argument. Ernst Mach (1905) is one empiricist who did take them seriously and his naturalized account moves in right direction. Thomas Kuhn did see thought experiments as playing a central role in conceptual change and characterized them as "one of the essential analytical tools which are deployed during crises and which then help to promote basic conceptual reform" (1964, p. 263). The historical record does indeed show the preponderance of thought experiments in periods of conceptual change in science. But, to understand why, i.e. how thought experiment functions, requires a fundamental recasting of the problem of conceptual change that rejects both traditional and Kuhnian accounts of it.

I agree with the positivists' conclusion that there is no *logic* (classical) of scientific discovery, but disagree with equating reasoning with applying logic to sets of propositions. Rather, I propose to start by acknowledging that such uses of analogy as Newton's are instances of quite powerful *reasoning* and attempt to determine how this kind of reasoning works. The essence of my point is that the positivists were right that history and psychology are essential components to understanding the processes of conceptual change and they were right that there is no logic of discovery, but their notion of reasoning was too narrowly constrained and that led to the mistaken view there are no issues of importance to philosophy in the context of discovery. In the case at hand, the problem of the nature of the constructive processes does bear on the philosophical problem of the nature of reasoning. Further, discovery processes have implications for other traditional philosophical issues that will not be addressed here, such as, realism, objectivity, rationality, and progress.

So, the problem thus far posed is whether is it possible to articulate a notion of reasoning that includes forms that other than deductive and inductive argument and yet fruitful in that they generate potential solutions to scientific problems. Can we ascertain the kind that will come up with good results, advance new and promising solutions? Is there a way we can evaluate as good or productive, reasoning that which can lead to incorrect or no solutions? With respect to the problem of conceptual change, asking these questions shifts focus from the products, or conceptual structures, to the processes of constructing concepts, which, in turn, puts the focus on the practices and practitioners of science. Although I cannot rehearse in this paper the various arguments and justifications for philosophical naturalism, my own brand calls for the necessity of a multidisciplinary analysis of conceptual change in science: an account that is informed by the constructive practices scientist actually use and by the best scientific accounts of how human cognitive capacities and limitations produce and constrain these practices.

3. THE COGNITIVE SCIENCE NOTION OF REASONING BY MENTAL MODELING

The first task is to provide a cognitive foundation for taking the modeling practices employed by scientists seriously as forms of reasoning productive of conceptual change in science. In cognitive psychology there is a controversy about the nature of human reasoning that parallels the philosophical issue. This is not surprising, since philosophers of psychology who hold the traditional view have played a significant role in the shaping the contemporary debate. On the traditional psychological view, reasoning consists of applying a mental logic to propositional representations. Critics of this view contend that a purely syntactical account of reasoning cannot account for significant effects of semantic information in reasoning. An early and now classic experiment is the Wason card task (cf. Johnson-Laird, 1983, for a discussion of it and subsequent research). In the task, the subject is presented with four cards, two displaying a letter (one consonant and one vowel) and two displaying a number (one even and one odd). Such a sequence might be: A G 2 5. The subject has been told that each card has a number on one side and a letter on the other. The task is to determine how many cards have to be turned over in order to make a given generalization true, e.g. "if a card has a vowel on one side then it has an even number on the other side." Most subjects failed to make the correct inference with this form of the test. However, when meaningful information and relations were provided on the cards, the number of correct inferences increased dramatically. For example, simply substituting destinations and modes of transportation or postage stamp costs and sealed or unsealed envelopes for the letters and numbers increased people's ability to reason correctly.

This result suggested that semantic information plays a more salient role in human reasoning than the traditional view allows. It sparked numerous investigations in different domains that have led many cognitive scientists to conclude that much of human reasoning is by means of "mental modeling" rather than through a process of applying a mental logic to propositional representations. The initial hypothesis of mental modeling as a significant form of reasoning derives from a proposal made by Kenneth Craik (1943). Craik proposed that, in general, people reason by carrying out thought experiments on internal models. Since he made this proposal at the height of the behaviorist approach in psychology, the hypothesis did not receive much attention. The development of a cognitive psychology in the 1960's created a more hospitable environment for its articulation and exploration. Though not uncontroversial, the centrality of mental modeling to cognition is a hypothesis under investigation by many domains. The main impetus for the resurgence of the hypothesis is experimental outcomes, such as the Wason experiments described above, that demonstrate the inadequacy of syntactic accounts of reasoning. Mental modeling has been investigated in a wide range of phenomena, including: reasoning about causality in physical systems (cf., e.g., deKleer and Brown, 1983); the role of representations of domain knowledge in reasoning (cf., e.g., Gentner and Stevens, 1983); analogical reasoning (cf., e.g. Gentner and Gentner, 1983); deductive and inductive inferencing (cf., e.g., Johnson-Laird, 1983); and comprehending narratives (cf., e.g., Perrig and Kintsch, 1985). Because the potential range of application is so extensive, some have argued that the notion of mental models can provide a unifying framework for the study of cognition (Gilhooly, 1986). For our problem, too, the hypothesis is attractive, in part, because it opens the possibility of furnishing a unified analysis of the widespread modeling practices implicated in conceptual change.

There are several distinct theoretical aspects of mental modeling that tend to be conflated in the literature. The most significant distinction for our purposes is between those investigations that treat mental models as structures that are stored in long term memory and are then called upon in reasoning and those that treat them as temporary structures constructed in "working memory" for a specific reasoning task. This distinction itself may be artificial given that it rests on the questionable notion that long-term memory is like a storage bin and is a different structure from working memory, where both may be dynamic in nature and related processes in the brain. Our analysis of modeling practices focuses on the processes of constructing models and reasoning through manipulating them. Philip Johnson-Laird's account (1983) remains the best articulated of those analyses that focus on the temporary reasoning structure, and it informs my analysis. However, Johnson-Laird's own focus has been on mental modeling in deductive and inductive reasoning tasks and not in creative scientific reasoning. Thus, my account provides an extension of the hypothesis into this domain.

Broadly construed, for Johnson-Laird, a mental model is a structural analog of a real-world or imaginary situation, event, or process that the mind constructs in reasoning. What it means for a mental model to be a structural analog is that it embodies a representation of the spatial and temporal relations among, and the causal structures connecting the events and entities depicted and whatever other information that is relevant to the problem solving task. Most of Johnson-Laird's published work has investigated mental modeling in deductive and inductive reasoning. To accommodate scientific reasoning, we need to include models that are dynamical in nature and to expand the notion to include mental models that are functional analogs to specific dimensions and behaviors of real-world systems. Although Johnson-Laird has not investigated such cases himself, he has hypothesized that since mental models need to be causally coherent, it should be possible to carry out simulative reasoning about the behaviors of a model for those tasks that are dynamic in nature. Such models would behave in accord with constraints that need not be stated explicitly.

Advocates of mental modeling argue that the original capacity developed as a way of simulating possible ways of maneuvering within the physical environment. It would be highly adaptive to possess the ability to anticipate the environment and possible outcomes of actions, so it is likely that many organisms have the capacity for mental modeling. Given the linguistic abilities of humans, it is likely they can create models from both perception and description. This hypothesis receives support from the research in narrative comprehension noted earlier. Although the original ability to perform simulative reasoning through mental modeling may have developed as a way of anticipating possible courses of action in the world, it is highly plausible that, as human brains have developed, this ability has been extended to more esoteric reasoning contexts, such as science. Further, differences in novice and expert reasoning skill in solving scientific problems (cf., e.g., Chi, et al., 1981) can be taken to provide evidence that skill in modeling is something that develops with learning (Ippolito and Tweney, 1995; Nersessian, 1995). The nature and richness of models one can construct and one's ability to reason develops as one learns domain-specific content and techniques.

Just what *format* a mental model takes, i.e., whether it is propositional, perception-based, or non-propositional but still amodal, is a matter of some debate, as is the issue of what are the *generative processes* in the brain for creating and operating on mental models. However, these issues do not have to be resolved before it is possible to make progress on an account of model-based reasoning in science. The essential points are that a mental model can be non-linguistic in form and the mental mechanisms are such that they can satisfy the model-building and simulative constraints necessary for the activity of mental modeling. My own cognitive-historical hypothesis is that the models are analog and non-propositional (though possibly including labeling, i.e., "mixed-mode") in nature, though at present my account is agnostic between perception-based and amodal accounts. To allay possible objections it should be stressed that mental modeling, even if it were to make use of perceptual mechanisms, would not need to be identical to the process of constructing a picture in the "mind's eye". That great thought experimenters, such as Bohr, have claimed not to be able to visualize well does not undermine non-propositional forms of mental modeling. This form of mental modeling would only require the ability to reason by means of an analog model. The relationship between a mental model and what has been called "mental imagery" is something that still needs to be worked out by cognitive scientists, but is not necessary for our purposes.

To carry out an analysis of model-based reasoning in conceptual change requires only that we adopt a "minimalist" version of a mental modeling hypothesis: that in certain problem solving tasks humans reason by constructing an internal model of the situations, events and processes that in dynamic cases provide the basis for simulative reasoning. Whatever the format of the model itself, information in various formats, including linguistic, formulaic, visual, auditory, kinesthetic, can be used in its construction. Although there is still much to learn about mental modeling, this minimalist hypothesis is attractive because it provides a cognitive foundation for taking seriously the modeling practices of scientists as the reasoning through which new conceptual structures are constructed. Further, it provides a basis for constructing a unified account of the various forms of model-based reasoning we will consider. Without going into the details, Figure 1 exemplifies why a unified account is needed. The figure is taken from a paper published in *Philosophi-*

cal Magazine by James Clerk Maxwell, in which, I have argued (Nersessian, 1992), he was trying to get his colleagues to understand the new representation of electromagnetic forces by leading them through processes he felt were salient to his constructing the mathematical representation of the electromagnetic field concept. As I interpret it, it is a *visual* representation of an *analogical* model that is accompanied with instructions for *animating* it correctly in thought: "Let the current from left to right commence in AB. The row of vortices *gh* above AB will be set in motion in the opposite direction to a watch [...]. We shall suppose the row of vortices *kl* still at rest, then the layer of particles between these rows will be acted on by the row *gh* on their lower sides and will be at rest above. If they are free to move, they will rotate in the negative direction, and will at the same time move from right to left, or in the opposite direction from the current, and so form an *induced* electric current" (1890, v. 1, p. 477, italics in original).

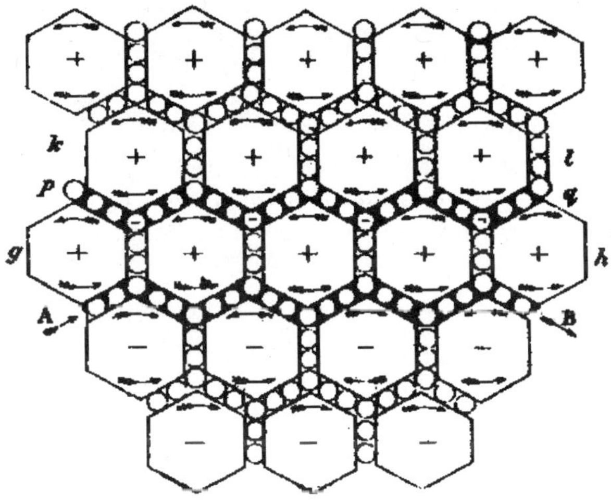

Figure 1. Maxwell's drawing of the vortex-idle wheel model (Maxwell 1890, Vol. I, Plate VII).

4. MODEL-BASED REASONING

The traditional account of reasoning has supported the belief that conceptual change cannot be the outcome of reasoned processes. By and large, philosophers consider the processes of conceptual change as mysterious and unanalyzable. Conceptual innovation is held to occur in sudden flashes of insight, with new concepts springing forth from the head of the scientist like Athena, fully grown. This does accord with retrospective accounts of some scientists, but if one examines their deeds – their papers, diaries, letters, notebooks – these records support a quite different interpretation in most cases. As I have been arguing for some years, conceptual change results from extended problem-solving processes. The records of these processes display extensive use of practices that I hypothesize constitute forms of model-based reasoning: analogical, visual, and simulative modeling. Further, the psychological theory of mental modeling provides a basis in human cognition for taking the external traces of modeling displayed in scientific practice as indicative that model-based reasoning is generative of concept formation and change.

The argument developed above in support of my hypothesis is that mental modeling is a fundamental form of human reasoning. It evolved as an efficient means of navigating the environment and solving problems in matters of significance to existence in the world. Humans have extended its use to more esoteric situations, such as constructing scientific representations. That is, the cognitive resources scientists call upon on in creative problem solving are not different in kind than those humans use in more ordinary circumstances. Mental modeling is applied by humans across a spectrum of problem solving situations and in numerous domains, ranging from solving the problem of how to get a chair through a doorway to problems traditionally classified as deductive and inductive logic problems. Additionally, there is significant experimental protocol evidence collected by cognitive psychologists to support it as a fundamental form of problem solving employed by contemporary scientists (cf. e.g., Chi, et al., 1981; Clement, 1989). These records of "think-aloud" reasoning processes provide additional support for the claim that the traces of modeling practices exhibited in the historical records of conceptual change are indicative that mental modeling played a central role in the historical process. Understood from the perspective of human cognition, these practices utilize and engage internal modeling processes that are highly effective means of problem solving and effective means of transmitting novel conceptions through a community. In the more mundane cases the reasoning performed is usually successful, e.g., one figures out how to get the chair through the door, because the models and manipulative processes embody largely correct assumptions about every-day real-world events. In the case of science where the situations are more removed from experience and the assumptions more imbued with theoretical assumptions, there is less assurance that a reasoning process, even if correct, will yield "success". In the evaluating process, a major criterion for success remains the goodness of fit to the phenomena, but success can also include such factors as enabling the construction of a viable mathematical representation.

The centrality of model-based reasoning practices in episodes of conceptual change, a creative form of problem solving, lends support to the position of several contemporary philosophers (cf., e.g., Cartwright, 1989;

Giere, 1988) that the basic units for the scientists in reasoning more generally are most often not axiom systems and propositional networks, but models. As in the theory of mental modeling, the term "model" is used here not in the logical sense of an abstract mapping of things to terms, but in the analogical sense of a structure intended as isomorphic to some aspect of a physical system. In using and constructing scientific theories, no matter how they may in principle be represented, models are the mental representations with which a scientist carries out much reasoning and by means of which she thinks and understands through the lens of a conceptual structure. In the constructive processes of conceptual change, specifically, one important lesson we should take from the historical records is that models come first, then further abstraction takes place to create formal expression in laws and axioms of theories (Nersessian, 1995, 1999). Modeling practices are employed both in experimental and in theoretical settings. The model is the mode of representation between the phenomena and expression in a language (including mathematics) and it is working with this intermediate form of representation that facilitates conceptual change. Further, there is significant evidence that much of the training of the practitioners of science takes place through learning the models of a community and developing facility with model manipulation (Clement, 1989; Giere, 1988; Nersessian, 1995). A cognitive account, then, raises the modeling practices from their traditional status of ancillary, inessential aids to reasoning to the actual forms of reasoning through which concept formation and change take place. We will now consider, briefly, how such model-based reasoning functions in conceptual change. The practices will be considered separately here, but they are all related forms of reasoning and they most often are used together in reasoning episodes I have elsewhere called "constructive modeling" (Nersessian, 1995; Griffith, et al., 1996).

5. THREE FORMS OF MODEL-BASED REASONING

To engage, specifically, in analogical modeling one calls on knowledge of the generative principles and constraints for physical models in a source domain. These constraints and principles may be represented mentally in different informational formats and knowledge structures that act as tacit assumptions employed in constructing and transforming models during problem solving. Inter- or intra-domain analogies may be retrieved and applied as models wholesale with suitable adaptation, but often, and especially in cases of conceptual change, no direct analogy exists and construction of an initial source model is required. In these cases the analogical domain serves as the source for constraints to be used in interaction with those provided by the target problem to create imaginary analogs. Evaluation of the analogical modeling process is in terms of how well the salient constraints of a model fit the salient constraints of a target problem.

Generic abstraction is a key reasoning process in analogical modeling, which often requires recognition of potential similarities across, and integration of information from, disparate domains. In viewing a model generically, one takes it as representing features common to a class of phenomena. This way of viewing the model can, of course, only take place in the mind. In reasoning, e.g., about a triangle, one often draws or imagines a concrete representation. However, to consider what it has in common with all triangles, one needs to imagine it as lacking specificity in the angles and the sides. That is, the reasoning context demands that the interpretation of the concrete polygon as generic. It was only through generic abstraction, e.g., that Newton could reason about the commonalities among the motions of planets and of projectiles, which enabled his formulating a unified mathematical representation of their motions.

The analogical model, understood generically, represents what is common among the members of specific classes of physical systems, viewed with respect to a problem context. Newton's inverse-square law of gravitation abstracts what a projectile and a planet have in common in the context of determining motion. The inverse-square-law model served as a generic model of action-at-a-distance forces for those who tried to bring all forces into the scope of Newtonian mechanics. In Maxwell's analogical model, represented visually in Figure 1, the dynamical relations among the idle wheels and vortices must be viewed in generic form. That is, the vortex-idle wheel system is understood to represent the class of such dynamical systems and the class includes electric and magnetic interactions on the assumptions of Maxwell's model. Although constructed for the purposes of showing how contemporary scientists use theories, I think the chart constructed by Ronald Giere (1994), shown in Figure 2, can also be interpreted as illustrating how the process of generic abstraction functioned in the process of constructing the modern system of classical mechanics. Starting from specific instances of pendula and springs, generic categories are created and are further abstracted to kinds of "harmonic motion", which then fall into generic categories of "conservative" or "non-conservative" models, and so forth.

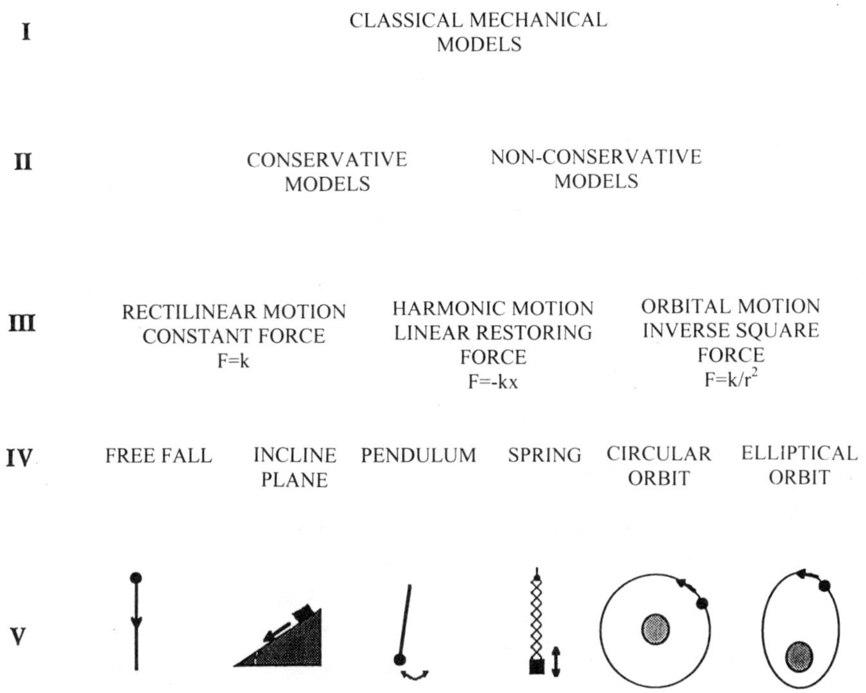

Figure 2. Illustration of the process of abstraction through generic modeling in classical mechanics.

Scientists use a variety of perceptual resources in modeling. These vary with the science, e.g., as Cameron Shelley (1996) has shown, an archeologist would make more use of kinesthetic information in model construction than a physicist. Both internal and external visual representations figure prominently in conceptual change across the science. A possible reason for this is that employing the visual modality may enable the reasoner to bypass specific constraints inherent in the linguistic and formulaic representations of existing conceptual structures. External visual representations provide support for the processes of constructing and reasoning with a mental model. They aid significantly in organizing cognitive activity during reasoning, such as fixing the attention of the salient aspects of a model during reasoning, enabling retrieval and storage of salient information and exhibiting salient interconnections, such as structural and causal, in appropriate co-location. Further external visual representations, such as the figure by Maxwell (Figure 1), facilitate the construction of shared mental models in a community and the transportation of a model out of the local milieu of its construction. The mental models perspective hypothesizes that the external visual representations support construction of an internal model. Internal visual representations need not be "pictorial" in format, but can be highly schematic.

Thus this modality may be operative even in the reasoning of scientists, such as Bohr, who claim not to experience imagery in reasoning.

As used in modeling in physics, external visual representations tend to be in diagrammatic form. These representations can model phenomena in several ways, including providing idealized representations of aspects of phenomena and embodying aspects of theoretical models. For example, early in Faraday's construction of a field concept the visual model represented in Figure 3 provided an idealized representation of the lines of force surrounding a magnetic bar.

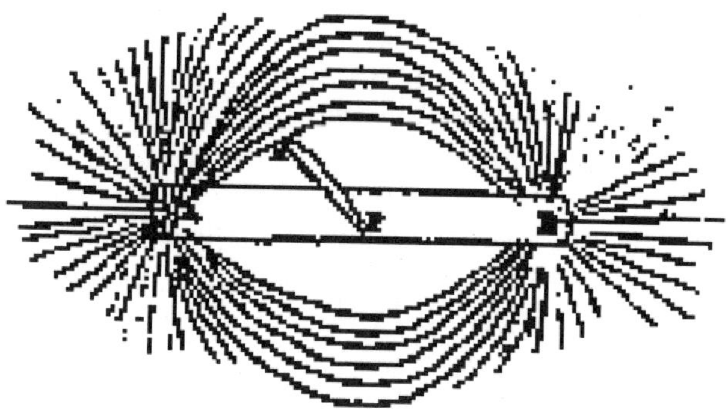

Figure 3. Faraday's drawing of the lines of force surrounding a bar magnet (Faraday 1839-1855, Vol I, Plate I).

Later in his problem solving, the visual model of lines of force functioned as the embodiment of a dynamical theoretical model of forces generally (Gooding, 1990; Nersessian, 1984, 1992). But, the visual model represented by Maxwell in Figure 1 is an embodiment of an imaginary system and not a theoretical model of electromagnetic actions.

There is a vast literature on mental imagery (Kosslyn, 1994) that provides evidence that humans can perform simulative imaginative combinations and transformations that mimic perceptual spatial transformation. These simulations are hypothesized to take place using internalized constraints assimilated during perception. Cognitive research also indicates that people use various kinds of knowledge of physical situations in imaginary simulations. For example, when objects are imagined as separated by a wall, the spatial transformations exhibit latency time consistent with having simulated moving around the wall rather than through it.

Thought experimenting is a specific form of the simulation that can occur in various forms of model-based reasoning. Because the thought-experimental narratives are what we have access to and because they are a central form of effecting conceptual change within a scientific community, my analysis begins with examining how these function. From that analysis one can infer that the original experiment involves a similar form of reasoning. To explicate the notion that thought experimenting is simulative model-based reasoning, we need to discuss: (1) how a narrative facilitates the construction of a model of an experimental situation in thought and (2) how one can reach conceptual and empirical conclusions by mentally simulating the experimental processes.

From a mental modeling perspective, the function of the narrative form of presentation of a thought experiment would be to guide the reader in constructing a structural analog of the situation described by it and to make inferences through simulating the events and processes depicted in it. So, extending from research on other forms of discourse models (cf., e.g., Perrig and Kintsch, 1985), the operations carried out in executing the thought experiment are performed not on propositions but on the constructed internal model. Unlike the fictional narrative, however, the context of the scientific thought experiment makes the intention clear to the reader that the situation is one that is to represent a potential real-world situation. That a thought experiment is presented in a polished form should make it an effective means of getting comparable mental models among the members of a community of scientists. The narrative has already made significant abstractions that aid in focusing attention on the salient dimensions of the model and in recognizing the situation as prototypical, so that the experimental consequences are understood to go beyond the specific situation of the thought experiment.

Although some kinds of mental modeling may employ static representations, those derived from thought-experimental narratives are inherently dynamic. The narrative delimits which are the specific transitions that govern what takes place. Constructing and conducting the experiment makes use of inferencing mechanisms, existing representations, and scientific and general world knowledge to make realistic transformations from one possible physical state to the next. Much of the information employed in these transformations is tacit. Thus, expertise and learning play a crucial role in the practice; as does what Gooding (1990) has called "embodiment". The constructed situation inherits empirical force by being abstracted both from our experiences and activities in the world and our knowledge, conceptualizations, and assumptions of it. In this way, the data that derive from thought experimenting have empirical consequences and at the same time pinpoint the locus of the needed conceptual reform. This understanding forms the basis of further problem-solving efforts to construct an empirically adequate conceptualization.

Thought experimenting plays a crucial role in conceptual change by showing that existing systems of constraints cannot be integrated into consistent models of the physical world. Thought experimenting may facilitate recognizing the undesirable consequences of a conceptualization in much the way that experimenting by computer simulation exposes undesirable consequences of the constraints of a scientific representation. By creating a simulative model that attempts to integrate specific systems of constraints, thought experimenting enables the scientist to grasp essential points of conflict and infer their consequences more readily than would reasoning through the logical consequences of a representation. Once the initial experimenter understands the implications of a thought experiment, she can guide others in the community to see them as well by crafting a description of the experiment into a narrative.

6. CONCLUSIONS

Assuming that analogy is at best a form of inductive argument leads to the conclusion of Carnap that "reasoning by analogy can yield only weak results" (1950, p. 589). Looking at examples of productive reasoning by analogy in science shows it to yield powerful and creative results. The way to resolve the discrepancy is to see that reasoning by analogy is not argument but model construction. So, although analogical modeling enables arguments, the heart of analogy is employing generic abstraction in the service of model construction, manipulation, and evaluation. In this, as opposed to the standard "argument by analogy," it is a powerful form of reasoning.

Visual representations may indeed have the potential to lead a reasoner astray. However, visual modeling appears to be highly developed and effective form of human reasoning in a wide variety of circumstances. Duhem was a particularly vocal opponent of visual representation in science. He went so far as to use the late 19th Century British proclivity for using visual representation and analogy as demonstrating the inferiority of the mind of the British scientist as compared with the analytical mind of the French scientist. The great irony, of course, is that the British produced the desired unified theory of electromagnetism, and not the French. Visual representation is powerful tool for science when sufficient constraints are incorporated into the reasoning process. Although many thought experiments can often be reconstructed as arguments, their modeling function cannot be supplanted by an argument. The argument is not evident until after the thought experiment has been constructed and executed. Exhibiting the soundness of a thought experiment by reconstructing it as an argument can perform an important rhetorical function. However, real-world experimental outcomes can be recast in argument form as well, but no one would argue that the experiment can be replaced by the argument. In similar fashion, we need to differentiate between the reasoning that is done with the thought experiment and that which is done with the reconstruction of it. The thought-experimental process, by linking the conceptual and the experiential dimensions of human cognitive processing, demonstrates the undesirable real-world consequences of a representation, thereby compelling representational change.

All three forms of model-based reasoning are complex forms of reasoning that integrate various forms of information – propositions, models, and equations – into mental models. There are several key common ingredients to the various forms of model-based reasoning we have considered. They are semantic reasoning processes in that the models are intended as interpretations of a target physical system, process, phenomenon, or situation. The models are retrieved or constructed on the basis of potentially satisfying salient constraints of the target domain. In the modeling process, various forms of abstraction, such as limiting case, idealization, generalization, generic modeling, are utilized. Evaluation and adaptation take place in light of structural, causal, and/or functional constraint satisfaction and enhanced understanding of the target problem through the modeling process. Simulation can be used to produce new states and enable evaluation of behaviors, constraint satisfaction, and other factors.

REFERENCES

Cartwright, N., 1989, *Nature's Capacities and Their Measurement*, Clarendon, Oxford.

Chi, M.T.H., Feltovich, P.J., and Glaser, R., 1981, Categorization and representation of physics problems by experts and novices, *Cognitive Science* 5:121-52.

Campbell, N.R., 1920, *Physics, the Elements*, Cambridge University Press, Cambridge.

Clement, J., 1989, Learning via model construction and criticism, in: *Handbook of Creativity: Assessment, Theory, and Research*, G. Glover, R. Ronning, and C. Reynolds, eds., Plenum, New York, pp. 341-81.

Craik, K., 1943, *The Nature of Explanation*, Cambridge University Press, Cambridge.

de Kleer, J., and Brown, J.S., 1983, Assumptions and ambiguities in mechanistic mental models, in: D. Gentner and A.L. Stevens, pp. 155-190.

Duhem, P., 1902, *The Aim and Structure of Physical Theory*, Atheneum, New York (reprinted 1962).

Faraday, M., 1839-1855, *Experimental Researches in Electricity*, Dover, New York (reprinted 1965).

Gentner, D., and Gentner, D.R., 1983, Flowing waters or teaming crowds: mental models of electricity, in: D. Gentner and A.L. Stevens, pp. 99-130.

Gentner, D., and Stevens, A.L., 1983, *Mental Models*, Lawrence Erlbaum, Hillsdale, NJ.

Gilhooly, K.J., 1986, Mental modeling: A framework for the study of thinking, in: *Thinking: Progress in Research and Teaching*, J. Bishop, J. Lochhead, and D. Perkins, eds., Lawrence Erlbaum, Hillsdale, NJ, pp. 19-32.

Giere, R.N., 1988, *Explaining Science: A Cognitive Approach*, University of Chicago Press, Chicago.

Giere, R.N., 1994, The cognitive structure of scientific theories, *Philosophy of Science* 61:276-296.

Gooding, D.C., 1990, *Experiment and the Making of Meaning: Human Agency in Scientific Observation and Experiment*, Kluwer, Dordrecht.

Griffith, T., Nersessian, N., and Goel, A., 1996, The role of generic models in conceptual change, in: *Proceedings of the Cognitive Science Society* 18, Erlbaum, Hillsdale, NJ, pp. 312-317.

Hempel, C.G., 1952, *Fundamentals of Concept Formation in Empirical Science*, University of Chicago Press, Chicago.

Hesse, M., 1966, *Models and Analogies in Science*, University of Notre Dame Press, Notre Dame, IN.

Ippolito, M.F., and Tweney, R.D., 1995, The inception of insight, in: *The Nature of Insight*, R.J. Sternberg and J.E. Davidson, eds., MIT Press, Cambridge, MA.

Johnson-Laird, P.N., 1983, *Mental Models*, Harvard University Press, Cambridge, MA.

Kosslyn, S.M., 1994, *Image and Brain*, MIT Press, Cambridge, MA.

Kuhn, T.S., 1964, A function for thought experiments, in: *The Essential Tension: Selected Studies in Scientific Tradition and Change*, University of Chicago Press, Chicago, pp. 240-265

Kuhn, T.S., 1991, The road since structure, in: *PSA 1990*, A. Fine, M. Forbes, and L. Wessels, eds., Vol. 2, PSA, East Lansing, MI, pp. 3-16.

Maxwell, J.C., 1890, *The Scientific Papers of James Clerk Maxwell*, W.D. Niven, ed., Cambridge University Press, Cambridge.

Nersessian, N.J., 1984, *Faraday to Einstein: Constructing Meaning in Scientific Theories*, Martinus Nijhoff, Dordrecht.

Nersessian, N.J., 1988, Reasoning from imagery and analogy in scientific concept formation, in: *PSA 1988*, A. Fine and J. Leplin, eds., Vol. 1, PSA, East Lansing, MI, pp. 41-48.

Nersessian, N.J., 1992, How do scientists think? Capturing the dynamics of conceptual change in science, in: *Cognitive Models of Science, Minnesota Studies in the Philosophy of Science 15*, R. Giere, ed., University of Minnesota Press, Minneapolis, pp. 3-44.

Nersessian, N.J., 1993, In the theoretician's laboratory: thought experimenting as mental modeling, in: *PSA 1992*, D. Hull, M. Forbes, and K. Okruhlik, eds., Vol. 2, PSA, East Lansing, MI, pp. 291-301.

Nersessian, N.J., 1995, Should physicists preach what they practice? Constructive modeling in doing and learning physics, *Science & Education* 4:203-226.

Nersessian, N.J., 1999, Abstraction via generic modeling in concept formation in science, in: *Correcting the Model: Idealization and Abstraction in Science*, N. Cartwright and M.R. Jones, eds., Editions Rodopi, Amsterdam, in press.

Norton, J. 1991, Thought experiments in Einstein's work, in: *Thought Experiments in Science and Philosophy*, T. Horowitz and G. Massey, eds., Rowman and Littlefield, Savage, MD, pp. 129-148

Perrig, W., and Kintsch, W., 1985, Propositional and situational representations of text, *Journal of Memory and Language* 24:503-518.

Shelley, C., 1996, Visual abductive reasoning in archaeology, *Philosophy of Science* 63:278-301.

Tracing the Development of Models in the Philosophy of Science

Daniela M. Bailer-Jones
Universität Gesamthochschule Paderborn, Germany, bailer-jones@t-online.de

Abstract: An overview is provided of how the concept of a scientific model has developed and changed in the philosophy of science in the course of this Century. I identify three shifts of interest in the treatment of the topic of scientific models. First, only from the 1950s did models begin to be considered relevant to the scientific enterprise, motivated by the desire to account for issues such as theory change and creativity in scientific discovery. Second, I examine how philosophers then increasingly concentrated on the analysis of the functions of models, e.g. for explanation or for guiding and suggesting new experiments. Finally, I show how an analysis of the functions of models could lead to the consideration of their function not just within science, but also in human cognition, so that models are now sometimes viewed as tools of actual (rather than logically reconstructed) scientific thinking.

1. INTRODUCTION

Are scientific models a topic discussed as part of the standard canon of philosophy of science? Well, even if they are today, this has certainly not always been the case. During the first half of this Century, theories and not models were the exclusive focus of investigation. Then, in the early fifties, interest in models first arose and began to accumulate. Not only did models undergo a period of intensive study in the sixties, there are also recent signs of continuing or renewed interest. This is reflected in conference sessions, e.g. at the PSA 1996 (Darden, 1997), in article collections, such as Herfel et al. (1995) and Morrison and Morgan (1999), and in journal issues dedicated to the topic, e.g. *Philosophia Naturalis* (35, 1998).

However, the motives for the study of models have changed over time; in fact, they did so in tune with three shifts of interest which I shall identify. Philosophy of science itself has progressed over the last decades and its development has, not surprisingly, also had repercussions on the treatment of models. Rising awareness of the issue of theory change and that of creativity in scientific discovery was of outstanding importance for such changes. I shall, in this paper, trace the "career" of scientific models from almost total

Model-Based Reasoning in Scientific Discovery, edited by L. Magnani, N.J. Nersessian, and P. Thagard, Kluwer Academic/Plenum Publishers, New York, 1999

disregard to the point where topics such as "model-based reasoning in scientific discovery" have become prominent in the philosophy of science.

I identify three shifts of interest regarding models under the following headings: (1) "from disregard to popularity"; (2) "from formal accounts to a functional characterisation of models", and (3) "from the role of models in science to their role in human cognition". These shifts describe tendencies in research interests and opinions not necessarily received at a time. The shifts have resulted from changes in emphasis and extensions of interest. In other words, if I claim, for instance, that there has been a development away from disregard to increased interest in models due to such-and-such an issue or insight, then this is not to say that such a shift has been universal. It is not to deny that there may be some (or many) philosophers who continue to think that models are either of no interest whatsoever or are of interest for completely different reasons from those suggested by the shift. Not everybody follows a shift, and views on models do not cease to exist merely because a (no matter how relevant) group of researchers abandons them. The shifts I identify here are not exclusive; they address tendencies and aim to trace roots of the current discussion. Moreover, they did not happen consecutively in the strict sense, but overlap in time. The order into which I put them is roughly historical, although, on many occasions, different developments took place synchronically and were intertwined with each other.

2. FROM DISREGARD TO POPULARITY

A negative attitude towards scientific models has a long tradition, going back to the beginning of this Century. Pierre Duhem ([1914] 1954) depicts the relation between models and theory as follows: "The descriptive part has developed on its own by proper and autonomous methods of theoretical physics; the explanatory part has come to this fully formed organism and attached itself to it like a parasite" (Duhem, ([1914] 1954, p. 32). The latter models feature as no more than appendices to theories; they have no crucial role and are redundant as far as any relevant aspect of scientific development is concerned. Duhem sharply distinguishes theories from models. For him, "[a] physical theory is not an explanation. It is a system of mathematical propositions, deduced from a small number of principles, which aim to represent as simply, as completely, and as exactly as possible a set of experimental laws" (Duhem, [1914] 1954, p. 19). Duhem considers this type of theory as the driving force of scientific progress and the only means of scientific development. Such theories are the "descriptive part" and are only subsequently supplemented by models – he has mechanical models in mind – to serve explanatory purposes.

It is nonetheless interesting to observe that Duhem, while he goes to great pains to disclaim the role of models in the scientific enterprise, succeeds, in spite of himself, in spelling out the immense potential of scientific models. As is well known, Duhem makes a distinction between the "abstract" minds of the French and the "ample" minds of the English. He supports this classification with the claim that, in the English scientific tradition, models almost invariably accompany the exposition of theories (Duhem, [1914] 1954, p. 69). While Duhem is full of contempt for this approach, he stresses the importance of models for the English-type mind: "Understanding a physical phenomenon is, therefore, for the physicist of the English school, the same thing as designing a model imitating the phenomenon; whence the nature of material things is to be understood by imagining a mechanism whose performance will represent and simulate the properties of the bodies" (Duhem, [1914] 1954, p. 72). The object of the English physicist is "to create a visible and palpable image of the abstract laws that [the Englishman's] mind cannot grasp without the aid of this model" (Duhem, [1914] 1954, p. 74). Duhem maintains categorically, however, that "only abstract and general principles can guide the mind in unknown regions and suggest to it the solutions of unforeseen difficulties" (Duhem, [1914] 1954, p. 93). Despite all this, Duhem admits in the end that, even in the face of any superiority of scientific rigor, we are never quite free from the need of some imagination: "At the bottom of our most clearly formulated and most rigorously deduced doctrines we always find again that confused collection of tendencies, aspirations and intuitions. No analysis is penetrating enough to separate them or to decompose them into simpler elements" (Duhem, [1914] 1954, p. 104). Duhem, against his own interest, provides an insightful and rather modern-sounding characterisation of models. Although he is highly adverse to any benefits of models, he embodies the dichotomy of theory bias and model need which is to shape struggles to come.

However, if a model need was grudgingly acknowledged on occasion, it still took many years until models were taken seriously as a topic of philosophical investigation. An important reason for this lies in the outlook and the philosophical interests of the very influential movement of logical empiricism associated with the Vienna Circle. Rudolf Carnap, for instance, attributed only a very minor role to models: "It is important to realize that the discovery of a model has no more than an esthetic or didactic or at best a heuristic value, but it is not at all essential for a successful application of the physical theory" (Carnap, 1939, p. 68). In Carnap's account, the meaning and interpretation of the theory is in no way dependent on the application of models. His whole project was not geared towards considering "heuristic" or "didactic" values of anything. This general outlook is made explicit in Reichenbach's (1938) distinction between the context of discovery and the context of justification, a distinction which aims at the separation of logic and psychology. For him, it is not "a permissible objection to an epistemological construction that actual thinking does not conform to it" (Reichenbach, 1938, p. 6). The interest is *not* in thinking and the context of discovery, but in "a relation of a theory to facts, independent of the man who found the theory" (Reichenbach, 1938, p. 382). Logic, correspondingly, the chosen tool of logical empiricism, is promising with regard to questions of justification, not of psychology; it supports the notions of confirmation in the form of deductive inferences, of clearly stated laws and of universal applicability. Carnap also thought that in modern theoretical physics, such as relativity theory and quantum physics, approaches based on intuition would play a lesser and lesser role.

Once the project of the logical empiricists is identified (see also Giere, 1996), their limited interest in modeling is no longer surprising, because, as Reichenbach put it succinctly, the *actual* thinking process is, for them, not a legitimate subject of epistemology. This disinterest in models is especially plausible in the light of the assumption that it is the "actual thinking", imagining and cognition (the "psychology") and their role in scientific discovery, precisely what the logical empiricists are *not* interested in, that becomes central to later uses of scientific models, and it is those uses that turned scientific models into a widely discussed topic in the philosophy of science. The logical empiricist project is more prescriptive than descriptive in any modern science studies sense. With the chosen outlook of rational reconstruction of science in terms of logic, there did not appear to be any space for the consideration of scientific models.

When, in the early fifties, scientific models began to take centre stage at least for a few philosophers of science, these philosophers were motivated by concerns and interests that were not predominantly those of the logical empiricists of the Vienna circle. The questions posed concerning science moved noticeably closer to scientific practice and to the *use* of scientific theories rather than their semantics or logical reconstruction. For instance, the question of experimental testing arises and has the air of a concrete, applied, practical problem, and similarly with the issues of how new terminology develops, how to explain creativity in scientific discovery or how theories change. (Ironically, this choice of issues and the style of approaching them, emerging mainly in Britain, is again "British" in the Duhemian, though not in his negative, sense.)

The British move towards models seems to have been formulated as a reaction to Russell's views of theories and Ramsey's criticism thereof. Critical issues were, for instance, whether "electron", being unobservable, yet displaying observable behaviour, is a theoretical or an empirical concept and how the empirical and the theoretical component in it are linked.[1] Russell furnishes an answer suggesting that it is possible "to exhibit the way in which electrons are logical constructions out of observable entities" (Braithwaite 1954, p. 35). This is criticized by Ramsey, and also by Braithwaite, because "[t]o treat theoretical concepts as logical constructions out of observable entities would be to *ossify* the scientific theory in which they occur: [...] there would be no hope of extending the theory to explain more generalizations than it was originally designed to explain" (Braithwaite, 1954, p. 36). Thus, the need for a development from a static to a dynamic understanding of theory, that allowed for theory development, was felt strongly in the British context of discussion.

Let me now review the motivations of the early proponents of the uses and benefits of scientific models in Britain. Richard Braithwaite, Mary Hesse and Ernest Hutten evidently developed their views in a similar research context. They all regarded scientific theories as hypothetico-deductive systems (Hesse, 1953, p. 198; Braithwaite, [1953] 1968; Hutten, 1954, p. 297) and they addressed some of the same issues concerning scientific models, though with different emphases.

Of the three, Braithwaite ([1953] 1968, 1954) was most committed to a formal reconstruction of scientific theorizing. He considers scientific theorizing as a task of deduction, whereby a calculus formally represents the deductive system of the theory. The calculus itself is uninterpreted which has the practical advantage that the calculus is clearly laid out and deductions from it are not confused by individual examples, but can be carried out merely in the form of symbolic manipulations (Braithwaite, [1953] 1968, p. 23). On the other hand, the difficulty arises of how the calculus can be interpreted. Interpretation means that the symbols of the calculus are given meaning in the light of the empirical data the occurrence of which the theory needs to explain. The underlying image Braithwaite employs is the following: imagine the development from premises (of the calculus representing a deductive system) to inferred conclusions as a movement from top to bottom. At the bottom, one finds "directly testable lowest-level generalizations of the theory" (Braithwaite, 1954, p. 38). In other words, observational data stands at the bottom of the logical chain of reasoning. The actual construction of theories, however, takes place precisely in the opposite direction, namely from bottom to top: it starts with observational data at the bottom from which the premises or hypotheses at the top of the logical chain need to be found. Although the premises are "logically prior", they are "epistemo-

[1] That Braithwaite proposed models as a promising strategy to address this question is not to imply that this was the only viable strategy, nor that others had completely overlooked this problem. Considerations similar in content to Braithwaite's led Carnap to explore the notion of correspondence principles, as he mentions in his ([1966] 1995, p. 236 and 237ff.).

logically posterior" (Braithwaite, [1953] 1968, p. 89), i.e., in the actual development of the theory, observed events are known before any higher-level hypotheses can be known. The theory formulation confronts an epistemological problem, because the logically posterior consequences (the observational data) determine the meaning of the theoretical terms, i.e. of the logically prior hypotheses in the calculus representation of the theory (Braithwaite, [1953] 1968, p. 90).

To be able to work from the logically prior to the logically posterior, i.e. from "top" to "bottom", provisional or hypothetical interpretations of the calculus and of the premises in particular are required. These, according to Braithwaite, can be provided in full by models, because models have a different epistemological structure from theories. A model, in contrast to a theory, is an interpreted calculus; in the model, the interpretation of the premises is fixed, even if hypothetically, while the model can still have the same structure as the theory. To illustrate the epistemological difference between model and theory, Braithwaite uses the metaphor of a zip-fastener: "the calculus is attached to the theory at the bottom, and the zip-fastener moves upwards; the calculus is attached to the model at the top, and the zip-fastener moves downwards" (Braithwaite, [1953] 1968, p. 90). Because a model is fully interpreted, whereas a theory is not, the model is a more accessible way to think about the structure represented by the calculus which makes the model an alternative way of thinking about the theory.

For Braithwaite, epistemological advantages established the role of models: the need to provide an interpretation of a calculus, at least hypothetically. The framework of Braithwaite's argument – and his conception of model – is largely formal, very much in the logical empiricist tradition, and his envisaged epistemology is akin to Reichenbach's of which "actual thinking" is not a subject. Hesse and Hutten departed from this. They developed their conceptions of scientific models more closely from scientific practice and from the actual needs of scientists (also actual needs for thinking), even though they remained committed to theories as hypothetico-deductive. For Hutten, it is no good if accounts of scientific method differ hugely between scientists and philosophers, and "model" is just such a term that features in scientists' understanding of scientific method, yet is totally neglected by philosophers (Hutten, 1954, p. 285). Hutten advises: "It is obviously best to follow the scientists here as closely as possible, at least in the first instance; we may hope in this way to avoid forcing science into a pre-conceived scheme, as philosophers have so often done" (Hutten, 1956, p. 81). Part of this endeavour is to "take as an example a modern theory and discuss actual laws, instead of illustrating scientific method by means of old-fashioned and very simplified examples" (Hutten, 1954, p. 284). He himself discusses, among others, the model of an oscillator applied to the specific heat of solids and other areas of physics. With this, Hutten anticipates the importance of case studies and examples for an exploration of scientific models that takes its lead from scientific practice. Hesse's (1953) paper does in fact contain a case study to illustrate her claims about models. The case Hesse discusses is the development of various 19th Century models of the transmission of light in the aether. Use such a case study introduces a new element within the philosophical discussion and carries much weight in demonstrating the need for the study of models in the philosophy of science.

Another important change is that Hutten views scientific models as partial interpretations of theories in the sense that models do not aim to be a *copy* of the theory nor of reality: "there is always some element that is changed or left out in the model as compared to the thing of which it is a model" (Hutten, 1954, p. 286). This means that models can be misleading; moreover, their status is that of being neither true nor false (Hutten, 1954, p. 296), and there can even exist multiple models of the same thing: "we may have many auxiliary models within a single theory; usually, they overlap and are mutually compatible though, on occasion, the models are alternatives" (Hutten, 1954, p. 298). In Hesse's article, Hutten's notion of partial interpretation is paralleled by the claim that models cannot be regarded as "*literal descriptions of nature, but as standing in a relation of analogy to nature*" (Hesse, 1953, p. 201). The point is here that the model only describes certain aspects of something in nature, but not others. The model may even misdescribe certain aspects (e.g. disanalogies to nature). This is why Hutten says that models can mislead and why they are neither true nor false.

Hutten, in particular, makes several more different points about scientific models. While he may not elaborate them in great detail, they are precursors of later areas of investigation, and he is a valuable source when looking for mentioned upcoming issues concerning the study of models. I shall omit the details of his philosophical position here in favour of listing general issues that motivate the discussion about scientific models. For instance, Hutten is one of the first to talk explicitly and positively about a psychological function of models. This "heuristic" or "pragmatic" use of models is based on the fact that models provide a *visual representation* of something. They do so either in three dimensions or in two dimensions in the form of pictures or diagrams (Hutten, 1954, p. 285).

A further practical issue in constructing scientific theories is making available appropriate vocabulary: "The model prescribes a context, or gives a universe of discourse" and it "supplies primarily a terminology" (Hutten, 1954, p. 295). Being confronted with a new situation, orientation comes from comparing this new situation with a familiar situation: "In science, we merely want to explain a new and unfamiliar phenomenon, and so we try to account for it in terms of what we already know, or to describe it in a language that is familiar to us. That is, the model is used to provide an *interpretation*" (Hutten, 1954, p. 286). Referring back to the familiar can be viewed as a strategy of "actual thinking". Hutten also compares models with metaphor. Models function like metaphor because they are used "when, for one reason or another, we cannot give a direct and complete description in the language we normally use" (Hutten 1954, p. 298; see also p. 293), i.e. when common, ordinary terminology fails us.

There are some hints that models are supposed to serve as a link between theory and experiment. According to Hutten, theories are explained and tested *in terms of* models (Hutten, 1954, p. 289), although he does not specify how. Slightly obscurely, he states that the model is not an application of the theory, although the theory is applied with its help. Hesse, in turn, endorses the link between models and experiment with a slightly different emphasis. She says: "Progress is made by devising experiments to answer questions suggested by the model" (Hesse, 1953, p. 199).

One aim of Hesse's article is to expand the concept of model to go beyond purely mechanical 19th Century models. She argues that mathematical formalisms can also be scientific models and that no sharp line should be drawn between the two because they function in essentially the same way (Hesse, 1953, p. 200). This is an important step towards a wider use of the concept of model, later taking effect in the notion of a theoretical model (e.g. Achinstein, 1965).

Like Braithwaite, Hesse considers theories as hypothetico-deductive, "that is, they consist of hypotheses which may not in themselves have any reference to immediate observations, but from which deductions can be drawn which correspond to results of experiments when suitably translated into the experimental language" (Hesse, 1953, p. 53). While this indicates a framework of discussion similar to Braithwaite's, Hesse highlights an entirely different point arising from a situation where inferences are required from "bottom" (empirical data) to "top" (hypotheses of the theory): "The main point that emerges from such a description of theories is that there can be no set of rules given for the procedure of scientific discovery – a hypothesis is not produced by a deductive machine by feeding experimental observations into it: it is *a product of creative imagination, of a mind* which absorbs the experimental data *until it sees them fall into a pattern*, giving the scientific theorist the sense that he is penetrating beneath the flux of phenomena to the real structure of nature" (Hesse, 1953, p. 198; my italics). Hesse here anticipates her own future philosophical concerns, exploring procedures for scientific discovery and creative imagination, in which models become central players. Moreover, she makes explicit reference to *actual thinking*, the mental activity of scientists.

To summarize, the move away from approaches in the philosophy of science that purely rely on the concept of theory was sparked by the need to tie in theory language with observation language (see also Sellars, [1956] 1997, p. 94ff). Attempts to fasten the link between theory and observation or experiment are found in all the "pro-model" authors discussed. Besides this, they had additional motives, some shared, some individual: Braithwaite aims to address the issue of theory change, as I shall discuss in the next section. Hutten studied theory construction and emphasizes that models serve to develop new scientific terminology in analogy to existing accounts. Moreover, he mentions a psychological function of models using visualization, just like Hesse emphasizes the heuristic role of models in that she identifies them as pointers towards further progress. One of Hesse's major motivations to explore models is the issue of scientific discovery inducing her to bring up the issue of creative imagination. So, in a nutshell, the factors motivating the inclusion of scientific models in an account of scientific method are the issues of theory construction and theory change as well as scientific discovery. With the appreciation of these factors, the shift away from disregard of models became firmly established. The shift was accompanied by a turn towards the study of the actual scientific practice, rather than purely to reconstruct scientific method logically.

Hutten's, Hesse's and Braithwaite's papers of the early to mid-fifties turned scientific models into a topic worthy of study. The question of linking theory with observation language was soon to be revolutionized by Hanson (1958) elaborating the theory-ladenness of observation. The discussion of theory change was further sparked by Kuhn's (1962) concept of scientific revolution, and the related issues of scientific discovery and creative imagination continue to be important concerns. The argument that models, as analogies or as metaphors, bridge the gap from the unfamiliar to the familiar has also remained prominent until more recent days, e.g. in Hesse (1966), Harré (1988), Gentner (1982, 1983).

By voicing such central themes, the discussed papers from the fifties prepared the ground not only for the second and third shift, but also for models to enter a phase of great popularity, as documented by an enormous proliferation of articles, beginning in the early sixties, e.g. Harré (1960), Apostel (1961) and Suppes (1961) in Freudenthal (1961), Achinstein (1964, 1965), Hesse (1966). By then, a distinct shift of interest from theories to models had taken place for growing numbers of philosophers of science.

3. FROM FORMAL ACCOUNTS TO A FUNCTIONAL CHARACTERISATION OF MODELS

Duhem pointed to the dichotomy of theory bias and model need. Even if models could no longer be disregarded, theory bias could still take the form of a formal, "theoretical" treatment of models. Mathematical model-theory, for instance, proved to be an attractive candidate to provide a formalized account of what scientific models are (proposed, e.g., in varying forms by Suppes, 1961; van Fraassen, 1980; Giere, 1988). Others, as we have seen, favoured characterizations of models inspired by scientific practice. This resulted in the tendency to characterize models in terms of their functions and the role they play in science and, more specifically, for creativity, discovery and theory development. Positions formulated in the literature on scientific models can consequently be categorized as pursuing two competing goals:

a) establishing, within a formal framework, what scientific models are, much in the logical empiricist tradition;
b) assessing the pragmatic role, the function, of models in the scientific enterprise.

Goal (a) points towards a general, universally applicable conception of model, while goal (b) leaves the option open for a diversity of conceptions corresponding to diverse functions of models. It will become evident that proponents of either goal could not entirely ignore the other, competing goal. Departing from a tradition of theory bias, this helped to establish the function of models as something that had to be taken into consideration. Some examples from either side of the divide will illustrate this.

Patrick Suppes (1961) promotes a formal, model-theoretic account of models. This model-theoretic position allows one to accommodate models in a systematic manner and to give them a central role with regard to theories. The model consists of a relational structure satisfied by the sentences of which the theory consists. The important point is that the structure of the model(s) is such that it does not lead to contradictions or inconsistencies within the theory. Thus, models are integrated with a formal concept of theory, in contrast to some early logical empiricist positions, e.g. Carnap, where models were seen as of little relevance, let alone be accommodated systematically.

According to Suppes, "a theory is a linguistic entity consisting of a set of sentences, and models are non-linguistic entities in which the theory is satisfied" (Suppes, 1961, p. 166). This is the formal account of models which,

Suppes admits, can have its difficulties. The reason for the conceptual difficulty is that it is an attempt to investigate models disregarding their function, almost *despite* their function. For Suppes, however, this is merely the difficulty of "how one is to explain the meaning of a concept without referring to its use" (Suppes, 1961, p. 165). The aim here is explicitly to overlook potential functions of models and consider them from a purely formal point of view. Suppes is nonetheless convinced that the standard notion of model, as defined in mathematical logic, can be applied without distortion to models in disciplines as various as particle physics, electrodynamics, mathematical statistics or social sciences. Suppes believes that there can be one universal conception of model underlying all the different uses of them, and that this conception coincides with the model-theoretic conception of a model. He is aware of the fact that many physicists treat models as "a very concrete physical thing built on analogy" (Suppes, 1961, p. 166), but in his opinion this view is not incompatible with his view based on mathematical logic: "To define formally a model as a set-theoretical entity which is a certain kind of ordered t-ple consisting of a set of objects and relations and operations on these objects is not to rule out the physical model of the kind which is appealing to the physicists, for the physical model may be simply taken to define the set of objects in the set theoretical model" (Suppes, 1961, pp. 166-167). In other words, while Suppes is fully in favour of (a), he thinks the requirements of (b) can be satisfied automatically.

Leo Apostel, in turn, is fully aware "that we cannot hope to give one unique structural definition for models in the empirical sciences" (Apostel 1961, p. 36), i.e. that carrying out (a) presents severe difficulties. In other words, a formal account of models, generally applicable to empirical models (such as the model-theoretic one), is hardly feasible. Apostel reasons that there is a multitude of different functions that models address, as a (b) approach suggests, because scientists use models in different ways; each model is "ambiguous" in that it can aid the scientific progress in a whole variety of ways (cf. Apostel, 1961, p. 5). Interestingly, despite this diversity, Apostel attempts an "adequate rational reconstruction" of models taking full account of the diverse uses scientists make of models. He expresses the hope that a unification of the different types of models could be achieved if these are studied in terms of their function (Apostel, 1961, p. 36), thus, in effect, aims for a compromise between goals (a) and (b).

Like Suppes, Braithwaite ([1953] 1968, 1954) provides a formal account following (a), but feels that he needs to allow for theory change in his account because theory change is observed in the practice of science, thus accepting a consideration from (b). In Braithwaite's terms, if one thinks about the theory in terms of a model, one needs not think about the language in which the theory is expressed, or rather the interpretation of the symbolism.

This is why Braithwaite reasons that the use of models positively affects theory change. In a model, all symbols are given direct meaning and the calculus can be interpreted in one piece. Because the model is hypothetical, the definitive interpretation of the theory is suspended, and, by simply swapping one model for another, the theoretical entities of the theory become easily adjustable to new interpretations. In his attempts to accommodate theory change, Braithwaite remains restricted, however, by the formal requirements of his account and only permits theory changes that preserve the structure of the theory, i.e. no radical changes in the Kuhnian sense. Thus, a compromise between (a) and (b) is here achieved at the cost of not accommodating radical theory changes.

Hesse (1966) belongs to the other side of the divide, (b), aiming to assess the contribution of models to creativity in scientific discovery. She argues that formal, hypothetico-deductive accounts of theories lack the tools to accommodate this important issue. This is why she suggests that scientific models are metaphors, viewing "theoretical explanation as metaphoric redescription of the domain of the explanandum" (Hesse, 1966, p. 157). This claim supports the hypothetical character of models and their propensity for suggesting further theoretical development. When a primary system, the domain of the explanandum, e.g. an ideal gas, is viewed in terms of a secondary system, e.g. billiard balls, then this inspires "creative imagination" and provides perfect grounds for an extension of the model. It points future research in directions of further investigation and experiments which researchers may not have thought of without the model. Clearly, analogy has an important share in devising a model; billiard balls and ideal gases have to have something in common for the model to be fruitful. Neither linguistic metaphors nor scientific models are chosen at random. Yet, if creativity can be claimed for metaphors, then, according to the analogy between linguistic metaphors and scientific models, metaphorical features of models can be held responsible for guiding research and supplying researchers with creative ideas for future development.

Hesse points out correctly that any metaphor approach to models evidently depends on one's view of metaphor. She adopts and elaborates Black's interaction view (Black, 1962). According to Black, the primary and the secondary object, A and B, of the metaphor, A = B, interact in our mind: each is viewed in the light of the other. In the metaphor "Our granddaughter is the sunshine of our lives", the notion of "granddaughter" is, for instance, *filtered through* the notion of "sunshine". Our notion of the granddaughter adopts some features of sunshine, and the reverse. Consequently, the claim goes, A and B shift in meaning because they have been applied to each other. Interpreting models as interactive metaphors aims to explain the creative potential of models for scientific development, a (b) project. This involves, however, employing a fairly formal account of metaphor, i.e. a (a) strategy (the interaction view), for these purposes. In other words, Hesse's (b) project remains dependent on a (a) strategy.

In sum, the struggle of various philosophers to strike a balance between (a) and (b), i.e. between formalistic and pragmatic, functional approaches, indicates that even when one of the goals, (a) or (b), is pursued single-mindedly, traces of the competing goal remain. During this struggle, however, it became increasingly accepted that models were *needed* in scientific methodology. Thus, functional approaches effectively gained ground, even if they are hard to systematize.

4. FROM THE ROLE OF MODELS IN SCIENCE TO THEIR ROLE IN HUMAN COGNITION

Of the diverse functions models can serve one stands out: explanation (e.g. Harré, 1960; Hesse, 1966; Achinstein, 1968). The explanatory advantages of theoretical models are frequently linked to the use of analogy. Achinstein states: "Analogies are employed in science to promote understanding of concepts. They do so by indicating similarities between these concepts and others that may be familiar or more readily grasped" (Achinstein, 1968, pp. 208-209). His examples are analogies between an atom and a solar system, between waves of light, sound and water, nuclear fission and the division of a liquid drop, between the atomic nucleus and extranuclear electron shells, etc. (Achinstein, 1968, pp. 203-205). The metaphor approach to scientific models also relies on the potential of analogy. With metaphor, a *cognitive* perspective on scientific modeling was introduced according to which models (or metaphors) creatively mould the ways of thinking about an object or phenomenon and therefore have a lasting effect on the ways in which their users think (Black, 1962; Hesse, 1966). The cognitive, already anticipated in Hesse's (1953) article ("a mind which absorbs the experimental data until it sees them fall into a pattern", Hesse, 1953, p. 198), is a natural companion to explanation, if explanations are viewed as providing understanding (Salmon, 1993) where understanding is a cognitive activity. In this context, it is interesting to observe that, more recently, analogy, which is viewed as central for explanation, has become a closely investigated, crucial candidate for patterns of human reasoning in cognitive science (Gentner, 1982, 1983; Gentner and Markman, 1997; Holyoak and Thagard, 1997).

It is not just analogy, however, that contributes to the cognitive function of certain scientific models, even if analogy is central. Rom Harré also focuses on the illustrative and creative functions models have for the development of scientific theories. Creativity is particularly needed when we lack a detailed account of the scientific problem and when there are "gaps in our knowledge of the structures and constitutions of things" (Harré, 1970, p. 35). While models fill the gaps in theories and are "putative analogue[s] for the real mechanism" (Harré, 1970, p. 35), modeling is also a crucial asset in the process of developing a *picture* of mechanisms. Harré claims that "[s]cientists, in much of their theoretical activity are trying to form a picture of the mechanisms of nature which are responsible for the phenomena we observe. The chief means by which this is done is by the *making or imagining of models*" (Harré, 1970, p. 34-35; my italics). Models have a creative function as "progenitors of hypothetical mechanisms" (Harré, 1970, p. 39), but their importance does not cease after discovery. Models of known objects continue to be used, serving a cognitive role: "Generally speaking, making models *for* unknown mechanisms is the creative process in science, by which potential advances are initiated, while the making of models *of* known things and processes has, generally speaking, a more heuristic value" (Harré, 1970, p. 40). It is one thing to think of the atom in terms of the solar system in order to develop a model of the atom, and an entirely different thing to use the image of the solar system in order to reason about or to teach ideas about the atom. Both *creating* and *using* a model are processes involving cognition, but not necessarily in the same way, and having a *picture* of a model may be crucial to *using* it.

The issue of *using* a model even for known things and processes and of "forming a picture" of how these things and processes might work points in the direction of mental model research. This type of research has developed quite independently of philosophical concerns, even though it slots in well where the relevance of models is not just considered in a discovery-related context, but where models are thought to have a more permanent function for human reasoning. In Gentner's and Stevens' (1983) collection of articles on mental models, the emphasis is on knowledge representation: "A typical piece of mental models research is characterized by careful examination of the way people understand some domain of knowledge" (Gentner and Stevens, 1983, p. 1). While it has practical advantages (e.g. for teaching and instruction) to know how people represent and process their knowledge, the research presented in the collection of papers has a more fundamental concern, and that is "understanding human knowledge about the world" (Gentner and Stevens, 1983, p. 1). Obviously, the inquiry into human knowledge needs to be restricted to domains that are easily tractable. This is why the investigated examples are mostly simple physical systems because they stem from such domains. An impressive illustration of a mental model of a simple physical system is Gentner's and Gentner's (1983) who explore people's reasoning about electrical circuits. Two central analogies are employed, water flow and moving crowds, which are suitable for solving battery and resistance problems respectively. The success of solving specific reasoning tasks can be shown to depend on whether batteries or resistors are involved and on the choice of mental model. Other related work has been done by de Kleer and Brown (1981, 1983) and Williams, Hollan, and Stevens (1983), just to mention a few. In view of such studies in various areas of cognitive science, it is not surprising that drawing from mental model research has become an attractive path towards examining the role of scientific models for human cognition (e.g. Giere, 1988; Nersessian, 1993; Bailer-Jones, 1997).

Another line of argument for the more permanent, not purely discovery-related use of models in science that relies on a cognitive component comes from science teaching. According to Thomas Kuhn, certain paradigmatic "patterns of thought" guide scientific thinking and determine the direction of scientific reasoning and investigation at a time. This idea is particularly prominent in the formulation of the Kuhnian concept of a paradigm, and, according to a later specification, of an exemplar (Kuhn, 1977, pp. 293-295, 297-298). Exemplars, according to Kuhn, are "concrete problem solutions accepted by a group as, in quite a usual sense, paradigmatic" (Kuhn, 1977, p. 298). A student of physics learns these problem solutions and an important step in his or her cognitive development is to be able to solve *other*, new problems by recognizing their similarity to the paradigmatic case. Kuhn claims: "The student discovers a way to see his problem as like a problem he has already encountered. Once that likeness or analogy has been seen, only manipulative difficulties remain" (Kuhn, 1977, p. 305). The skill involved is that of *thinking about something* (the problem to be solved) *in terms of something else* (the known, paradigmatic problem solution). Kuhn argues from some common experience of learning physics; Giere (1988) takes a similar line and studies undergraduate textbooks. Both approaches have their problems, but independent research from physics teaching exists (Halloun and Hestenes, 1987; Wells, Hestenes, and Swackhamer, 1995) that suggests that actively and deliberately encouraging the use of modeling strategies can greatly improve the success of physics teaching.

In the context of mental model research it is commonly assumed that scientific modeling is subject to the same cognitive processes and ramifications as all other areas of human reasoning. This is why it is tempting to turn around the question and ask, instead of what can role cognition plays in scientific modeling, what scientific modeling tell us about human cognition. In a way, this is just what has happened with the advent of mental model research. Moreover, looking back on philosophy of science, the shift towards the cognitive importance of models adds an interesting twist to the current discussion because it circumvents the dilemma between a tidy, formal account and a functional, pragmatic characterization of scientific models. It does so by no longer viewing models exclusively in terms of their role in science, but in terms of their role for human cognition. This is a significant shift in perspective.

5. CONCLUSION

Taking the three shifts together, the study of scientific models has developed from disregard under logical empiricism to a situation where cognitive aspects of model use have become a well-established topic of research. Scientific models can now not only be considered an important tool of cognition, they can even be used as a source of information of how scientists reason. The move towards *actual thinking* in the exploration of science involved more than one step: first, it needed to be established that models are *used*, and, for this, the actual scientific practice needed to become a criterion for the study of science. Second, it needed to consider what models are *used for*. The second shift of interest illustrated that investigating the function of models led to tension with more formal, "theoretical" approaches to models aiming at very systematic and generally applicable accounts. The third shift towards cognitive considerations regarding models, keeping in mind the function of explanation and the provision of understanding, led to encouragement from independent developments in cognitive science. Although not simple in detail, it has now become possible to challenge or to support philosophical claims about modelling by employing empirical results from cognitive psychology (mental models research and research into human reasoning, into visualization, into diagram use) and science teaching.

In sum, only through the shift from formal to functional approaches in the study of scientific models has become possible to approach at the question of "actual thinking" which Reichenbach so fiercely rejected as a subject of epistemology. The result is a study of scientific models as part of a "humanized" (or naturalized) epistemology centring on the cognizing scientific researcher.

ACKNOWLEDGEMENT

Many thanks to Andreas Bartels for commenting on, questioning and advising me on many issues of this paper.

REFERENCES

Achinstein, P., 1964, Models, analogies, and theories, *Philosophy of Science* 31:328-350.

Achinstein, P., 1965, Theoretical models, *British Journal for the Philosophy of Science* 16:102-120.

Achinstein, P., 1968, *Concepts of Science*, John Hopkins Press, Baltimore/Maryland.

Apostel, L., 1961, Towards the formal study of models in the non-formal sciences, in: *The Concept and the Role of the Models in Mathematics and Natural and Social Sciences*, H. Freudenthal, ed., Reidel, Dordrecht, pp. 1-37.

Bailer-Jones, D.M., 1997, Scientific models: a cognitive approach with an application in astrophysics, Ph.D. Thesis, University of Cambridge (unpublished).

Black, M., 1962, *Models and Metaphors*, Cornell University Press, Ithaca/New York.

Braithwaite, R., [1953] 1968, *Scientific Explanation. A Study of the Function of Theory, Probability and Law in Science*, Cambridge University Press, Cambridge.

Braithwaite, R., 1954, The nature of theoretical concepts and the role of models in an advanced science, *Revue Internationale de Philosophie* 8/1-2:34-40.

Carnap, R., 1939, Foundations of logic and mathematics, *International Encyclopedia of Unified Science*, Chicago University Press, Chicago.

Carnap, R., [1966] 1995, *An Introduction to the Philosophy of Science*, M. Gardner, ed., Dover Publications Inc., New York.

Darden, D., ed., 1997, *PSA 1996*, Philosophy of Science Association, East Lansing/Michigan.

de Kleer, J., and Brown, J. S., 1981, Mental models of physical mechanisms and their acquisition, in: *Cognitive Skills and Their Acquisition*, J.R. Anderson, ed., Lawrence Erlbaum Associates, Hillsdale/New Jersey, pp. 285-309.

de Kleer, J., and Brown, J.S., 1983, Assumptions and ambiguities in mechanistic mental models, in: *Mental Models*, D. Gentner and A.L. Stevens, eds., Erlbaum, Hillsdale, NJ, pp. 155-190.

Duhem, P., 1954 [translated from the French 2nd edition, 1914], *The Aim and Structure of Physical Theory*, Princeton University Press, Princeton/New Jersey.

Freudenthal, H., ed., 1961, *The Concept and the Role of the Models in Mathematics and Natural and Social Sciences*, Reidel, Dordrecht.

Gentner, D., 1982, Are scientific analogies metaphors?, in: *Metaphor: Problems and Perspectives*, D.S. Miall, ed., Harvester Press, Brighton, pp.106-132.

Gentner, D. 1983. Structure mapping: a theoretical framework for analogy, *Cognitive Science* 7:155-170.

Gentner, D., and Gentner, D., 1983, Flowing waters and teeming crowds: mental models of electricity, in: *Mental Models*, D. Gentner and A.L. Stevens, eds., Erlbaum, Hillsdale, N.J., pp. 99-129.

Gentner, D., and Markman, A.B., 1997, Structure mapping in analogy and similarity, *American Psychologist* 52:45-56.

Gentner, D., and Stevens, A.L., eds., 1983, *Mental Models*, Erlbaum, Hillsdale. N.J.

Giere, R., 1988, *Explaining Science: A Cognitive Approach*, University of Chicago Press, Chicago.

Giere, R., 1996, From wissenschaftliche Philosophie to philosophy of science, in: *Origins of Logical Empiricism*, R. Giere and A. Richardson, eds., Minnesota Studies in the Philosophy of Science, Vol. XVI, University of Minnesota Press, Minneapolis, MN, pp. 335-354.

Halloun, I.A., and Hestenes, D., 1987, Modeling instruction in mechanics, *American Journal of Physics* 55:455-462.

Hanson, N.R., 1958, *Patterns of Discovery*, Cambridge University Press, Cambridge.

Harré, R., 1960, Metaphor, model and mechanism, *Proceedings of the Aristotelian Society*, 60:101-122.

Harré, R., 1970, *The Principles of Scientific Thinking*, Macmillian, London.

Harré, R., 1988, Where models and analogies really count, *International Studies in the Philosophy of Science* 2:118-133.

Herfel, W.E., Krajewski, W., Niiniluoto, I., and Wojcicki, R., eds., 1995, *Theories and Models in Scientific Processes*, Poznan Studies in the Philosophy of the Sciences and the Humanities, Rodopi, Amsterdam, pp. 137-149.

Hesse, M., 1953, Models in physics, *British Journal for the Philosophy of Science* 4:198-214.

Hesse, M., 1966, *Models and Analogies in Science*, University of Notre Dame Press, Notre Dame/Indiana.

Holyoak, K.J., and Thagard, P., 1997, The analogical mind, *American Psychologist* 52:35-44.

Hutten, E.H., 1954, The role of models in physics, *British Journal for the Philosophy of Science* 4:284-301.

Hutten, E.H., 1956, *The Language of Modern Physics*, George Allen & Unwin Ltd., London.

Kuhn, T.S., 1962, *The Structure of Scientific Revolutions*, University of Chicago Press, Chicago.

Kuhn, T.S., 1977, Second thoughts on paradigms, in: *The Essential Tension*, University of Chicago Press, Chicago, 293-319.

Morrison, M. and Morgan, M., eds., 1999, *Models as Mediators*, Cambridge University Press, Cambridge, in press.

Nersessian, N.J., 1993, In the theoretician's laboratory: thought experimenting as mental modeling, in: *PSA 1992*, D. Hull, M. Forbes and K. Okruhlik, eds., Philosophy of Science Association, East Lansing, MI, pp. 291-301.

Reichenbach, H., 1938, *Experience and Prediction*, University of Chicago Press, Chicago.

Salmon, W., 1993, The value of scientific understanding, *Philosophica* 51:9-19.

Sellars, W., [1956] 1997, *Empiricism and the Philosophy of Mind*, Harvard University Press, Cambridge/Mass.

Suppes, P., 1961, A comparison of the meaning and uses of models in mathematics and the empirical sciences, in: *The Concept and the Role of the Model in Mathematics and Natural and Social Sciences*, H. Freudenthal, ed., Reidel, Dordrecht, pp. 163-177.

van Fraassen, B., 1980, *The Scientific Image*, Clarendon Press, Oxford.

Wells, M., Hestenes, D., and Swackhamer, G. 1995. A modeling method for high school physics instruction, *American Journal of Physics* 63:606-619.

Williams, M.D., Hollan, J.D., and Stevens, A., 1983, Human reasoning about a simple physical system, in: *Mental Models*, D. Gentner and A.L. Stevens, eds., Erlbaum, Hillsdale, N.J., pp. 131-153.

Using Models to Represent Reality

Ronald N. Giere
Department of Philosophy, Center for Philosophy of Science, University of Minnesota, Minneapolis, MN, USA, giere@maroon.tc.umn.edu

Abstract: In this paper I develop a unified interpretation of the nature and role of models in science. Central to this interpretation is an understanding of the relationships between models and other elements of an understanding of science, particularly theories, data, and analogy. I begin by criticizing a standard interpretive or instantial view of models, derived from mathematical logic, as not being adequate for empirical science. I then go on to develop a representational view of models which, I argue, is much more adequate to the needs of empirical science. I conclude that scientific reasoning is to a large extent model-based reasoning. It is models almost all the way up and models almost all the way down.

1. INTRODUCTION

There has recently been an increase in interest in the role of models in science, of which the Pavia workshop on model-based reasoning is a manifestation. One result of this increased attention has been a proliferation of views on what models are and how they are used in science. In this presentation I will develop a unified interpretation of the nature and role of models in science. Central to this interpretation is an understanding of the relationships between models and other elements of an understanding of science, particularly theories, data, and analogy. My conclusion will be that models play a much larger role in science than even the most ardent enthusiasts for models have typically claimed. Modeling, on my view, is not at all ancillary to doing science, but central to constructing scientific accounts of the natural world.

When I say I seek an *interpretation* of the nature and roles of models in science, I allow that other interpretations are possible. There is no unique essence to the nature of models that might be revealed by philosophical analysis. Nevertheless, I think my interpretation is better than some others, and I will attempt to convince you that this is so.

Model-Based Reasoning in Scientific Discovery, edited by L. Magnani, N.J. Nersessian, and P. Thagard, Kluwer Academic/Plenum Publishers, New York, 1999

2. MODEL THEORY

The claim that an understanding of models is central to an understanding of science is not new. Almost forty years ago, Patrick Suppes (1960) published a much cited paper with the title: "A comparison of the meaning and uses of models in mathematics and the empirical sciences". The thesis of this paper was that the meaning and use of models can be interpreted as being the *same* in the empirical sciences as it is in mathematics, and, more particularly, in mathematical logic.

At the time Suppes wrote this paper, the theory of models was closely tied with logic. Thus Suppes wrote: "A theory is a linguistic entity consisting of a set of sentences and models are non-linguistic entities in which the theory is satisfied". More specifically, a model for Suppes is a set-theoretical structure consisting of a set of objects together with properties, relations, and functions defined over the set of objects. The important point is that when the specific objects, properties, relations, and functions are coordinated with terms in the axioms of a theory, the axioms all come out to be true, relative, of course, to our prior understanding of the domain of objects considered. So a model, on this account, provides an *interpretation* of a set of uninterpreted axioms. For this reason such models are often called "interpretive models." They might also be called "instantial models" since they instantiate the axioms of a theory understood as comprised of linguistic (including mathematical) statements.

For logicians, most of the models considered consist of *abstract* entities such as numbers or geometrical points and lines. *In principle*, however, the objects considered could be *physical* objects, such as the Earth and the Moon. This is the basis for Suppe's claim that the concept of a model is the same in empirical science as in mathematics. Later I will consider whether any interesting scientific theories, suitably reconstructed, might *in fact* have physical models.

I should note that over the past forty years, as the study of abstract models moved from the province of philosophers and logicians to that of mathematicians, the connection between model theory and logic has become rather attenuated. Current treatises on model theory, such as that by Hodges (1993), focus directly on what are called structures, which are abstract, nonlinguistic entities. For example, the groups of group theory and vector spaces are structures in this sense. It would, therefore, be misleading to refer to the models of contemporary mathematical model theory exclusively as interpretive or instantial models.

It is Suppes' view, however, that became, and largely remains, *a*, if not *the*, standard view of models within the philosophy of science. Let me highlight the point that Suppes' view of models incorporates a quite specific relationship between a *theory* (a set of axioms) and a *model* (a set of objects satisfying the axioms).

3. INSTANTIAL MODELS AND ANALOGY

The instantial conception of models supports a quite specific understanding of the nature of *analogy* in science. Uninterpreted logical formulae may be interpreted using many different instantial models. These models will all be *isomorphic*, that is, there will be a one-to-one correspondence between the elements of the different models. This provides a basis for saying that the corresponding elements of the models, as well as the models themselves, are *analogous*. Of course the models in question must be *physical*, and not merely mathematical.

$$a\, d^2u/dt^2 + b\, u = 0$$

$$m\, d^2x/dt^2 + k\, x = 0 \qquad L\, d^2q/dt^2 + 1/C\, q = 0$$

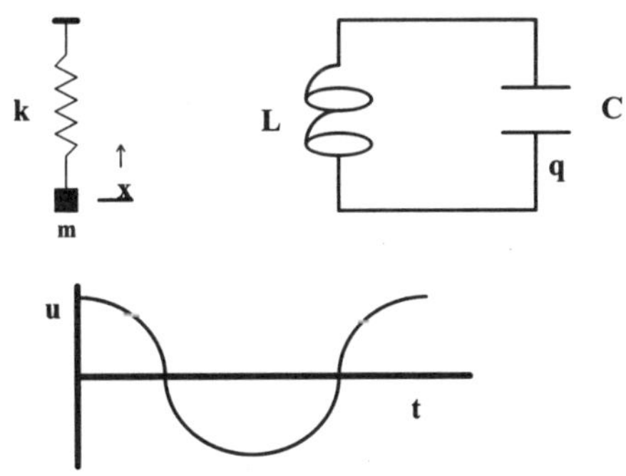

Figure 1.

A classic example of such an analogy is that between, on the one hand, an electrical circuit consisting of an inductance coil and a capacitor (an L-C circuit) and, on the other hand, a mechanical oscillator, such as a bouncing spring. Here the inductance coil is said to be analogous to the mass on the oscillating spring while the capacitance is said to be analogous to the spring constant. The current at any point in the circuit is then analogous to the position of the mass on the spring. Both exhibit sinusoidal change over time. The differential equations describing both change in current and change in position have exactly the same abstract form. These relationships are shown in Figure 1.

4. REPRESENTATIONAL MODELS

The instantial conception of models is a well-defined conception of considerable value, particularly in the study of formal logic and the foundations of mathematics. Nevertheless, contrary to Suppes' claims, I do not think it is the best conception of models for understanding models as used in the practice of the empirical sciences. I will not directly criticize the instantial conception of models as a means for understanding scientific practice. Rather, I will simply point out the difficulties I see from my own, alternative, perspective. For contrast, I call my understanding of models *representational* because it takes models not primarily as providing a means for interpreting formal systems, but as tools for *representing the world.* That is not their only function, but it is, I think, the central function of models used in empirical science. For the moment, then, forget about logic and concentrate on scientific practice, indeed, on the practice of a lowly science, *cartography.*

Maps. Figure 2 is a standard tourist map of the central area of Pavia. Let us explore some relevant properties of maps. First, maps are not linguistic entities. They are *physical objects*, for example, a piece of paper with lines on it. It does not, therefore, strictly make sense to ask whether a map is *true* or *false*. Those designations are usually reserved for linguistic entities. Moreover, maps are not usually thought of as *instantiations* of any linguistic forms. Of course, one can create a language-like version of any map by creating a digitized bit-map, as was done to produce Figure 2. One could then say that the map of Figure 2 is an instantiation of a long, uninterpreted binary code. But this is an extreme way to secure a conception of maps as instantial models. Such an interpretation plays absolutely no role in understanding the nature or function of maps, which were well-established long before anyone even had the idea of a bit-map. Nevertheless, even though they are neither linguistic entities nor instantiations of linguistic entities, maps are *representational.* Just *how* they are representational is another question, one I will take up shortly. First let us consider some further characteristics of maps.

Maps are *partial.* Only some features of the territory in question are represented. For example, the map of Figure 2 represents very few buildings. Moreover, even the indicated features are not fully specified, such as the height of the Palazzo Università. Maps are of *limited accuracy* regarding included features. Relative distances on the map, for example, will not correspond exactly to relative distances on the ground. This could not be otherwise. No real map could possibly indicate literally all features of a territory with perfect accuracy. In the limit, the only perfect map of a territory would be the territory itself, which would no longer be a map at all. Here one may recall the story by Borges (1954) in which the cartographers of a fictional land set out to construct a map of their land on a scale of one to one. As they complete their project, the people of the land begin to move onto the new territory. The map of Figure 2, by contrast, is a *representational model* of Pavia. It *represents* Pavia in its own special way.

Figure 2.

Let us now return to the question: *How* does this map represent Pavia? The answer is: by being *spatially similar to* aspects of Pavia. For example, the lines on the map have similar spatial orientations to some streets in Pavia. In using a map we are using features of one two dimensional surface (the map) to represent features of another two dimensional surface (the surface of the city). For example, some lines on the map represent streets in the city. To generalize: a map represents the region mapped in virtue of shared spatial similarities between the map and the region mapped. Here one *object* (a map) is used to represent another *object* (a geographic region). This notion is explicitly opposed to that of a statement representing a state of affairs.

Similarity versus isomorphism. Philosophers tend to be suspicious of appeals to similarity. A standard objection is that, since anything is similar to anything else in some respects or other, claims of similarity are vacuous. One might be tempted to invoke isomorphism here, saying that the map is isomorphic, or partially isomorphic, to aspects of the city. But this just cannot be right. No reasonably detailed map can be accurate enough to exhibit a literal isomorphism with identifiable features of a real geographical surface. So the best one could do is invoke something like "approximate isomorphism". Absent some account of what "approximate" might mean in this context, such talk only gives the appearance of clarity. It offers no real conceptual advantages over talk of similarity. And it may disguise problems that need to be faced directly.

Here, by the way, is a basis for questioning the idea that there can *in fact* be *physical* instantiations of the statements of a linguistically formulated theory. It seems easy enough to imagine the objects of an instantial model being physical objects, such as the Earth, Moon, or the planets. But as soon as one adds quantitative functions, such as the mass of the Earth or the distance between the Earth and the Moon, one is in great danger of ending up with false statements, which is to say, no model at all.

Charges of vacuity for claims of similarity can be met by specifying 1) the respects in which the map is said to be similar to the region mapped, and 2) the degree of similarity regarding these respects. Thus, a map might be highly accurate with respect to relative linear distances, but contain very little information about relative elevations. Here an important general point is that the respects and degrees of similarity must be specified from the outside, so to speak. They are not intrinsic to any map or geographical region. Thus maps necessarily reflect the interests of map makers and map users. Maps are *interest relative*, and necessarily so.

Not only have philosophers been suspicious of the concept of similarity, they typically claim there is no way to give a satisfactory *general* account of the notion of similarity. But there is no need to look for a general account of similarity between a model and what is modeled. Similarity is *context dependent*. In any particular context, what is said to be similar to what, in what ways, and to what degrees, can be specified. Of course, there is no unique specification. There are many possible specifications depending on the particular interests of those doing the modeling.

These points may be reinforced by considering a somewhat more abstract kind of map, a subway map, as exhibited in Figure 3. Here spatial locations are indicated only very schematically. The important information is *topological*. One gets the ordering of stations on individual lines together with indications where two lines meet and thus where transfers from one line to another are possible. So the important similarities are those between these topological features of the map and of the whole metro system.

Figure 3.

5. OTHER MATERIAL MODELS

Diagrams. There are many types of diagrams. I will restrict my comments to two dimensional line drawings, such as the circuit diagram shown in Figure 4. The similarity between maps and diagrams is obvious. One could call this a map of the electrical circuit. It shows the pathways electricity can follow. Here I want to say: the diagram is a *representational model* of the circuit. Again we have *one thing*, the diagram on paper, being used to represent *another thing*, an electrical circuit.

In this diagram, the spatial locations of the wires are not important. There need be no strong similarity between the relative positions of the wires in the diagram and in the physical circuit. What matters is only what is connected to what. So what is modeled are connections, not spatial locations. Connections are more abstract than locations. The locations of lines representing wires in the diagram should be organized so as to make it easy for the human eye and brain to perceive the connections. How things are actually wired is a matter of convenience or efficiency in the physical wiring process.

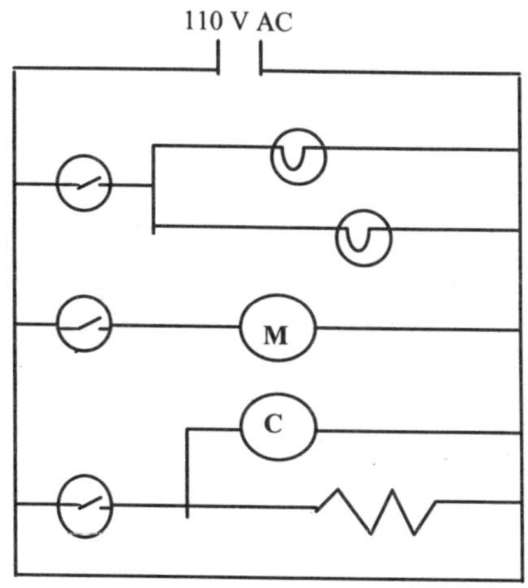

Figure 4.

Scale models. There are many sorts of scale models, from model houses to models of the solar system. A canonical example in twentieth century science is the three dimensional scale model Jim Watson built in the process of discovering the double helical structure of DNA molecules. This scale model was a representational model of DNA molecules. And it was representational in virtue of three-dimensional spatial and structural similarities between the scale model and real DNA molecules. The base pairs of DNA were claimed to be arranged in a helical structure similar to the pieces of tin and cardboard in Watson's scale model. Here again we have a physical object being used to represent other physical objects.

6. ABSTRACT MODELS

Consider a simple linear relationship between two variables, x and y, expressed by the equation

$$y = ax + b \qquad (1)$$

This equation is a linguistic object, but also a physical object, letters on paper. But the relationship described is some sort of abstract object, more abstract than any written equation, which could use different letters or be written in another form, such as:

$$y - ax - b = 0 \qquad (2)$$

For the purposes of this exposition, I will take the existence of such abstract objects as unproblematic. We might call them *pure mathematical models* to distinguish them from what are more commonly called mathematical models, which I would then call *applied* mathematical models. More about these shortly.

This relationship can also be presented graphically, as in Figure 5. What are we to say about this graph? It is, I would say, a physical counterpart of the abstract model of the same linear relationship, that is, a physical model of a linear relationship. Like all physical models, of course, it is imperfect, and thus at best only similar to the abstract model.

Beginning with the *pure* mathematical model, we can construct an *applied* mathematical model by replacing its mathematical elements with models of real objects and relations. For example, we can create a general model in which the variable y is distance from a fixed origin, x is time t from an arbitrary starting time, which can be zero, a is the velocity v of a moving point, and b the initial distance d_0 of the moving point from the origin.

We can then create a still more specific model, say of an auto moving away in a straight line from an intersection at velocity v having started at time zero a distance d_0 away. Here we are talking about *models* of an auto and of an intersection. In the model, the auto travels in a perfectly straight line at a perfectly constant velocity. Its distance from the idealized intersection at any time is then given by the equation

$$d(t) = vt + d_0 \qquad (3)$$

$$\mathbf{y = a\,x + b}$$

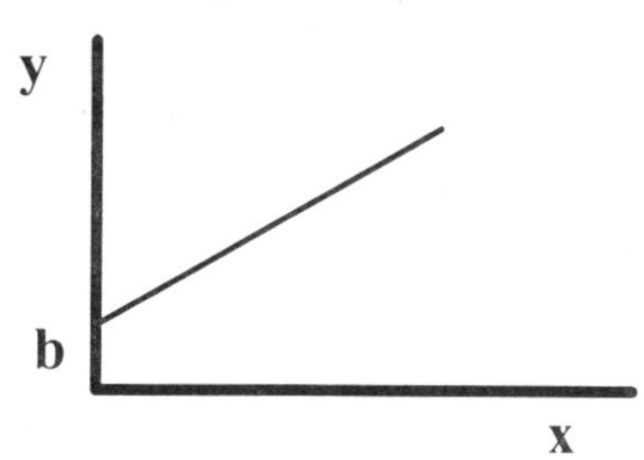

$$\mathbf{d(t) = v\,t + d_0.}$$

$$\mathbf{d \pm \delta_d d = (v \pm \delta_v v)\,(t \pm \delta_t t) + (d_0 \pm \delta_d d_0)}$$

Figure 5.

One may say that, in the model, this equation is *true*. What one cannot say is that the equation is true of the position of a real auto. No real auto can maintain a genuinely constant velocity in a perfectly straight line. The question, as always, is how similar the real situation is to the model of the situation.

Here one may complain that I am creating models beyond necessity. One must judge this complaint in light of the traditional way of handling the undeniable fact that no real objects exactly satisfy any simple mathematical relationships. The traditional way is to introduce *margins of error* into equation. Thus equation (3) becomes

$$d \pm \delta_d d = (v \pm \delta_v v)(t \pm \delta_t t) + (d_0 \pm \delta_d d_0) \quad (4)$$

and this equation may indeed be true of the real auto. In this way one can preserve the idea that representation in science is to be understood solely in terms of the truth of statements.

While technically correct, this is not necessarily the best way of interpreting the actual use of abstract models in the sciences. The margins of error rarely appear in the descriptions or calculations until one gets to the point of comparing theoretical predictions with actual measurements. This practice strongly supports interpreting the original equations, without explicit margins of error, as referring not to actual things but to abstract models of which they are true by definition. When it comes time to compare the abstract model with reality, the deltas may then be understood as specifying the degree of similarity (either expected or actual) between the abstract model and the real system.

On this view, mathematical modeling is a matter of constructing an idealized, abstract model which may then be compared for its degree of similarity with a real system. The tendency to identify the model with the equations used to define it are then seen as a holdover from an excessively positivistic view of science that attempted to avoid abstract entities and identified underlying structures with their observable manifestations, such as minds with behavior, probability with relative frequency, and theories with their linguistic formulations.

7. HYPOTHESES

There is a use for statements like (4) above, although it is a little different from what is usually thought. Equation (4), as I understand it, is a *hypothesis* about a particular system in the real world. As such it may be judged true or false based on evidence obtained by examining the real system. In more general terms, statements like (4) say that some particular real system is similar to the proposed model, to the degree specified by the various error terms. It is also possible to generalize statements like (4) to cover classes of systems. In this case, the error terms might be explicitly specified, or they may be place-holders for the specific error terms in the various individual hypotheses included in the generalization.

8. THEORETICAL MODELS

In a slight departure from my earlier (Giere, 1988) uses of the term *theoretical model*, I now want to reserve it for a special class of abstract models, those constructed with the use of what I shall call *theoretical principles*. Examples of such principles are Newton's Laws, the Schrödinger equation in quantum mechanics, the Principle of Relativity, the Principle of Natural Selection, and the laws of Mendelian Genetics.

There has long been a debate in the philosophy of science as to whether Newton's Laws, for example, should be understood, on the one hand, as definitions or conventions or, on the other hand, as empirical claims, either universal generalizations or claims of natural necessity. My view is that the initial question is defective because it presupposes that Newton's Laws are genuine statements which must be either true or false. The question is then one of the source of the truth or falsity. If one rejects this presupposition, one is free to answer, "both".

The issue may be posed as one of the status of Newton's term *body*. Does this term refer to empirical objects such as cannon balls and planets, or does it refer to abstract objects. I suspect Newton himself thought it referred to real physical objects. But we can make better sense of what he was doing, I claim, if we take him instead to be referring to abstract objects. These objects are then *defined* as things satisfying the three Laws of Motion plus the Law of Universal Gravitation. So, on this interpretation, Newton's Laws are definitions. One can then explore, as a mathematical exercise, the characteristics of various systems of bodies, such as two bodies moving in three dimensional space subject only to the three laws plus gravitation. Here I would say one is exploring features of a particular type of theoretical model. No empirical claims are being made, only claims about the model, which, if true, are true by definition.

However, if one identifies various real objects, such as the Earth and Moon, as bodies, then one has a model of this particular real system. In this case, one can formulate empirical claims as theoretical hypotheses about how the real system should behave if it is indeed similar to the model in the requisite respects. So the laws can be used to make empirical claims even though, taken by themselves, they imply no such claims.

Newton's gravitation law is usually rendered as "All bodies attract..." In this form it sounds like an empirical generalization. And perhaps that is how Newton himself conceived of it. But a slight change to "Bodies attract..." sounds like it could be part of a definition of bodies, on the order of "A circle is a plane figure..." That is how I suggest we think of the law of universal gravitation. There are independent reasons for preferring this interpretation.

For one, Newton's bodies are said to be mass points. No real object can be a mass at a point. Any real thing with (classical) mass must be somewhat extended. To apply Newtonian models to real objects one must treat their mass as being concentrated at their "center of mass," which ideally is a point. This supports the interpretation of Newton's laws as *defining* idealized abstract objects rather than as *describing* real objects.

Second, there can be no direct evidence for the gravitation law understood as an empirical generalization. Even for as few as three bodies subject only to their mutual gravitation, the equations of motion admit of no exact solution. One must make simplifying assumptions, which, in my view, is to create simplified models known to be less than perfectly similar to a real three-body system. What we do have is a great number of one and two-body models that have been shown to be very similar to real systems. Textbooks in mechanics are devoted to developing a variety of such models and exploring their mathematical properties.

9. MATHEMATICAL MODELING

The practice of mathematical modeling in various areas illuminates the contrast between theoretical models and merely abstract models. Much mathematical modeling proceeds in the absence of general principles to be used in constructing models. Rather, one has a number of different *mathematical techniques* useful for constructing models, such as differential equations and systems of linear equations. These are deployed as the situation requires.

For example, in modeling the growth of organic populations, as in ecology, two sorts of models one may employ are exponential models and logistic models. The former go to infinity with time; the latter level off to a finite limit, as shown in Figure 6. The latter better represent most real populations over extended periods of time. Here there are no over-arching theoretical principles at work.

If, however, one wishes to model changes in gene frequency in a population, then one employs principles of genetics, such as Mendel's laws, and not merely mathematical techniques, in the construction of well fitting models. One such model is characterized by the well-know Hardy-Weinberg Law.

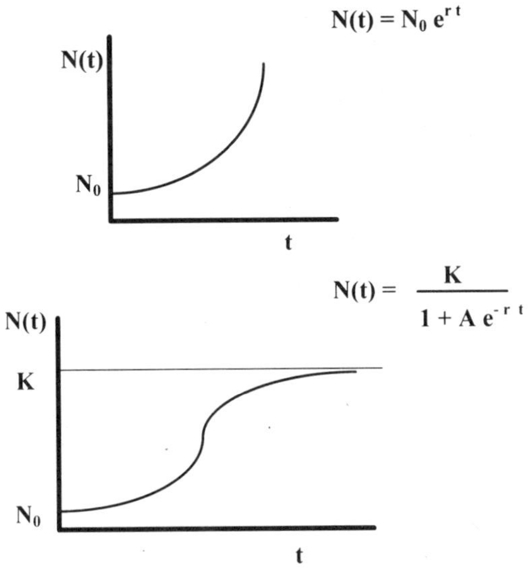

Figure 6.

Here, by the way, it is obvious that one has models and not literal descriptions. The mathematics of population growth and change typically employs continuous variables. But populations consist of discrete individuals, so their growth cannot literally be continuous. Nevertheless, for even just moderately large populations, continuous models may fit very well. In population genetics one finds people saying that they are assuming an *infinite* population. Here it is abundantly clear that they are talking about models and not about any real populations.

10. MODELS AND THEORIES

In the introduction and several papers in a forthcoming collection of papers entitled *Models as Mediators*, Mary Morgan and Margaret Morrison (1999) argue that models should be though of as somewhat autonomous agents operating the region between data and theories. Much of what they say about the *uses* of models in science strikes me as quite illuminating. But they say very little about what models *are* and even less about what theories might be. They give the impression of presuming the traditional view of theories as sets of statements. They also seem to take for granted that models are typically quite inexact and oversimplified in known ways while theories are much less so. They even suggest that as models are refined, they may *become* theories. This seems to imply that models are also linguistic entities. How could something not a statement become a statement?

I think most of these latter ideas are mistaken. On my interpretation, the model/theory distinction is mainly a reflection of the extent to which a branch of inquiry is guided by broad general principles. Where there are such principles, as in many areas of physics and biology, the models employed often, though not always, embody these principles. Where such principles are lacking, the models employed derive principally from various mathematical techniques. In both cases, however, reasoning about the world is primarily reasoning with models. It is models almost all the way up.

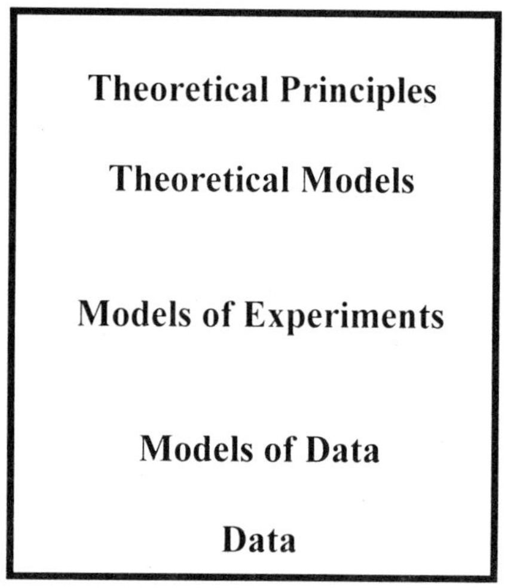

Figure 7.

11. MODELS AND DATA

Two years after publishing his paper on the meaning of models in empirical science, Suppes published an equally influential paper titled: "Models of data" (1962). A central message of this paper was that higher level models are not compared directly with data, but with models of data which are lower down in a hierarchy of models. A similar point has been made in recent years by James Woodward (1989) who insists that is not data but phenomena that theories explain and that are used to test theories. And phenomena are first constructed from the data. Statistical techniques, for example, are among the primary means of constructing models of data from data.

On this view, when testing the fit of a model with the world, one does not compare that model with data but with *another model*, a model of the data. Thus one reasons from a high level model not to predictions about data, but to predictions about a model of possible data. The actual data are processed in various ways so as to fit into a model of the data. It is this latter model, and not the data itself, that is used to judge the similarity between the higher level model and the world. In between, as Suppes already insisted, must be a *model of the experiment*. A version of Suppes' hierarchy of models is shown in Figure 7. It is models almost all the way down.

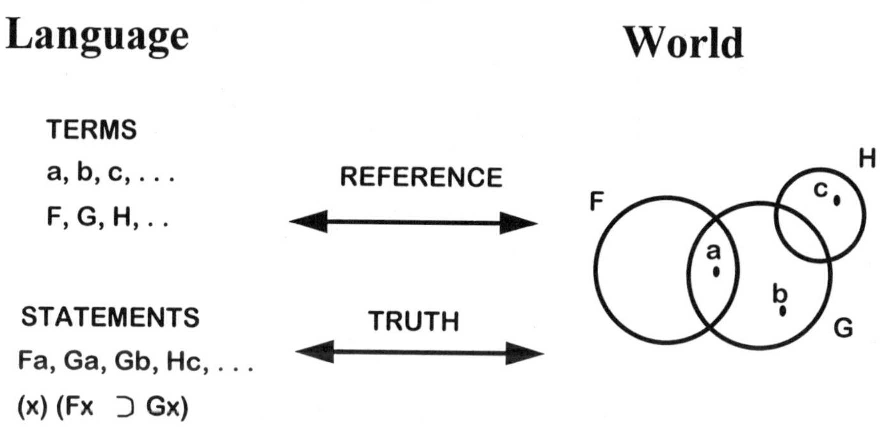

Figure 8.

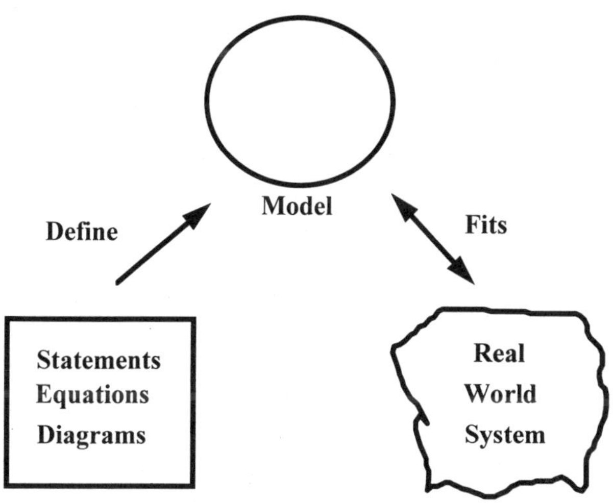

Figure 9.

12. MODELS AND ANALOGIES

Finally, one can also provide an account of the role of analogy in science based on representational models rather than on instantial models. Scientists have at their disposal an inventory of various known phenomena and the sorts of models that fit these phenomena. When faced with a new phenomenon, scientists may look for known phenomena that are in various ways similar to, which is to say, analogous with, the new phenomenon. Once found, the sorts of models that successfully accounted for the known phenomena can be adapted to the new phenomenon. In the process, features of the old models may suggest unknown features of the new phenomenon. So reasoning by analogy can be a fruitful means to new discoveries. I would say, however, that such analogies are suggestive only, and provide very little reason actually to believe that the suggested features will in fact be found in the new phenomenon. That, I think, always requires independent confirmation.

13. CONCLUSION

By way of conclusion, let me present one final contrast between the instantial view of models and the representational view. On the instantial view of models, there are direct relationships of reference and truth between linguistic expressions and objects, as pictured in Figure 8. This conception works fine for well-defined abstract objects. It does not work so well for physical objects, whose natures and relationships are not so well-defined.

On a representational conception of models, language connects not directly with the world, but rather with a model, whose characteristics may be precisely defined. The connection with the world is then by way of similarity between a model and designated parts of the world, as shown in Figure 9.

In sum, scientific reasoning is to a large extent model-based reasoning. It is models almost all the way up and models almost all the way down.

REFERENCES

Borges, J.L., 1954, *Historia Universal de la Infamia*, Emece, Buenos Aires.

Giere, R. N., 1988, *Explaining Science: A Cognitive Approach*, University of Chicago Press, Chicago.

Hodges, W., 1993, *Model Theory*, Cambridge University Press, Cambridge.

Morgan, M., and Morrison, M., 1999, *Models as Mediators*, Cambridge University Press, Cambridge, forthcoming.

Suppes, P., 1960, A comparison of the meaning and uses of models in mathematics and the empirical sciences, in: *Studies in the Methodology and Foundations of Science*, P. Suppes, ed., Reidel, Dordrecht, 1969.

Suppes, P.,1962, Models of data, in: *Logic, Methodology and the Philosophy of Science: Proceedings of the 1960 International Congress*, E. Nagel, P. Suppes and A. Tarski, eds, Stanford University Press, Stanford, CA, pp. 252-61.

Woodward, J., 1989, Data and phenomena, *Synthese* 79:393-472.

Models and Diagrams within the Cognitive Field

Kenneth J. Knoespel
Ivan Allen College, Georgia Institute of Technology, Atlanta, GA, USA, kenneth.knoespel@iac.gatech.edu

Abstract: My objective in this paper is to argue that research on modeling within science and technology should be cautious about approaching models in isolation but should regard them as part of a complex, generative field. While there certainly may be rhetorical purposes for isolating a particular model to bring into relief a particular "way of seeing" in the history of science, caution is essential if such a presentation of models doesn't become more a caricature of science. The plethora of books representing the history of science through a series of icons or great figures is an example. Others examples can be found in the misleading assimilation of Kuhn's idea of paradigms (Hoyningen-Huene, 1993; Nersessian, 1998). After making several brief comments about model-based reasoning, I will notice what might be learned from *reading* about the generative cognitive field in which models emerge. I will conclude my comments by making several comments on the relation between models and diagrams.

1. MODELS AND CONTEXT

Whether one considers the papers presented at a recent conference devoted to "Model-Based Reasoning in Scientific Discovery" at the University of Pavia or surveys recent literature on scientific modeling, one is struck by the multiple vantage points from which models in science are approached. In his book on magnets, the seventeenth-century polymath Anasthasius Kircher describes a thought experiment in which a statue, located on a stage, is placed on movable pedestal operated by magnet hidden from the audience. When a magnet is turned beneath the floor of the stage, the statue moves. To complicate his experiment, Kircher notices that multiple mirrors behind the statue generate a seemingly infinite number of reflected images from the vantage point of individuals seated in the audience. Kircher's splendidly baroque experiment reminds the reader that there are multiple ways for viewing phenomena but unless one comprehends the fundamental elements of cause and effect defined by the magnet the audience will be kept guessing (Kircher, 1643). The multiple images of the turning statue offer an interest-

Model-Based Reasoning in Scientific Discovery, edited by L. Magnani, N.J. Nersessian, and P. Thagard, Kluwer Academic/Plenum Publishers, New York, 1999

ing analogy to the multiple ways for approaching the use of models in science. In the broadest sense, a distinction may be made between research directed toward to experimental deployment of models (Gooding, 1999; Wimsatt and Schank, 1999), the increasing connection between modeling and visualization technologies within computer environments (Thagard, 1999), to work that would identify what might be called the logics of modeling and the pedagogical role of models. The range includes patterning within biological system (Pennock, 1999) to the administrative enforcement of models within scientific projects (Freedman, 1999)

The observation that there are models "all the way up and all the way down" (a comment made by Ronald Giere at the Pavia conference) deserves consideration especially since the large majority of studies have looked at modeling within the context of particular case studies from the history of science. The comment is suggestive and even a bit enigmatic. While there may be models "all the way down," it is not simply a matter of general pluralism. I think it is important to emphasize the continuum of modeling that takes place in cognition. Furthermore, it may be useful to recognize how reading provides an important analog for approaching such a continuum of modeling. Before I turn to reading in particular, it is useful to ask whether the discussions about cognitive modeling are not imitating discussions which have already taken place in the study of poetics?

Aristotle's analysis of genres in poetry can hardly be separated from his analysis of animals. The question we are asking must be posed not on the level of the individual work but as a question of the field. The interest in ordering the written forms of science really marks the development of a poetic of science. The desire to establish a hierarchy of such forms points as well to an idea of development and the assembly of more complex forms of communication.[1] For example, while just as we might speak of a poetic of science that would create a taxonomy of forms, so it might be possible to speak of a genealogy of scientific forms that presupposes an idea of evolving complexity and purpose. Indeed, here it is appropriate to underscore that the very effort to establish such orders within literature are regarded as part of *scientia* through the early modern period and continue to be embraced as such within the heritage of *Geisteswissenschaften* (Bambach, 1995; Gadamer, 1987). It is possible to see the interest in modeling in philosophy of science as an impulse to develop a typology of models that parallel literary typology.

The implications are important: for it suggests that models in science point to a desire to analyze scientific discourse. Do we have literary analysis

[1] Russian interest in the genres of scientific and technical communication as defined in a Euro-American setting provide an intriguing site for the location of such taxonomies. See Knoespel (1999).

turned inside out? It is hardly surprising that historical surveys of such forms provoke ideas of evolving order.[2] It is valuable to make a comparison between the taxonomic structures identified in literature and the impulse to create taxonomies of writing in science because in both cases the historical field provides examples that lead to the description of models. Beyond the taxonomies of the written forms of science, there have been multiple impulses to consider the topoi, or figures, that have been used in science. For example, efforts to identity the use of metaphor in science are part of an evolving impulse to formulate a tropology for science. The recognition that it is possible and even necessary to assume a "rhetoric" of science is another element of such a tropology (Gross, 1990; Bazerman, 1988). One of the most significant components of such work has been the recognition that the elements that comprise such a tropology may also emerge within the non-natural or created languages associated with science. Modeling in literature is displayed by a developed practice of collecting of topoi or figures which are assumed to contribute to the evolving constitution of literature.[3]

From the vantage point of modeling, efforts to identify taxonomies or tropologies within science as it has been practiced reminds us that scientific models are hardly isolated but continually in play with one another. Kuhn's work on Copernicus offers a good example (Kuhn, 1957). In contrast to broad overviews of the history of science that often give the impression that the change between Ptolemaic and Copernican views of the universe occurred quickly, Kuhn's work reminds one that the shift from one model to another took place innumerable times if viewed from the vantage point of the individuals who constituted to the scientific community. The modeling practice associated with taxonomies and tropology pertains fundamentally to our own acts of interpreting historical evidence rather than defining the actual use of models in the practice of science. From the vantage point of practice, it is important to make a distinction between status of a model as a vehicle that reveals something about how science *was* practiced in contrast to the model as an instrument that shows how science *is being* practiced.[4] The distinction is important because it shows that models may assume different roles as the mediate between interpreting the past and the future. It is pre-

[2] See the interesting history of mathematics included in the introductory essay of Wallis (1685). Wallis's history provides a rudimentary tropology for mathematics that bears comparison to Vico. Although Vico receives much credit for delineating ideas of evolving order within natural languages, he does not take up the idea of evolving orders within symbolic languages.

[3] A substantial example of topoi in literature is found in Curtius (1963).

[4] See Latour (1987). While Latour's discussion of modalities is in part indebted to Greimas, his discussion of inscription devices does not really take up the physical intervention of different forms of writing or of graphic representation within techno-scientific literature.

cisely for this reason that models can function both as rituals that represent the way something has been done or as projective forms that may lead to reorientation.[5] Just as the idea of modeling in science may function in historical analysis of science or in the practice of science, so the idea of modeling within the field of literature has a function that may be associated either with interpretation or with creation. But while such a comparison is certainly appropriate, reading itself provides an even better analog to the binary function of the model in science.

2. THE MODEL OF READING

I want to consider a model of reading which provides a synthetic structure for approaching modeling within science. The model is proposed by Cesare Segre and in itself combines several components from formalist linguistics (Segre, 1979, pp. 1-56). As we read we are continually in the process of generating and evaluating hypothesis. On a fundamental level each word we read is the consequence of a learned ability to recognize letters and form words. The recognition of individual words leads to the formation of sentences and the mental practice that we associate with meaning. The act of reading involves not simply the activation of meaning on the level of the sentence but an evolving formulation of hypothesis about what we are reading which are ever under revision as we continue through a text. If we wish, we think about reading as the movement of an arrow that discards possible meanings at the same time that it makes projections about what will be encountered next on the horizon of reading. The process which I have described is truly multilayered because it involves both the chronological assimilation of sentences and their reformulation into the ideational or narratival structure of the text. Even though the text has been read it may receive multiple interpretations in as much as comprehension of the plot or fabula may not at all be synonymous with an understanding or interpretation of the narrative. There is a sense in which the process of reading is never completed.

Although the process described by Segre may be augmented (Iser, 1974, 1978), even its simplest formulation provides a vehicle that may be used to approach modeling within science. As I noticed above, the act of reading inevitably functions as a moment of mediation on an evolving ho-

[5] Kurz and Martignon (1998) observed the place ritual within mathematical practice in a paper presented at the Pavia Conference. I have also benefited from discussions with Elke Kurz on her recent visit to Georgia Tech.

rizon. Such mediation is hardly passive but involves a process of formulation and reformulation.

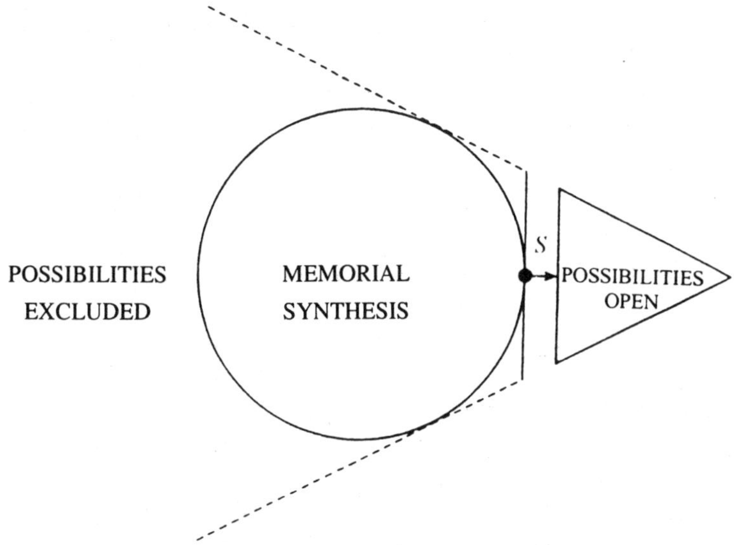

Figure 1. Segre model of reading.

Even though it may seem that such a process ceases when the text has been completed, it is evident that the text either may be read again or that it comes under interrogation as one reads other texts. A similar practice occurs in the practice of science for the construction of model is first grounded on one's previous experience and second, continually under examination given one's evolving experience. While there are broad similarities, differences also appear in the referential assumptions associated with the model of reading and the deployment of the model in scientific practice. Although the act of reading may have multiple referents (i.e., the landscape in a novel by Jane Austen or of a planet in a work of science fiction), the referent of the scientific model is assumed to be a symbolically mediated link to nature. Another difference appears in condensed forms of representation found in scientific modeling. Rather than being narrative structures alone, the scientific model is often formed in a metalinguistic setting.

Isaac Newton's "Trinity College Notebook" provides remarkable examples of the formulation of a working model or diagram (I will return to the distinction between the two of them below) (Newton, 1992). In addition to serving as an important document within the history of science, the notebook provides a rich collection of cognitive problems in the practice of science. From an array of problems that range from comets to quantity, I would like to consider Newton's effort to describe turbulence.

~~Of Reflection Refraction & Undulation~~

Of violent motion

Violent motion is ~~made~~ [continued] either by ye aire or by ~~motion~~ [a force] imprest. Or by the naturall gravity in ye body moved. Not by ye aire since ye aire ~~since ye air~~ crowds more uppon ye thing projected before, yn behind, & must therefore rather hinder ~~it but if ye aire helps it why doth ye thing at last fall~~ for you may observe in water yt a thing moved in it doth carry ye same water behind it along wth it as in a cone or at least ye water is moved from behind it wth but a small force as you may observe by ye motes in ye water supose (a) to be ye boddy moved: (b. d. e. f.) to be ye water moving behind (a) to give it place. (c) ye water behind (a) followinig it & going along wth it. The if ye water at (f) ran so violently against ye backside of (a) it would beate away ye water at (a) wth violence but yt water is moved very slowly from behind a.~~as you~~ if it be moved away: as you may perceive by

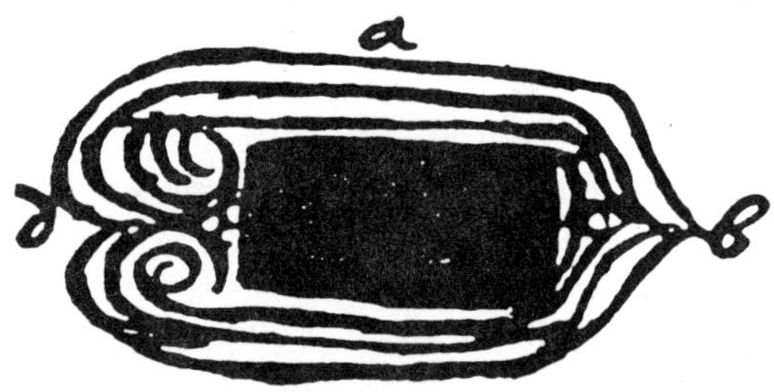

Figure 2.

ye motes in ye water. ~~Againe this wil~~ the like must happen in aire if you say no I answer must yn move (a) forwards in water. So if hot leade drop into water yt pte wch is behind ~~(?)~~ will be pointed. Ye fore pte round wch would be otherwise if ye aire pressed as much on it ~~before~~ hind as ~~behind~~fore. Thirdly how can ye aire continue ye motion of a globe on its axis [...] (pp. 367-68)

Figure 3.

Newton's text reminds us that correction and clarification is an aspect of formulation. The words that are crossed out are important as a record of Newton's thought process. Newton's drawings, however, serve as the focal point and orient the way that we approach the text. The drawings are remarkable because they show Newton providing a mathematical framework for the movement of water around a small obstruction. The emended text accompanies the two drawings as a commentary and should be distinguished both from the drawing of turbulence and the symbolic inscriptions that impose a schema of potential interpretation on the phenomena. Significantly, it is the symbolic mathematical language used to describe the phenomena that provides the link with the commentary. It is precisely Newton's effort to formulate the problem through symbols that is so interesting. The formulation of the problem, however, appears neither in the text or the drawings alone but in their interaction. Once the problem is figured out, however, the drawing may become the representation of the problem. While diagrams or models participate in the mediation of the narrative, they also stand out and serve as a focal points or registers provoking further investigation. In effect, drawings or diagrams inevitably function as memorial devices as well for they seize the strains of the possible just as they let others go. By simultaneously opening and excluding possibilities, they anticipate the course of a projected exploration. Every drawing becomes a potential site for discovery. The drawings organize future discourse.[6]

Our efforts to understand Newton's problem are also shaped by our knowledge of Newton and our understanding of the development of science. Each shapes a horizon of expectation. The fact that the problem appears in a notebook of person recognized as a icon of science shapes our expectation as we read. The more we know about Newton, the more detailed our expectations may be. Another set of expectations comes from the way we situate the problem within the history of science. Who else sought to represent turbulence mathematically in the seventeenth century? Although there still may be historians of science who believe they can recreate Newton's own expectation, it is more prudent to recognize that our efforts to understand the past are inextricably shaped by our own experience.

A conception of reading as a process of continuous hypothesis formulation and evaluation especially when it includes the creation of meaning through metalinguistic forms provides a useful ground for approaching modeling within science. Just as models are constructed in the practice of reading

[6] An analogous function is found in literature where the figurative becomes posed as an interpretive problem that needs to be explained. Drawings, however, do not speak entirely for themselves but invite written or oral elaboration. The drawings – and here we may also say diagrams – are analogous to figures of speech such as metaphor: both are intuitive and provoke elaboration on another level.

they are formed in the practice of science. While scientific models may be articulated through natural language, they are defined through drawings or diagrams. What is of utmost importance, however, is that the practice of modeling is not static but involves a continuous conversation with other possibilities. Above all, reading provides a ground for approaching the generative field in which modeling occurs in science.

3. DIAGRAMMATICS AND THE FIELD OF GRAPHIC REPRESENTATION

I have suggested that the discussion of models within science cannot take place without recognition of the prevalence of modeling within cognition. Furthermore, I have suggested that reading provides a vehicle for approaching the omnipresence of models in experience. In the following paragraphs, I want to suggest that rather than using the term model too broadly, it is useful to distinguish between different forms of graphic representation in science. In particular, it is useful to identify a progression within the field of graphic representation. Such a progression includes multiple symbolic forms. (The study of these symbolic forms and their relation might be referred to as *diagrammatics*).

In turning my attention to diagrammatics, I want to refer to my earlier comments on reading. However, at this moment, I also want to register the differences between reading and writing. As an extension of writing, modeling in science should be regarded as an active process which is fundamentally projective. As in the case of reading, modeling looks not only to the future but carries an implicit rejection of other possibilities. It is significant that the modeling process in science at an earlier stage may be accurately described as process of reading especially when the observer thought of himself as copying out patterns that had been written by God (Bono, 1995). In fact a distinction between the modeling process used in reading (interpretation of meaning) and the modeling process used in writing (the construction of meaning) is crucial in as much it indicates a particular relation to nature. It is possible to view western science as fundamentally shaped by hermeneutical strategies that would "read" natural structures and then suppress the very technological strategies that show the ways nature is also being written.[7] By

[7] Vulcan provides mythographic analog especially if one considers his contradictory identity in the Middle Ages. While being viewed as a discoverer and fabricator as well as his association with ingenium. This identity is contradicted by another that portrays him as someone who is ashamed when he ejaculates prematurely when he is in the presence of Venus. See Wetherbee (1972).

emphasizing the centrality of diagrammatics, we are calling attention not to an act of interpretation but to an act of intervention.

In his recent book, *Les enjeux du mobile: mathématique, physique, philosophie*, Gilles Châtelet surveys the ways diagrams have been used in the invention of geometric space.[8] The following are among the functions that may be associated with diagrams:

- *Diagrams constitute technologies that mediate between other technologies of writing*. The physicality of the diagram is also an extension of the literal technologies used in their construction. "J'ai donc ainsi en quelque sorte propulsé une main par la pensée et on serait tenté de dire que la pince ou le compas donnent un *point de vue* à la main, en associant à la saisie un angle sou lequel l'intervalle est "vu." (p. 221) "Nous pouvons apprécier tout ce qui sépare la règle du compas et de l'angle. La règle implique une mise bout à d'unités étalons, une manipulation d'un morceau de direction. La règle est un transfert qui a été gélé – un barre" (pp. 221-222).

["Hence I have thereby as it were propelled a hand by thought and one might be tempted to say that the pincers or compasses give a *point of view* to the hand, by associating an angle in which the interval is 'seen' with the grasp" (5.3). "We get the full measure of what distinguishes the ruler from the compasses and the angle. The ruler implies an end-to-end placement of standard measuring units, a manipulation of a piece of direction. The ruler is a transfer that has been frozen – a bar" (5.3)].

- *Diagrams create space for mathematical intuition.* "On pourrait parler à son propos d'une technique d'allusions" (p. 33). "Une philosophie du physico-mathématique ne saurait ignorer cette pratique symbolique en amont du formalisme, pratique de condensation et d'amplification d'intuition" (p. 34).

["We could describe this as a technique of allusions" and "A philosophy of the physico-mathematical cannot ignore this symbolic practice which is prior to formalism, this practice of condensation and amplification of the intuition" (Introduction, p. 17)].

- *Diagrams are not static but project virtuality onto the space which they seek to represent.* "Faraday bouleverse la physique en décidant de ne plus les voir comme des figures subsidiaires, mais comme des

[8] All quotes are from Gilles Châtelet (1993). Translations are from the forthcoming English translation to be published by Kluwer in 1999 under the title *Configuring Space*. Page reference are to preliminary page-proof in which each chapter is paginated separately.

incitations à provoquer l'espace, comme des diagrammes, comme des *expériences en pointillé* faisant allusion à des expériences 'réelles' qui rendent flagrantes des actions latentes. Les lignes de force ne sont ni 'réelles' ni 'artificielles'; elles ne mathématisent pas les forces; elles ne plaquent pas du géométrique sur du physique, elles, substitualité de reprendre ses droits" (pp. 241-42).

["Faraday turns physics on its head by deciding no longer to see them (lines of force) as subsidiary figures, but as *incitements to provoke space*, as diagrams, *as dotted-line* experiments alluding to 'real' experiments that manifest latent actions. Lines of force are neither 'real' or 'artificial'; they do not mathematize forces; they do not pin geometrical onto physical, rather they substitute an idea, but elastic and tough, network, which allows virtuality to reclaim its rights" (5.21)].

Diagrams represent a visual strategy for entailment. "L'horizon n'est ni une borne qui interdit ou sollicite un dépassement, ni une barre inscrite dans le ciel en pointillé. Une foi qu'il a été décidé, on emporte toujours son horizon avec soi. C'est le côté exaspérant de l'horizon: décapant comme le visible, tenace comme une odeur, compromettant comme un tact, il n'habille pas les chose d'apparences, mais imprégne tout ce que nous nous sommes résolus à saisir" (p. 92). "Saisir une dimension, c'est inventer un diagramme où s'équilibrent la pure dispersion de ce que Hegel appelait *aussereinandersein* (l'étant l'un à côté de l'autre simplement extérieur') et la pétulance de l'intensif: le déploiement de celui-ci nourrit la mobilisation et le redressement de celui-là, et réceproquement" (pp. 172-173).

["The horizon is neither a boundary marker that prohibits or solicits transgression, nor a barrier drawn in a dotted line across the sky. Once it has been decided, one always carries one's horizon away with one. This is the exasperating side of the horizon: corrosive like the visible, tenacious like a smell, compromising like touch, it does not dress things up with appearances, but impregnates everything that we are resolved to grasp" (2.22). "To grasp a dimension is to invent a diagram where the pure dispersion of what Hegel called *aussereinandersein* (the 'being outside one another') and the exuberance of the intensive balance one another out: the unfolding of the latter nourishes the mobilization and straightening of the former, and vice versa" (4.19)].

Diagrams are mediating vehicles which means that they cannot only be recovered but rediscovered. While they mediate between other technologies of writing, they may also mediate or reveal links between problems. "Rien n'est plus fécond, tous les mathématiciens le

> savent, que ces obscures analogies, ces troubles reflets d'une théorie à une autre, ces furtives caresses, ces brouilleries inexplicables; rien aussi ne donne plus de plaisir au chercheur. Un jour vient où l'illusion se dissipe; le pressentiment se change en certitude [...]. Heureusement pour les chercheurs, à mesure que les brouillards se dissipent sur un point, c'est pour se reformer sur un autre" André Weil as cited by Châtelet (p. 28).

["Nothing is more fertile, all mathematicians know, that these obscure analogies, these murky reflections of one theory in another, these furtive caresses, these inexplicable tiffs; also nothing gives as much pleasure to the researcher. A day comes when the illusion vanishes: presentiment turns into certainty [...]. Luckily for researches, as the fogs clear at one point, they form again at another" (André Weil as cited by Châtelet, Introduction, p. 11)].

> *Diagrams have a pedagogical force that could be integrated into mathematical education.* "Les lignes de force de Faraday réussissent à s'approprier ce que nous proposons d'appeler des *expériences diagrammatiques* et à donner un statut à cette langue du 'parler avec les mains', du 'sens physique' dont les plus grands électrogéomètres ont deviné la puissance d'entraînement et qui sait saisir l'instant de la solidification de la métaphore, au moment où l'opération lui dispute la souveraineté. Cette langue se nourrit de gestes bien plus que de formes mathématiques déjà disponibles et possède une puissance allusive propre, transversale aux chaîne déductives de la physique mathématique et capable d'illuminer des pans entiers de la géométrie. Il y a une vocation expérimentale intrinsèque du champ qui le soustrait aux prétentions de l'entendement (en ce sens il est irréductible à une saisie transcendantale) et aux appétits des causalités. Il n'y a pas, au sens propre, d''application' des lignes de champs, mais une étonnante pédagogie par allusions qui introduisent d'emblée à un espace élastique et à jouer de son aptitude à rententir" (p. 244).

["Faraday's lines of force succeed in appropriating what we are proposing to call *diagrammatic experiments* and in giving a status of this language of 'speaking with the hands', of the 'physical sense' whose driving power was sensed by the greatest electrogeometers and which can grasp the instant when the metaphor solidifies, when the operation contends with it for sovereignty. This language feeds on gestures much more than on already available mathematical forms and possesses a particular allusive power, which can traverse the deductive chains of mathematical physics and illuminate whole sections of geometry. The field has an intrinsic experimental mission that protects it from the pretensions of the understanding (in this sense it is irreducible to a transcendental grasp) and from the appetites of causalities. There is, in the literal sense, no 'application' of the lines of fields, but rather an astonishing pedagogy by allusions which straight away introduces an elastic space and place with its capacity to resonate" (5.23)].

Although I cannot do justice to Châtelet's discussion in the space that I have here, even the identification of these functions demonstrates what I have referred to above as the generative field of graphic representation of science. Their resonance may be demonstrated by noting that each has particular relevance to Newton's representation of the turbulence problem cited above. Rather than spelling each out in detail, I would make several further comments applicable both to Châtelet's discussion and the example from Newton. Diagrams should not be thought of only within a rigid and controlled terrain of mathematics, but should also be approached as ludic forms. In effect, diagrams may be approached as aspects of "play" in the sense that Piaget described the ways children explore space through gestures and drawings (Piaget, 1962, 1973). As the example of Paul Klee and Cy Twombley show, ludic aspects of diagrams are also found in art (Barthes, 1985a, 1985b). It is hardly misleading to view Newton's drawings as a form of playful and puzzling interaction with his experience.

In my earlier discussion of reading, I observed that graphic representations were complemented, at an initial stage, by a written text. The narrative setting for graphic representation may also be determined by speech. The question of how we move between or integrate different symbolic codes – the problem of translating between the narrative and diagrammatic matter of mathematical discourse – has been raised by many scholars (Rotman, 1993, 1995). I want to suggest that we learn something important about these negotiations by paying attention to speech. Although we readily acknowledge the place of oral tradition in the transmission of Homer or of speech in the shift from philology to linguistics, the importance of orality in the transmission of mathematics and science has remained oddly bracketed. Reminding ourselves of the orality of our inquiry may sound like a truism. I don't think it is. The integrative function that relates the diagram to the text is one that occurs through speech. The centrality of speech in performing these translations is exhibited everyday in classrooms and laboratories around the world. We must be prepared to conceive of orality not simply as a phenomenon that may help us understand a particular experiment but as a constant phenomenon that contributes substantially to science.[9] There is a sense in which the texts

[9] The function of orality in the practice of science is the subject of a current project entitled *Deaf Science*.

assembled by the history of science amount to cues or a score waiting to be performed or enacted.

Diagrams are controlled by modalities. These modalities must be approached as having aspects that are both written and oral. If we approach the broad range of graphic representation within science and technology as a generative field, we see that technology may be thought of as a progression of states which involve the continuous interaction of diagrams or models. Rather than viewing technology as jump from an idea to an artifact it would be constituted by a process of increasingly complex sign systems. We need to think of technology as a continuous set of interactions with signs that become increasingly reified. One way to enter this zone is to think of diagrams as devices that mark a process of becoming. Diagrams might also be thought of as providing access to a trace of thinking – somewhat analogous to efforts to describe the transformation of matter from one stable state to another.[10] From such a vantage point diagrams are interstitial structures. The architectural blue-print would be an example. There are multiple mediating drawings in the development of most structure and artifacts.

Diagrammatics pertains not simply to the limited use of drawings in scientific texts but identifies a cognitive field concerning with the use of figures in the mediation of meaning. From such a vantage point, figures are not neutral but reveal matrices through which meaning is negotiated on an individual and social level. What the Renaissance regarded as "thinking in pictures" is not simply allegorical but an indication of the cognitive field in which meaning emerges (Gombrich, 1989; Fletcher, 1964). Allegory is not only an interpretive strategy for dealing with a hierarchy of meanings (the control and moderation of meaning) but a practice that permits us to see the coordination of emergent models and emergent meaning. Allegory assumes commentary whether associated with a written or graphically portrayed narrative. In effect, allegory creates a setting for the emergence of meaning with symbolic languages. It is important, of course, that Renaissance painters were also mathematicians (Baxandall, 1972). Painting in effect also works as a symbolic language that provokes commentary. Nicholas of Cusa's use of a self-portrait by Roger van der Weyden to demonstrate the omnipresence of God is an example (Cassirer, 1972). Paintings, like diagrams, identify the cognitive space in which meaning is generated.

Somewhat more abstractly, we may say that technologies of representation simultaneously have the capacity to reveal and transform space. While oriented by language, our thought process is shaped in fundamental ways by

[10] The notion of a "trace" within natural languages has been shaped by Jacques Derrida in various works. The idea of a "trace" or within non-phonetic symbolic languages deserves further consideration. See Derrida (1974, 1981).

pattern recognition and pattern manipulation. Models and diagrams embody strategies of writing that function as vehicles for constituting idiosyncratic coherence and meaning. But beyond such individual meanings, they also work as vehicles for the constitution of shared meaning within groups and finally, as mechanisms for the exploration and social constitution of transhistorical meaning that give coherence to the practices of science. Perhaps it is appropriate for us to think of ourselves as members of the Kircher-like audience commenting not only on the multiple angles from which "models" may be viewed but on the broad field of representations from which they emerge.

REFERENCES

Bambach, C.R., 1995, *Heidegger, Dilthey, and the Crisis of Historicism*, Cornell University Press, Ithaca.

Barthes, R., [1982] 1985a, Cy Twombly: works on paper, in: *The Responsibility of Forms*, Univ. of California Press, Berkeley, pp. 157-176.

Barthes, R., [1982] 1985b, The wisdom of art, in: *The Responsibility of Forms*, Univ. of California Press, Berkeley, pp. 177-197.

Baxandall, M., 1972, *Painting and Experience in Fifteenth-Century Italy,* Oxford University Press, Oxford.

Bazerman, C., 1988, *Shaping Written Knowledge: The Genre and Activity of the Experimental Article in Science*, University of Wisconsin Press, Madison.

Bono, J.J., 1995, *The Word of God and the Languages of Man*, University of Wisconsin Press, Madison.

Cassirer, E., [1927] 1972, *The Individual and the Cosmos in Renaissance Philosophy*, University of Pennsylvania Press, Philadelphia, pp. 7-45.

Châtelet, G., 1993, *Les enjeux du mobile: mathématique, physique, philosophie*, Seuil, Paris. (English translation to be published by Kluwer in 1999 under the title *Configuring Space*).

Curtius, E.R., [1948] 1963, *European Literature and the Latin Middle Ages*, Harper Torchbooks, New York.

Derrida, J., [1967] 1974, *Of Grammatology*, Johns Hopkins University Press, Baltimore.

Derrida, J., [1972] 1981, *Positions*, University of Chicago Press, Chicago.

Fletcher, A., 1964, *Allegory: The Theory of a Symbolic Mode*, Cornell University Press, Ithaca.

Freedman, E.G., 1999, An executive-control-process model of scientific discovery, Special Issue "Analogy and mental modeling in scientific discovery", L. Magnani, N.J. Nersessian, and P. Thagard, eds., *Philosophica*, forthcoming.

Gadamer, H.G., 1987, Geisteswissenschaften, in: *Religion in Geschichte und Gegenwart and Theodor Bodammer, Philosophie der Geisteswissenschaften*, Alber, Freiburg.

Gombrich, E.H., [1960] 1989, *Art and Illusion: a Study in the Psychology of Pictorial Representation*, Princeton University Press, Princeton.

Gooding, G., 1999, A simulation of model-based reasoning about disparate phenomena, this book.

Gross, A.G., 1990, *Rhetoric of Science*, Harvard University Press, Cambridge.

Hoyningen-Huene, P., 1993, *Reconstructing Scientific Revolutions: Thomas s. Kuhn's Philosophy of Science*, University of Chicago, Chicago.

Iser, W., 1974, *The Implied Reader: Patterns of Communication in Prose Fiction from Bunyan to Beckett*, Johns Hopkins University Press, Baltimore.

Iser, W., 1978, *The Act of Reading: A Theory of Aesthetic Response*, Johns Hopkins University Press, Baltimore.

Kircher, A., 1643, *Magnes sive de arte magnetica opus tripartitum*, London.

Knoespel, K.J., 1999, Russian discourse communities in the information age, in: *Publications of the Russian Academy of Science*, forthcoming.

Kuhn, T.S., 1957, *The Copernican Revolution: Planetary Astronomy in the Development of Western Thought*, Harvard Univ. Press, Cambridge.

Kurtz, E., and Martignon, L., 1998, Weighing, then summing: the triumph and tumbling of a modeling practice in psychology, presented at the Conference Model-Based Reasoning in Scientific Discovery, Pavia, 1998.

Latour, B., 1987, *Science in Action*, Harvard Univ. Press, Cambridge.

Nersessian, N.J., ed., 1998, *Configurations*, special Issue devoted to Thomas S. Kuhn, 6/1, 1998.

Newton, I., 1992, *Certain Philosophical Questions: Newton's Trinity Notebook*, J.E. McGuire and M. Tamny, eds., Cambridge University Press, Cambridge.

Pennock, R.T., 1999, Can darwinian mechanisms make novel discoveries? Learning from discoveries made by evolving neural networks, Special Issue "Model-based reasoning in science: learning and discovery", L. Magnani, N.J. Nersessian, and P. Thagard, eds., *Foundations of Science*, forthcoming.

Piaget, J., [1951] 1962, *Play, Dreams and Imitation in Childhood*, W.W. Norton, New York.

Piaget, J., [1970] 1973, *Main Trends in Inter-Disciplinary Research*, Harper Torchbooks, New York.

Segre, C., 1979, *Structures and Time: Narration, Poetry, Models*, University of Chicago, Chicago.

Rotman, B., 1993, *Ad Infinitum: The Ghost in Turings's Machine*, Stanford University Press, Palo Alto.

Rotman, B., 1995, Thinking Dia-Grams: mathematics, writing, and virtual reality, in: *Mathematics, Science, and Postclassical Theory, The South Atlantic Quarterly* 94:2, Duke University Press, Durham, pp. 389-415.

Thagard, T., 1999, Scientific discovery and technological innovation: ulcers, dinosaurs extinction, and the programming language Java, this book.

Wallis, J., 1685, *A Tratise of Algebra*, 1st edition, London.

Wetherbee, P., 1972, *Platonism and Poetry in the Twelfth Century*, Princeton University Press, Princeton.

Wimsatt, W.C., and Schank, J.C., 1999, Discovering emergence in evolutionary dynamics, presented at the Conference Model-Based Reasoning in Scientific Discovery, Pavia, 1998.

Theories, Models, and Representations

Mauricio Suárez
Department of Philosophy, Bristol University, Bristol, UK, m.suarez@bristol.ac.uk

Abstract: I argue against an account of scientific representation suggested by the semantic, or structuralist, conception of scientific theories. Proponents of this conception often employ the term "model" to refer to bare "structures", which naturally leads them to attempt to characterize the relation between models and reality as a purely structural one. I argue instead that scientific models are typically "representations", in the pragmatist sense of the term: they are inherently intended for specific phenomena. Therefore in general scientific models are not (merely) structures. I then explore some consequences of this pragmatist account of representation, and argue that it sheds light upon the distinction between theories and models. I finish by briefly addressing some critical comments due to Bas Van Fraassen.

1. THE STRUCTURALIST CONCEPTION OF REPRESENTATION

Although many of the tenets of logical empiricism have today been abandoned the logical empiricists' form/content distinction still runs deep in much of contemporary philosophy of science. In the logical empiricist syntactic, or linguistic view of scientific theories, a theory is a set of formal axioms, and a model is a structure that physically interprets those axioms. Thus theories can be seen to provide the logical form of scientific knowledge while models are responsible for filling in the content. By contrast, in the semantic, or structuralist view of theories once advocated by Patrick Suppes and currently defended by Bas Van Fraassen, Ronald Giere, and Rig Hughes among others, there is no fundamental distinction between theories and models: a theory is in fact nothing but a family of structures. The family in question is picked out by what Giere has called the *theoretical definition*. In and by themselves these structures provide no indication of their intended applications: they fail to demarcate any real-world system as their object. Defenders of the semantic view often refer to the theoretical structures as "representations" (Giere, 1988, p. 6 and pp. 78-86). And indeed in both the syntactic and the semantic view all scientific claims (high-level or low-level, realistic or idealized, general or particular, and so forth) are always properly representational.

Model-Based Reasoning in Scientific Discovery, edited by L. Magnani, N.J. Nersessian, and P. Thagard, Kluwer Academic/Plenum Publishers, New York, 1999

The "representations" of the semantic view are analogous to what William James (1976) called *flat surfaces* in his discussion of interface theories of perception. They are theoretical constructions devoid of any empirical content. But how can such theoretical structures be said to "be about" anything in the physical world? And where, then, is the empirical content of the theory to be found? The semanticists are well aware of the issue: at the very least, they acknowledge, a further *theoretical hypothesis*, stating the domain of application, is needed to fully specify a scientific theory. Hence according to the semantic view a scientific theory has two parts: a family of formal "representations", or structures, and a statement that links those structures to different phenomena. What is the nature of this link, or relation? In keeping with the spirit of the semantic view, we may suggest this: a theory accurately represents a phenomenon if it is possible to lay out an isomorphism between the theoretical structures and the phenomenon, i.e. if the theory and the phenomenon share their *form*.[1]

More specifically, theories represent phenomena by *embedding* phenomenological structures within theoretical structures. A phenomenological structure $P = <A, P_i>$ may be embedded in a theoretical structure $T = <B, T_i>$ (with $i \geq j$), if the former is isomorphic to an "empirical substructure" $E = <C, T_j')>$ of the latter. The domain of the substructure is strictly a subset of the domain of the theoretical structure ($C \subseteq B$). The relations T_j' that appear in the empirical substructure are the restrictions of some of the relations T_i to the smaller domain of objects that appear in the empirical substructure ($\{T_i'\} \subseteq \{T_i|_C\}$).

An isomorphism is a one-to-one mapping, a function that maps every element in C uniquely onto some element in A, and every relation in $\{T_j\}$ uniquely onto some relation in $\{P_j\}$. Formally, an isomorphism between two structures $P = <A, P_j>$ and $E = <C, T_j>$ is a function $f: A \rightarrow C$, such that if $x_1,\ldots,x_n$ form an n-tuple of objects in A and if $y_1,..,y_n$ is an n-tuple of C then: there is some P_i which holds of $x_1,\ldots,x_n$ if and only if there is a T_j' which holds of $f(x_1,\ldots,x_n)$; and there is some T_j' which holds of $f(y_1),\ldots,f(y_n)$ if and only if there is a P_i which holds of $f^{-1}(y_1),\ldots,f^{-1}(y_n)$. More concisely, an isomorphism between $P = <A, P_j>$ and $E = <C, T_j>$ is a function $f: A \rightarrow C$ such that $P_i[x_1,\ldots,x_n]$ *iff* T_j' $[f(x_1),\ldots,f(x_n)]$, for any $x_1,\ldots,x_n \in A$; and $T_j'[y_1,\ldots,y_n]$ *iff* $P_i[f^{-1}(y_1),\ldots,f^{-1}(y_n)]$, for any $y_1,\ldots, y_n \in C$.

It follows that the domain C ($B \supseteq C$) that the theory must put forward as candidate for representing the phenomena must be of the same cardinality as the domain of objects A in the phenomenological structure. The number of relations T_j' defined over this domain must be identical to the number of re-

[1] This is Van Fraassen's characterization, which I follow in this paper. Giere's is in terms of similarity, but the difference does not matter for the purposes of the present discussion.

lations P_j that appear in the phenomenological structure. And although the objects and their properties (represented by the n-place relations P_i on A) need not be preserved with the mapping, the properties of the relations that denote those properties must be so preserved.[2] The existence of an isomorphism between two structures is, so to speak, a statement that the *structures* have identical properties. In the context of the semantic view, isomorphism is *structural identity*.

2. CRITIQUE OF THE STRUCTURALIST CONCEPTION

Identity is a limiting case of resemblance, and it is important to emphasize that resemblance is neither sufficient nor necessary for representation. As William James noted, resemblance on its own, no matter how extensive, can never entail that a *flat surface* is a "representation" of an object. His memorable "Harvard Memorial Hall" example illustrates this vividly. A mental image that resembles Memorial Hall is not necessarily a representation of Memorial Hall, even if accurate to some degree. As he writes (James 1976, pp. 38-39):

> For instance, if you ask me what hall I mean by my image, and I can tell you nothing; or if being led by you I am uncertain whether the hall I see be what I had in mind or not; you would rightly deny that I had "meant" that particular hall at all, even though my mental image might to some degree have resembled it. The resemblance would count as coincidental merely.

In just the same way, a mathematical structure, even if isomorphic to a particular physical phenomenon, need not constitute a *representation* of that phenomenon. Mere formal resemblance, or identity, is neither sufficient nor necessary for representation.

[2] The relation T_j' put forward by the theory to match up the relation P_j in the phenomenological structure will not hold of the same objects because the phenomenological structure is not *part of* the theoretical structure. As a relation can always be defined extensionally by appeal to sets of n-tuples of objects in its domain, and as the domains of the phenomenological and the theoretical structure will typically be different, it follows that the relations themselves will be different. In other words, isomorphism between structures does not require that the *objects* in the structure be the same, nor that the *properties* of those objects be identical. However, if there is to be an isomorphism between the structures, the domains A and C must possess identical properties. For instance, an equivalence relation over a domain of cardinality x is mapped onto another equivalence relation over a domain of the same cardinality, etc.

It is instructive to consider here some examples from the history of art. The painter Francis Bacon was obsessed with Diego Velázquez's portrait of Pope Innocent X. Bacon imitated Velázquez's portrait in a series of his own paintings, some of which show remarkable similarity to Velázquez's original painting. There is indeed a sense in which these are all representations of the Pope; but it is not a very interesting sense. Suppose that the original Velázquez canvas has x number of flecks, while one of the Bacon's has y. Consider Pope Innocent X and the room that he was in when posing for Velázquez. Divide up the Pope and the room, at that instant, into x dots and then again into y dots. It is then possible to lay out two one-to-one mappings that correlate uniquely every dot to every fleck in each canvas. Neither will be exactly isomorphic, but they will both be good approximations (the Velázquez may be slightly more accurate).

Hence on the "flat surface" conception of representation, both paintings can be said to *represent*, and moreover to represent the very same thing. And this sounds absurd. For while Velázquez's painting was commissioned as a portrait of Pope Innocent X, Bacon's may at best count as representations of the Velázquez, and were never intended to portray the Pope in the first place.

Thus isomorphism is not sufficient for representation. Another example from the history of art serves to illustrate how it isn't necessary either. Picasso's *Guernica* represents in at least two different ways. It represents, first, the concrete pain of the murdered inhabitants of the Basque town of Guernica, bombed by Hitler's Condor legion in July 1937, during the Spanish Civil War. *Guernica* also represents the more abstract threat posed by the rising forces of Fascism in Europe to the progressivist ideals embodied in the Spanish Second Republic. (Picasso asked for the painting to be exhibited in the Spanish Republican Pavilion during the World Exhibition in Paris in 1938.) But the canvas cannot be placed into a one-to-one correspondence, not to mention an isomorphism, with those things that it represents. We can at best recognize a wild horse, the torso of a bull or perhaps a Minotaur, a crying mother holding up a baby, etc. Somebody entirely unaware of the history of *Guernica*, and of the events that preceded and motivated it, could never come to determine the representational use of this painting.

The analogy leads to a further thought, namely that a canvas need not be representational at all. Consider Mondrian's paintings, such as the *Diagonal Composition* on display at Chicago's Art Institute. Mondrian's work is often discussed in the context of the distinction between objectivist and non-objectivist art.[3] We have spent centuries trying to figure out who Leonardo's

[3] See for instance the essays by constructivist artists contained in Martin, Nicholson, and Gabo (1971), and in particular the paper by Mondrian (1971). Mondrian's original distinc-

Monna Lisa really was; perhaps she never existed, other than in Leonardo's imagination – but the question makes sense. It is a question that also makes sense about much of 20th Century art, cubism and surrealism included. But to insist that a Mondrian, a Kandinsky, or even a Rothko, *really* represents is to entirely miss the point of the art.[4] This type of abstract painting is capable of inducing esthetic, or emotive responses, but it does not do so by "representing" anything. Non-objectivist painting is concerned exclusively with pure abstract form.

3. THEORIES AND MODELS: A PROPOSAL

This final example from the history of art gives us a way of understanding the distinction between scientific theories and models, a way that interestingly alters the traditional order. It used to be assumed that while theories are propositional, models are merely heuristic.[5] But recently this distinction has come under heavy fire. As a matter of fact, that *there is* here any principled distinction to be drawn seems now dubious. Some types of *mediating* models are not merely heuristic but contain substantial local knowledge of the phenomenon that they *represent*.[6] The distinction between theories and models, therefore, seems possible only in use; and I would suggest that the following is a good guide: while a theory may function as a pure form, a model has its content "built in". There are no "flat surface" representations; and there is no need to postulate further external relations, or mappings, to fill in the content of a representation.[7] When a model is presented, its intended use is given by implicit or explicit reference to the object of its application. Thus, from a historical point of view, a model is a representation, as it is essentially intended for some phenomenon; its intended use is not an external relation that we can choose to add to the model, but an essential part of the model itself.

---

tion is between "figurative" and "non-figurative" art. I thank Joanna Huggett for bringing this collection to my attention.

[4] Rothko turns out to be a difficult case, for he himself liked to claim that his work was in some sense representational – it was intended to represent Rothko's own angst at the Holocaust. I would still want to include him in the list: it is always possible to suppose that there is a gap between the representational use of a work of art and the artist's own *post-facto* reconstruction of her intentions. See Baxandall (1985), especially pp. 41-42.

[5] See for instance Hesse (1966).

[6] See Suárez (1999a), and the other essays contained in the volume by Morgan and Morrison (1999).

[7] Hilary Putnam has argued a similar point in the Dewey lectures (1994, pp. 445-456); also in (1997, p. 173). I explore this theme further, including the connection with James's and Dewey's views, in a forthcoming paper (1996b).

By contrast a theory's intended uses are arguably not an essential part of it.[8] Theories are more general than models precisely in the sense that they are not so tightly linked to their intended applications: we don't typically construe the discovery of a new application of, say, quantum mechanics, as fundamentally changing the theory in any way. When a theory is presented we often are given a set of theoretical structures, with no further hypothesis that delimits their domain of application. Only some heuristic rules of thumb or rough guides of application will be given; but even those are constantly being revised, as new and surprising applications of the theory continue to be found. A theory is then used non-representationally if it is used *as if* it had no empirical content. By this I do not mean that the theory really has no content of its own and is constituted exclusively by pure form. There is no *pre-given* distinction between form and content: we construct this distinction for specific practical purposes, by abstracting certain formal features of a theory. My point is only that theories lend themselves to this practice of abstraction more readily than models.

An example: it is often useful to consider classical electromagnetism as characterized by the set of Maxwell's four formal equations (this is the content of the well-known slogan: "Maxwell's theory is Maxwell's equations"). But is it not possible in the same way to characterize Fritz and Heinz London's electromagnetic model of superconductivity as merely a set of formal equations. This model is not just a mathematical entity; the diamagnetic conception of superconductivity is also an important part, and hence the model cannot be fully understood without reference to the Meissner effect, for which it was intended.[9]

I am then suggesting that what makes a theory non-representational is the absence of intended applications, and the corresponding lack of empirical content. In the same way, Mondrian's *Diagonal Composition* is a non-representational painting because it is not intended of any real object. In William James's apt thought "it does not lead us towards its object" in the way a genuine representation must.

A different answer, suggested by the semantic or structuralist conception of scientific theories, would hinge on the degree of structural resemblance, or isomorphism. If any structure is *prima facie* a representation then both theories and models are representational, for both can be construed as structures. Of course a representation of a physical object is accurate only if it is possible, in principle, to map out the object onto the structure and preserve,

[8] For a different view, see Moulines (1991, pp. 279-293). The German structuralists tend to include the intended applications as part of the theory itself. But this, I think, and partly for the reasons that I give in the text, is a counterintuitive consequence of their programme.

[9] For the details of this episode see Suárez (1999a), and Cartwright, Shomar, and Suárez (1995).

at least in part, or to a good approximation, the formal features of the object's properties. The Velázquez, although not fully isomorphic, is pretty good in this respect: red dots are generally mapped onto red flecks, blue dots onto blue flecks, and so on. The Mondrian, on the other hand, is not a representation of Pope Innocent X because there is no mapping that will preserve to any reasonable degree the properties of the Pope.

This, however, seems to me a hopeless answer. Not only does it ignore the symbolic and metaphoric aspects of the representation, but most importantly, it presupposes both a *pre-given* distinction between form and content, and an inappropriate interface conception of scientific representation. In general the question "does *x* represent *y*" can never be fully answered, independently of the history of *x*'s intended uses, by considering all the logically possible hypotheses that may link the form of *x* to the form of *y*. For, as James's "Memorial Hall" example shows, mere structural agreement is no guide to representational power. It is besides the point that it is possible to rationally reconstruct the theory as a structure, and then to lay out a mapping from such structure to some phenomenon. For there is no doubt that it is always possible to do this; but then, this is not the issue.

The issue instead concerns the actual use of the theory in practice. For it is the addition of an intended use that turns a "flat structure" into a *representation*. Velázquez intended his painting to represent the Pope, as a person and as a symbol of the opulence and power of the Catholic Church. A model is typically representational because it is intended for a particular phenomenon or type of system. And yet, scientists often use theories very differently. If a theory operates *as if* it had no intended uses – if it operates as a pure form, devoid of empirical content –, then the theory is not working representationally.

4. REJOINDER

I have been arguing that the notion of scientific representation within the semantic conception of scientific theories is inadequate because it ignores the scientists' own representational intentions. Bas Van Fraassen (1995) agrees that an intentional notion is needed, but he sees room for it within the semantic conception. This unlikely combination may be possible, but not without some cost. In particular, I think, there is tension between an intentional notion of representation and Van Fraassen's *constructive empiricism*.

According to Van Fraassen a theory is empirically adequate if it represents all *observable* phenomena, whether actually observed or not. However, on the intentional account of representation a theory cannot "represent" a phenomenon that hasn't been observed. For a theory cannot be intended for a phenomenon that hasn't yet been established. Therefore the claim that a theory is empirically adequate loses its modal force: a theory will then be empirically adequate if it represents phenomena already observed, not if it represents all *observable* phenomena. By contrast, for a scientific realist – who I think cannot countenance an intentional notion of representation –, a theory is *true* if it accurately represents the whole world, including of course all phenomena, whether actually observed or not. Van Fraassen's constructive empiricist would then fail to be "sticking his neck out" as much as the realist: the claim that a theory is empirically adequate would not carry the same commitments regarding the theory's future predictive successes as the claim that it is true.

Moreover Van Fraassen continues to understand the notion of a model as a "flat structure", to which intended use must be added as an external relation. By contrast, I want to defend the more radical conception of models as historical entities, with their intended uses as "internal" components. The models themselves change, if and when their intended uses change. Van Fraassen (in personal correspondence) raises a question about this view: will this conception of models as historical entities not force a commitment to an ontology of intentional objects? For as he writes:

> There are only things which can be used in various ways, or not used as the case may be; and when we call something a representation, we make implicit reference to that use. But the entity denoted is just a structure, a thing, and to me it is either a use/mention confusion or a modal fallacy when more is asserted.

But the view of models as historical entities need not involve a commitment to an ontology of intentional *objects*. In proposing it, I am not indulging in revisionary metaphysics, but merely suggesting a more appropriate regimentation of our language. A scientific model is, on this view, not merely an object, or a structure. If it is to be a representation at all, a model must be thought of as a heterogeneous combination of a mathematical or physical object (a structure, or a structured object) *and* an intended use, or application. But the intended use is not an object, and the structure is not intentional. An example, which I owe to Todd Harris (1999), is a spiral staircase (i.e. a physical entity), which *could* be used to represent DNA. The very same staircase could also be taken, for instance, to represent a spring. It seems fair then to speak of two different representations, two distinct models. How do the DNA model and the spring model differ? Not in structure, which is the same in both cases – they differ only in intended use.

Therefore according to the intentional view that I have been exploring, and quite contrary to what happens in the semantic view, a model is more than a mere object or structure; a model is a *representation* of a particular physical phenomenon.

AKNOWLEDGEMENTS

Research for this paper was carried out while I was a postdoctoral research fellow at Northwestern University. I want to thank Francesca Bordogna, Arthur Fine and Thomas Ryckman for useful discussions on these topics. Thanks also to Nancy Cartwright and Adam Morton for comments on a previous draft.

REFERENCES

Baxandall, M., 1985, *Patterns of Intention*, Yale University Press, London, 1985.

Cartwright, N, Shomar, T., and Suárez, M., 1995, The tool-box of science, in: *Theories and Models in Scientific Processes*, W. Herfel, W. Krajewski, I. Niiniluoto, and R. Wójcicki, eds., Rodopi, pp. 137-149.

Giere, R., 1988, *Explaining Science: A Cognitive Approach*, University of Chicago Press, Chicago.

Harris, T., 1999, A hierarchy of models and electron microscopy, this book.

Hesse, M., 1966, *Models and Analogies in Science*, University of Notre Dame Press, Notre Dame, IN.

James, W., 1976, *Essays in Radical Empiricism*, Harvard University Press, Cambridge, MA.

Martin, J., Nicholson, B., and Gabo, N., eds., 1971, *Circle: International Survey of Constructive Art*, Praeger Plublishers, New York.

Mondrian, P., 1971, Plastic art and pure plastic art (figurative and non-figurative art), in: J. Martin, et al., pp. 41-56.

Morgan, M., and Morrison, M., 1999, *Models as Mediating Instruments*, Cambridge University Press, Cambridge, forthcoming.

Moulines, U.C., 1991, *Pluralidad y Recursión: Estudios Epistemológicos*, Alianza Universidad, Madrid.

Putnam, H., 1994, The Dewey lectures: sense, non-sense and the senses: an inquiry into the powers of the human mind, *Journal of Philosophy* 91:445-517.

Putnam, H., 1997, James's theory of truth, in: *Cambridge Companion to William James*, R.A. Putnam, ed., Cambridge University Press, Cambridge, pp. 166-185.

Suárez, M., 1999a, The role of models in the application of scientific theories: epistemological implications, in M. Morgan and M. Morrison, 1999, pp. 168-196.

Suárez, M., 1999b, Scientific representation: some pragmatist views, in: *Royal Institute of Philosophy conference on Pragmatism*, C. Hookway, ed., Cambridge University Press.

van Fraassen, B., 1995, The theory of tragedy and of science: does nature have narrative structure?, unpublished manuscript.

How Scientists Build Models InVivo Science as a Window on the Scientific Mind

Kevin Dunbar
Department of Psychology, McGill University, Montreal, Quebec, Canada,
dunbar@ego.psych.mcgill.ca

Abstract: How do scientists think, reason, and generate new models and theories? Over the past decade, I have been addressing these questions by investigating scientists in their own labs, reasoning about their research "live" and by conducting experiments on scientific thinking and model building. The labs are molecular biology and immunology laboratories in the U.S., Canada, and Italy. I have found that one place where much reasoning and new discoveries are made is at weekly lab meetings. We have performed extensive cognitive analyses of these meetings and have identified some of the key components of contemporary scientific thinking that are important in generating new models, modifying old models and solving difficult problems. In this paper I will outline four important activities that are important in model building: analogical reasoning, attention to unexpected findings, experimental design, and distributed reasoning.

1. INTRODUCTION: A NEW WAY OF INVESTIGATING SCIENCE. THE INVIVO APPROACH

What scientists do and how they think has been a surprisingly difficult issue to investigate. Researchers have used notebooks, interviews, diaries, historical reconstructions and colleagues' accounts to obtain a picture of the cognitive processes that underlie scientific thinking and model building. Using these approaches, researchers have built many important accounts of what happens in science. However, over the past decade we have been pursuing a different approach to investigating the ways that scientists think and reason. We have been investigating the thinking and reasoning that occurs "live" at laboratory meetings in molecular biology and immunology laboratories (Dunbar, 1995, 1997, 1999). What we have done is to videotape and audiotape scientists as they think and reason in their own labs. We then analyze, sentence by sentence, the types of thinking and reasoning that the scientists use when formulating theories, analyzing data, designing experiments, and building models. We have investigated leading laboratories in The United States, Canada, and Italy. Each laboratory we have followed

Model-Based Reasoning in Scientific Discovery, edited by L. Magnani,
N.J. Nersessian, and P. Thagard, Kluwer Academic/Plenum Publishers, New York, 1999

from three months to a year, taping the weekly laboratory meetings and supplementing the meetings with interviews and other documents such as grant proposals, drafts of papers, and one-on-one meetings. Using this approach we have been able to build detailed models of the cognitive underpinnings of the scientific mind. In particular, we have been investigating the roles of analogy, unexpected findings, and group reasoning in contemporary science.

Why use the lab meeting as a source of data on scientists? I found that the most representative cross-section of the ways scientists think and reason is the weekly laboratory meeting. It is in the weekly lab meeting that scientists present their data, reason about findings and propose new experiments and theories. Thus, by analyzing the reasoning that scientists use at their lab meetings it was possible to obtain a clear vista into how scientists really reason. Furthermore, much of what happens at meetings consists of spontaneous reasoning, and sometimes scientific discoveries and breakthroughs are made at these meetings. Using this method it is possible to directly monitor thinking and reasoning rather than uncovering reasoning through after-the-fact interviews, questionnaires or think aloud protocols. The scientists externalize much of their thinking through interactions with other scientists in the lab. Thus by recording laboratory meetings it is possible to gain access to "online" thinking and reasoning. This approach allows us to see science as it unfolds rather than the selective information that the scientists put in their notebooks or reconstruct when attempting to remember what happened in their lab. The lab meetings therefore provide a unique opportunity to see science "live." Borrowing terminology from biology I have called this the "InVivo" cognitive approach (Dunbar, 1993, 1995, 1997, 1999a). Another approach that we have used is the "InVitro" cognitive approach where we conduct experiments on scientific thinking in our own cognitive laboratory. By using a combination of InVivo and InVitro approaches it is possible gain a clearer picture of the cognitive processes involved in science.

2. A TYPICAL LAB MEETING

The laboratories that we have been investigating consist of a senior scientist, three or four post-doctoral fellows, five or six graduate students, and one or two technicians. Each week one of the members of the lab will present the results of their latest experiments. The presentations consist of a brief rationale, many data slides, and what they are going to do next. While this is the basic plan of a meeting, some tend to focus more on theory, others on method, others on data, and many on future experiments. Thus, the meetings provide a nice cross section of the things that scientists do and provide a window into the workings of the scientific mind. While the lab meeting does have a presenter, the format of the meetings is very informal. Much of the lab meeting consists of many scientists reasoning about all aspects of the enterprise, models are discussed, diagrams drawn, inductions made, deductions given, competing models discussed, experiments designed and dissected. In addition, the possibility of alternate models, methodological errors, and feasibility of various approaches is discussed. From the point of view of a cognitive scientist, this type of data is unrivalled in its richness. Here we can see all the basic reasoning processes at work and how different cognition really works in science. These meetings are supplemented by interviews both before and after the meetings that provide detailed background information.

Using this InVivo approach we have uncovered a number of basic features of the way scientists think, reason and build models. In the next sections I will provide an overview of four important aspects of building scientific models – Analogy, Unexpected findings, Distributed Reasoning, and the cognitive processes subserving experimental design. While each of these different aspects of model building can be discussed separately, it is important to note that scientists freely move between analogy, unexpected findings, distributed reasoning, and designing experiments. Thus, while I have four different sections on these aspects of scientific model building, I do not mean to imply that each process works alone, or moves along in a serial manner.

3. ANALOGY

Analogy has been regarded as an important cognitive component of scientific thinking for many decades. Furthermore analogical reasoning has been the focus of intense investigation over the last 15 years culminating in a number of detailed models of the cognitive processes involved in making an analogy (e.g., Forbus, Gentner, and Law, 1995; Holyoak and Thagard, 1995, 1997). A number of cognitive accounts of analogy have noted that many different types of analogy are involved in science (see Gentner and Jeziorski, 1993; Holyoak and Thagard, 1995). Accounts of analogy distinguish between two components of an analogy; the target and the source. The target is the concept or problem that the scientist is attempting to solve or explain. The source is another piece of knowledge that the scientist uses to understand the target, or explain the target to others. What the scientist does when he or she makes an analogy is to map features of the source onto features of the target. By mapping the features of the source onto the target new features of the target may be discovered, or the features of the target can be rearranged so that a new concept is invented, or the scientist can highlight a specific feature of the target for other people. I will illustrate this discussion of analogy with an analogy that Rutherford ostensibly used in his research. When Rutherford was attempting to understand the structure of the atom he made an analogy to the solar system. In this case, the target was the atom and the source was the solar system. Rutherford ostensibly mapped the idea that the planets revolve around the sun onto the atom and argued that the electrons revolve around the nucleus. Thus, a number of historians have argued that by drawing an analogy to the solar system, Rutherford was able to propose a new account of the structure of the atom. By mapping the feature of the planets revolving around the sun, Rutherford was able to align his data with those predicted by a solar analogy. According to this view, the analogy resulted in a major restructuring of his knowledge and a scientific discovery was made.

The Rutherford example and other examples of analogy in the history of science raise a number of important questions regarding the role of analogy in science. Do scientists use analogy at all in their day-to-day science? If they do, is it the distant analogies that people have talked about in the historical creativity literature? Do less distant analogies play any role in science, as the empirical psychological work would suggest (see Forbus, et al., 1995; Holyoak and Thagard, 1995)? Does analogy work alone, or does it work in conjunction with other mental operations? Finally, is analogy involved in scientific discoveries and model building in science?

We sought to investigate the role of analogy in science by coding the use of analogy in four laboratories. We coded each use of an analogy. Any instance where a scientist drew knowledge from one domain and used that knowledge to either fill in gaps in their knowledge of the issue they were investigating, or used their prior knowledge to help others understand an issue. What we have found is that scientists frequently use analogy when there is not a straightforward answer to their current problem. Their problem may be in formulating a theory, designing an experiment, interpreting data, or in going from theory to an experiment. We found that analogy is frequently used. There were 99 analogies made at 16 meetings. We coded the goals that the scientists had and found that the goals that the scientists had when they made the analogies could be grouped into four classes: formulate an hypothesis, design an experiment, fix an experiment, or explain a result. We found that the type of analogy that the scientists made were directly related to their goal. When the scientists were designing and fixing experiments, the analogies were made to very similar experiments or the same organisms. For example, the scientists might draw an analogy from one gene on the HIV virus to another gene on the virus. These types of analogies accounted for around 45% of the total number of analogies that the scientists made. However, when the goal was one of formulating hypotheses the scientists tended to make analogies to other organisms that have similar underlying structure, or might have similar underlying biological mechanisms. For example, from the Ebola virus to the HIV virus. These types of analogies were frequent, accounting for over 50% of the total number of analogies. When the goal is to explain a concept to other members of a lab or a more general audience, the scientists make analogies to a very distant domain. For example, one scientist attempted to explain the way a vacuole works by drawing an analogy to the pop song "Hotel California." These "distant analogies" were very rare accounting for 2% of the total number of analogies.

These results were very surprising; we did not expect to see so many analogies made to other organisms. Most research on analogy has shown that people tend to make analogies based on superficial features, rather than deep structural features. In the case of science, the scientists have both a knowledge of underlying biological mechanisms, and access to structural information such as the homology of a gene in one organism to a similar gene in another organism. This knowledge makes it possible for the scientists to go beyond superficial characteristics in making an analogy. Furthermore, despite the anecdotal accounts of distant analogies having a major role in science, our findings indicate that distant analogies are rarely used, and are primarily used to explain concepts to others, rather than as a source of novel hypotheses and experiments. In particular, we found that much analogy use revolved around accounting for unexpected findings. I will discuss this use of analogy in the next section.

Overall our research on analogy paints a very different picture of analogy use in science than what the anecdotal literature would suggest and is more consistent with the psychological literature (e.g., Holyoak and Thagard, 1995). Rather than new theories and hypotheses being generated by making analogies to very different domains, new theories are generated by making analogies to related domains. One final note on analogy is that we have found that scientists have little memory for the analogies that they use. When we go back and ask the scientists to remember how they generated a new concept or solved a problem, at a meeting that we recorded, they have little memory of how it occurred. Thus, biographies and anecdotes do not mention many of the analogies that went into making a scientific discovery.

4. THE UNEXPECTED: A COMMON OCCURRENCE!

One of the most frequently mentioned aspects of scientific discovery is that a finding was due to chance or was unexpected. The recent discoveries of Naked DNA and Buckey balls, not to mention penicillin, nylon, and gravity itself are among the many significant discoveries that have been attributed to the unexpected. Many scientists have claimed that their discovery was "fortuitous," "lucky," or due to "pure chance." However, when we analyze the strategies that scientists use in their research, we can see that they structure their research to take advantage of unexpected findings, and that they actually conduct experiments that lead to unexpected findings that they can then exploit. Thus rather than being the victims of chance, scientists are ready to use chance events. Thus we can modify Pasteur's famous phrase of "chance favors the prepared mind" to "the prepared mind favors chance."

To answer the question of how scientist deal with the unexpected, we have been conducting extensive analyses of every finding that scientists make and what happens to the finding. An unexpected finding occurs where the outcome of an experiment is different from that predicted by the scientist. The analysis of the 12 laboratory meetings yielded 28 research projects, with 165 experiments, and they reasoned about 417 results. We found that the first step that the scientists take is to classify their data. All findings – both expected and unexpected – were classified in one of two main ways. The first was to classify the finding without reference to the other conditions in the experiment. Here, the scientist says immediately what a particular result is by looking at that result alone. For example, a scientist might look at a newly obtained gel and say "This is P30." Clearly this way of classifying the data refers to knowledge that the scientists already have and can recognize relatively easily. The second way of classifying their data is by comparing one result to another result from a control condition and inferring what their result is by making this comparison. For example the scientists might say that "This band is larger than the control one so this is P30." Interestingly, only 15% of the findings were classified alone, whereas 85% of their findings were classified by comparing results from one condition with results from other conditions. This holds for both expected and unexpected findings. Thus, classification of data is not a simple process. Instead, it usually consists of comparisons between experimental conditions and the numerous other conditions that scientists use in their experiments.

Experimental results were coded as unexpected if the scientist claimed that the result was not what she or he expected to find. When we divided their results into expected and unexpected findings, we found that over half of their findings were unexpected (223 out of 417 results). Thus, rather than being a rare event, the unexpected finding was a regular occurrence that the scientists reasoned about. The large number of unexpected findings is important: it is not the case that scientists can take any unexpected finding and make a discovery merely by focusing on it. Rather, the scientists have to evaluate which findings are due to methodological error, faulty assumptions, chance events, or to new mechanisms.

Once a finding was classified, expected and unexpected results were treated in different ways. Expected results usually led to the next step in a sequence of experiments, whereas unexpected findings led to either replication, change in the protocol, or use of an entirely new protocol. Interestingly, scientists initially proposed a methodological explanation for their unexpected findings. For 223 unexpected findings 196 methodological and 27 theoretical explanations were offered. Thus, the first strategy that the scientists used was to blame the method. One reasoning strategy that they used to support their methodological explanations was to draw analogies to other experiments that have yielded similar results under similar conditions, usually with the same organism. For example, if one scientist obtained an unexpected finding, another scientist might draw an analogy to another experiment using the same organism in their lab and say "Oh. I also incubated the cells at 37 degrees and failed to get digestion of trypsin, but when I incubated the cells at zero degrees I got it, maybe you should do that". The types of analogies that the scientists used at this point in their research were what we term "local," that is, the analogies were to very similar experiments, usually with the same organism and within their own lab. Using local analogies is the first type of analogical reasoning that scientists use when they obtain unexpected findings and is an important part of dealing with such findings. Scientists rarely mention these types of analogies in their autobiographies as they appear to be uninteresting, yet these types of analogies are one of the first reasoning mechanisms that are used in dealing with unexpected findings.

The way that the scientists reason about their data changed when they obtained a *series* of unexpected findings. This usually occurred when the scientist continued to obtain unexpected findings despite having modified the method, attempted to replicate the finding, or when they obtained a whole series of unexpected findings in related conditions. It is at this point that a major shift in the reasoning occurs; the scientists begin to offer new more general models, hypotheses, or theoretical explanations. In fact, model building is most important and most common when a series of unexpected findings are made. Thus, for repeated unexpected findings 51 theoretical explanations were offered as compared with 33 methodological ones. These models usually covered anywhere from 2-5 unexpected findings. The types of reasoning that occur at this point were very different from the reasoning that scientists initially used when they obtained isolated unexpected findings. Three different types of reasoning processes were used to reason about a series of findings that were repeatedly unexpected. First, was to draw analogies to different types of mechanisms and models in other organisms rather than making analogies to the same organism. This also involves making analogies to research conducted outside their lab. The scientists switched from using local analogies to more distant analogies. For example, a scientist working on a novel type of bacterium might say that "*IF3* in *ecoli* works like this, maybe your gene is doing the same thing". Thus, the way that analogies are used dramatically changes now.

A second type of model building that occurs following a series of unexpected findings is that the scientists attempt to generalize over the series of findings. Often the scientists will search for common features of the unexpected finding and attempt to propose a general model that can explain these results. For example, in one immunology lab two different types of T-cells went to two different places that neither cell should have been able to enter. The scientists asked what was common to the two different types of cells and inferred that it was the cells had a particular property. Starting with this generalization, the scientists then attempted to build a causal chain of events that could provide a complete model. The causal models that the scientists build usually specify a starting state, a process that transforms the starting state to another state, what the components of this process are and what is the end state. Often this model building phase goes through a number of cycles where a number of small models are linked together to provide a larger model of the entire biological process that is being explained. This model building process makes reference to findings from the scientists' own lab, other people's labs, and general knowledge of the domain. Note that many reasoning processes are involved here: causal reasoning, generalization, analogy, deduction and visual reasoning are often all present. Inevitably, there are numerous gaps in the models that the scientists propose. Often the gaps are crucial. The scientists may think that some mystery protein is causing something to happen. This will lead to the proposal of new experiments to find the protein. New experiments will also be proposed to test the predictions of the model.

How do we understand the role of the unexpected in science? Again it is important to note that unexpected findings are common and that scientists expect the unexpected. Scientists use the unexpected in many different ways. First, as we will see in the next section, scientist build very complex experiments with control conditions that often capture unknown and unexpected processes. Second, scientists attend to both expected and unexpected findings. Initially, they gave methodological explanations for their findings, and support these explanations by making analogies to very similar results obtained in other experiments. Third, if they continue to see the same type of unexpected result despite changes in methodology, they form new models and theories using more distant analogies and group reasoning. Thus, it is only when the scientists pass a threshold in unexpected findings that they propose general models and discoveries begin to unfold.

5. MODEL BUILDING AND EXPERIMENTAL DESIGN

A crucial component of model building in science is the experiment itself. While some analyses have been conducted on the design of experiments over the past twenty years, relatively little is known on what really goes on when an experiment is designed (but see Friedland and Isakawi, 1985; Galison, 1987; Hacking, 1983; Kulkarni and Simon, 1988). Instead, many researchers have assumed that an experiment is virtually entirely predetermined by the scientists' theoretical stance, or that the apparatus that the scientists use is so permeated by the theoretical assumptions that the scientists hold, that the experiment is again merely an appendage a theory. This viewpoint leads to the conclusion that experimental design is of little interest. While these views of the experiment are certainly interesting, we wanted to go beyond them to see what scientists actually do when an experiment is designed. Lisa Baker and I have conducted a series of analyses, and yes, experiments to determine what the cognitive components of the experimental design process are (Baker and Dunbar, 1996, 1999). It turns out that while theory is usually the starting point for experiments, much of the decision making that takes place while designing an experiment takes place independent of the hypothesis being tested.

Our first investigations of the experimental design process consisted of a qualitative analysis of designing an experiment in an immunology laboratory (Baker and Dunbar, 1996). What we found was that experimental design consists of a number of basic processes. The first component is choosing an overall method for investigating the hypothesis or question that the scientist has. It is at this point that the relation between theory and experiment is strongest. After selecting an overall type of methodology, the scientist's concern is with getting the method to produce an interpretable result. While the subsequent steps can follow any order, I will describe them in a serial fashion. Following the choice of a particular methodology the scientists must fill in the details of the experiment. What they do is to unpack the design, asking questions about which components are needed and what their values should be. The scientists evaluate each component of the experiment locally while at the same time attempting to coordinate each local component of the experiment to see if it will really work. What we found was that four sets of criteria are used when they evaluate various components of their experiment: ensuring a robust internal structure to the experiment, optimizing the likelihood experiments will work, performing cost/benefits analyses on possible design components, and ensuring acceptance of results by the scientific community.

Our qualitative analyses of experimental design revealed that while the experiments that the biologists conduct are very complex, the scientists appeared to be working with a general model of experimental design that guided what they did. One key aspect of the experiments is the choice of conditions. Scientists design experiments that have both experimental conditions and control conditions. The experimental conditions usually consist of a manipulation of the variable that the scientist is interested in. For example if the scientists thinks that a particular protein is responsible for a cell having a particular function, the scientist might add this protein to cells and to see if the cells gain this function. However, the scientists also have numerous control conditions in their experiments. One type of control that they use is "known controls" (Baker and Dunbar, 1996, 1999).

"Known controls" consist of conditions that have previously been used and validated and are standardized. These known controls are important for comparisons. What these controls do is to ensure that the technique really works and can allow the scientist to conclude that they are conducting the experiment in an appropriate manner. In the complex experiments that the scientists conduct, many things can go wrong. By having numerous known controls the scientists can ensure that the various steps in the experiment really are working. Another type of control is a "baseline control" in which something is taken away from the experiment or not added to an experiment. This "baseline control" tests the hypothesis that doing something very similar to the experimental condition does not produce the result that the experimental condition produces. When we analyzed the types of experiments that scientists conducted we found that "known controls" were particularly important in controlling for error.

Following our initial analyses we decided to further probe the use of control conditions and the relationship of controls to error (Baker and Dunbar, 1999). We analyzed the design of experiments at four meetings in two Canadian immunology labs. We coded the goal that the scientist had (testing an hypothesis, or anticipating error) and we also coded the type of control conditions used ("baseline control" or "known control"). What we found was that most "baseline controls" were proposed when testing an hypothesis, and most "known controls" were used when ruling out potential errors in use of the techniques. This result is interesting as it shows that the use of controls, particularly "known controls" is one way that scientists attempt to determine whether methodological errors have occurred in their experiments. One important point to note is that these controls are usually put into the experiment before the experiment is conducted. The scientists are attempting to minimize error before they conduct an experiment. Thus the scientists assume that an experiment is an error prone process. We also saw this in the previous section on unexpected findings where scientists usually interpret the unexpected findings as being due to error. Clearly, the possibility of error underlies the models that the scientists use and their initial reaction to unusual data.

Having found that scientists spend much of their time both anticipating and dealing with error, we decided to go back into the cognitive laboratory and conduct our own experiments on the ways that scientists deal with error (Baker and Dunbar, 1999). In one experiment we asked immunology and molecular biology students to design experiments that would test an hypothesis regarding a particular gene function. Once the students had designed their experiments we told them that the professor who wanted the experiment done was afraid that there was an error in the experiment. Thus, initially we gave the students the goal of testing a hypothesis and then we switched the students to the goal of worrying about the possibility of error in their experiments. Would the science students, like the scientists, use "baseline controls" when testing an hypothesis and "known controls" when anticipating error? Yes. We found that both immunology and developmental biology students used "baseline controls" when testing hypotheses and "known controls" when anticipating error. These results indicate that using "known controls" is a common strategy that scientists use for anticipating error and interpreting results once they obtain data. Interestingly, non-science students did not use this "known controls" strategy.

Overall, our work on experimental design indicates that while experiments are usually designed in the context of theoretical questions, much of the experimental design process is constrained by methodological considerations. While it is also true that the methods and instruments that the scientists use are predicated on theory, these background assumptions are really background assumptions that are rarely considered. What is important here is that experimental design is a complex cognitive process in which the scientists must navigate through many obstacles to get a result. The process by which the scientists do this is appears to be a search in a problem space (see Klahr and Dunbar, 1988). One important side effect of the use of complex experiments with many controls is that these controls, particularly baseline controls, can often reveal hitherto unknown phenomena that are at the root of many discoveries attributed to unexpected findings. The use of multiple conditions thus serves a dual function – guarding against error and generating potentially important unexpected findings that can underlie later model building.

6. DISTRIBUTED REASONING: AN IMPORTANT SOURCE OF NEW MODELS IN SCIENCE

The image of science that we are all familiar with is one of the alone scientist toiling away under a naked light bulb for long hours. Suddenly, inspiration strikes and the scientist has made a discovery. This is the image of science that has motivated much research in cognitive science. Its key features are that the scientist works alone and that scientific discovery occurs in a flash of insight. How representative of contemporary science is this view? Our research and that of other cognitive scientists such as Paul Thagard (Thagard, 1997) is revealing that reasoning in science, particularly at the critical moments of hypothesis formation, experimental design, data interpretation, and discovery is by groups of scientists and not individual scientists. We call this type of reasoning distributed reasoning (Dama and Dunbar, 1996, 1999; Dunbar, 1995, 1997).

What is distributed reasoning? There are many differing uses of this term. We define distributed reasoning as reasoning in which more than one person is contributing to a reasoning episode. For example, when a scientist obtains an unexpected finding the scientist and other members of the laboratory may induce a new concept to explain the finding. The different members of the lab provide different elements to the induction of the concept. In other words, induction is distributed over individuals. We have analyzed the distributed reasoning that occurs in five laboratories and have found that over 50% of the reasoning that takes place at meetings is distributed. Clearly, distributed reasoning is an important component of science. Are there precipitating events for distributed reasoning to occur? What do scientists do during distributed reasoning? Does distributed reasoning do something that does not occur during reasoning alone? These are some of the questions that we are currently investigating.

While I have reported our general findings on distributed reasoning elsewhere (Dunbar, 1997), here I will focus on some recent analyses of distributed reasoning that Mike Dama and I have conducted. What we did was to analyze sentence by sentence, the way that distributed reasoning occurs in labs. We used a new statistical technique that allows us to see how social processes such as question asking influence the ways that scientists represent their knowledge. What we found is that the major event that occurs during social interactions is generalization. The scientists in the lab often take a finding and attempt to generalize to other findings, both within and outside the lab. This process is important, as scientists frequently force each other to consider other representations and explanations for the data being considered or theory being proposed. What we see happening is that large changes in representation occur at these moments.

One place where distributed reasoning is particularly important is when a series of unexpected findings occur. When a single unexpected finding occurs, little distributed reasoning is present. In this situation, social interactions are usually a straightforward suggestion about method that does not necessitate any major change in the scientists' representation of the issue. But when a series of unexpected findings occur there is a large amount of distributed reasoning. Thus, it is not one scientist shouting "Eureka," but a number of scientists building a new model by adding different components to the model. As noted above, an important feature of this process is that often the members of the lab propose alternate models and explanations for the unexpected findings as well as resulting in the use of more distant analogies. By looking at many different laboratories we also have found that when groups of scientists reason, the diversity of the group is very important (Dunbar, 1995). When all the scientists are from the same background it is difficult for them to generate multiple hypotheses, but when the scientists are from different backgrounds many different hypotheses can be generated. Clearly, these types of social-cognitive processes are at work in the peer review process underlying the publication of papers and evaluations of grant proposals.

Is distributed reasoning always beneficial? We believe that there are a number of important constraints on when and whether distributed reasoning will be successful. The first is the knowledge base that the group has. We have found that when the members of a group all have the same background, such as all having done graduate work on a particular organism, then the group performs no better than the individual. This can be seen in the analogies that the group makes. Groups from the same background will draw their analogies from that background. Groups with a varied background, but with common goals, will produce many different types of analogies and these analogies can be used to solve the problem that the lab is working on. Overall, our work on distributed reasoning indicates that it is an important component of contemporary science and is frequently an aspect of scientific discovery. Furthermore, our analysis reveals that distributed reasoning can help circumvent one of the main problems in human reasoning generating different representations and understandings of both theory and data.

7. IN VIVO SCIENCE: WHAT DOES IT REVEAL?

Overall, the investigation of the cognitive mechanisms involved during

"InVivo" scientific thinking and reasoning reveal a number of important mechanisms underlying model building in science. I have demonstrated that analogy, unexpected findings, experimental design, and distributed reasoning all play a crucial role in the building of scientific models. Thus, no single cognitive process is involved in model building. Furthermore, while scientists can subjectively point to key moments in the changes in their models, an analysis of the changes in models over an extensive period of time indicates that new models arise through a series of small changes produced by a variety of different cognitive mechanisms. The goal of our current research is to determine how these different aspects of scientific model building are integrated to produce these new models.

ACKNOWLEDGMENTS

This research was supported by grants from the Spencer Foundation, Office of Naval Research, NSERC, SSHRC, and FCAR. I would like to thank Laura-Ann Petitto for her comments on an earlier draft and Annie St. Laurent, Lisa Baker, and Mike Dama for their work on different aspects of this enormous project.

REFERENCES

Baker, L.M., and Dunbar, K., 1996, Constraints on the experimental design process in real-world science, in: *Proceedings of the Eighteenth Annual Conference of the Cognitive Science Society*, G.W. Cottrell, ed., Erlbaum, Mahwah, NJ, pp. 154-159.

Baker, L.M., and Dunbar, K., 1999, Experiment design heuristics: the use of known controls and baselines, manuscript submitted for publication

Dama, M., and Dunbar, K., 1996, Distributed reasoning. When social and cognitive worlds fuse, in: *Proceedings of the Eighteenth Annual Meeting of the Cognitive Science Society*, pp. 166-170.

Dama, M., and Dunbar, K., 1999, Distributed reasoning and representational change in real-world science laboratories, manuscript submitted for publication

Dunbar, K., 1999, Beyond the myth of the unexpected: are scientists the victims of chance?, manuscript submitted for publication

Dunbar, K., 1997, How scientists think: online creativity and conceptual change in science, in: *Conceptual Structures and Processes: Emergence, Discovery and Change*, T.B. Ward, S.M. Smith, and S. Vaid, eds., APA Press, Washington DC.

Dunbar, K., 1995, How scientists really reason: scientific reasoning in real-world laboratories, in: *The Nature of Insight*, R.J. Sternberg and J. E. Davidson, eds., MIT Press, Cambridge, MA, pp. 365-395.

Dunbar, K., 1993, In Vivo cognition: knowledge representation and change in real-world scientific laboratories, paper presented at the "Society for Research in Child Development", New Orleans, LA, March 26.

Forbus, K., Gentner, D., and Law, K., 1995, MAC/FAC: a model of similarity-based retrieval, *Cognitive Science* 19:141-205.

Friedland, P.E., and Iwasaki, Y., 1985, The concept and implementation of skeletal plans, *Journal of Automated Reasoning* 1:161-208.

Galison, P., 1987, *How Experiments End*, Chicago University Press, Chicago, IL.

Gentner, D., and Jeziorski, M., 1993, The shift from metaphor to analogy in western science, in: *Metaphor and Thought*, A. Ortony, ed., Cambridge University Press, Cambridge.

Hacking, I., 1983, *Representing and Intervening*, Cambridge University Press, Cambridge.

Holyoak, K.J., and Thagard, P., 1995, *Mental Leaps*, MIT Press, Cambridge, MA.

Holyoak, K.J., and Thagard, P., 1997, The analogical mind, *American Psychologist* 52:35-44.

Klahr, D., and Dunbar, K., 1988, Dual space search during scientific reasoning, *Cognitive Science* 12:1-48.

Kulkarni, D., and Simon, H.A., 1988, The processes of scientific discovery: the strategy of experimentation, *Cognitive Science* 12:139-175.

Thagard, P., 1997, Collaborative knowledge, *Noûs* 31:242-261.

Discovery Processes and Mechanisms

A Simulation of Model-Based Reasoning about Disparate Phenomena

David C. Gooding* and Tom R. Addis**

*Department of Psychology, University of Bath, UK, hssdcg@bath.ac.uk; **Department of Computer Science, Mercantile House, University of Portsmouth, Portsmouth, UK, addistr@sis.port.ac.uk

Abstract: This paper argues that an adequate theory of discovery must bring the formal approach of computer science and cognitive psychology closer to the empirical approach of history and the social studies of science. We argue that a computer simulation is analogous to a *theory* of scientific investigation in that it is comprised of a number of *models* of activities and entities. We describe a model of experimentation to illustrate an approach to discovery simulation that shows how formal and empirical approaches may be integrated.

1. INTRODUCTION

Scientific knowledge is produced by people, interacting sometimes with the natural world but always with each other. This paper describes an approach to modeling experimental science as a process to which both types of interaction are essential. We are interested in experimentation as a means of learning (gaining new information about the world) and as a way of establishing and rejecting hypotheses. Our approach has been to work from historical case studies of experimental scientists involved in the interpretation and development of novel, anomalous phenomena (Gooding, 1989, 1990a, 1990b). These studies provide information about the structure of investigative processes and the types of information exchanged between observers. The studies serve a dual role as sources for models which are implemented as part of a computer simulation of a network of interacting agents and to provide empirical benchmarks against which the behavior of the simulation can be validated.

The simulation is analogous to a *theory* of scientific investigation, in that it is comprised of a number of *models* of activities and entities such as:

a) *experiments* (experimental set-ups, procedures, physical and conceptual resources),

Model-Based Reasoning in Scientific Discovery, edited by L. Magnani, N.J. Nersessian, and P. Thagard, Kluwer Academic/Plenum Publishers, New York, 1999

b) *observation* (interpretation, the ability to discriminate between phenomena, representation of observation by images or numbers),
c) *beliefs* comprising degrees of confidence in a range of interpretative models or hypotheses which together make up a view of the world,
d) *investigators*, here modeled as *agents* which are defined by certain sets of beliefs that create dispositions to act (e.g., to make an experiment or to consult another agent),
e) *learning* (as the modification of the belief-profiles of agents in response to the outcomes of actions,
f) *agency* (defined pragmatically as a disposition to act on the basis of beliefs).

More precisely, the simulation is an approach to a theory of the interaction of experiments and their results with experimenters and the hypotheses they entertain.

An adequate theory of discovery must draw together the formal approach of computer science and cognitive psychology with the more empirical approach of history and the social studies of science. As one of us has argued elsewhere (Gooding, 1996a) neither approach can get very far without the other. The history of science contains many precedents for the synthesis of *a priori* and empirical approaches to a single domain. The example most relevant to this paper is the response to H.C. Oersted's discovery of electromagnetism in 1819. Some investigators such as A.M. Ampère applied *a priori* principles that were consistent with mathematical methods. This dictated their style of experimentation. Others, including Michael Faraday and Humphry Davy, took an experiential approach in which they looked for features in the phenomena from which they could generate process models of the interaction electric, magnetic and mechanical forces (Gooding, 1992a, 1996; Tweney, 1992). By 1826 two distinct approaches to electromagnetic phenomena had emerged, based on action at a distance and field concepts, respectively. Neither was complete, however. Subsequent developments showed that these two approaches contributed information and ideas that were eventually synthesized in the work of Maxwell, Lorentz and Einstein between 1860 and 1905 (Nersessian, 1984, 1985, 1992).

A major difficulty facing any attempt to produce a computationally tractable theory of discovery is the phenomenological complexity of discovery processes as presented in scientists' notebooks, letters and papers, and in the narratives of historians (Kulkarni and Simon, 1988; Gooding, 1996a, 1996b). Earlier approaches tend to emphasize formal models of inference and problem solving (Newell, 1977; Simon, 1992). While these have been successful within highly abstracted and simplified representations of some *solved* discovery problems this is a limitation (Gillies, 1992; Gorman, 1987; Hesse,

1992). They have also been criticized for their lack of realism (Gorman, 1992; Tweney, 1990). We have approached the needs of computational representation differently, developing an abstract model of the structure and process of experiment, analysing, modeling and simulating experiments by cognitive and social psychologists, and using the model of experimentation to guide our approach to more complex historical cases. Our exemplars include the 4-card test (Johnson-Laird and Wason, 1970, 1977) and the shadow-box experiments (Gruber, 1985, 1990). The simulation is therefore an interdisciplinary synthesis of a number of approaches to the investigation of scientific discovery. These include history of science, experimental and cognitive psychology (Gooding, 1990a, 1996a), knowledge elicitation, computational modeling, functional analysis (Addis, Gooding, and Townsend, 1993), logical (hypothetico-deductive) and Bayesian models of inference and games theory (Addis, 1985).

2. SIMULATING DISCOVERY

The process of discovery has several components. These include:

a) *innovation* (which may be driven by new phenomena or data, new experimental technologies, or new insights and concepts),
b) *exploration* of novel information (expansion of the range of phenomena or data in a new domain),
c) *validation* of the new phenomena or concepts (by the proving of the techniques of observation, measuring, derivation or proof – usually in response to criticism by other practitioners, and the replication of results) and finally,
d) *acceptance* (recognition of a novel phenomenon, quantity, concept or law as a natural fact).

Whereas most computational approaches concentrate on the search for patterns and relationships in numerical data that has already been through processes a) – c), we have concentrated instead on the first two. This is a matter of emphasis rather than exclusion. No aspect of discovery is conducted in isolation of the others. The familiar distinction between the contexts of discovery and justification obscures the inter-dependence of the construction and the validation of new results (Nickles, 1989a, 1989b). Even the earliest stages of the attempt to develop, say a new interpretative hypothesis are carried out with awareness of the need to articulate, communicate, explain and defend the new idea (Gooding, 1986, 1990b).

Simulation of a comprehensive theory of discovery should recognize and implement each of these components of the process. Similarly, a full treatment of the computer implementation of such a theory would deal with the following:

a) Identification of key features and processes to be included,
b) Derivation of model(s) of discrete processes, as sets of objects, relationships, constraints, etc.,
c) Justification of the assumptions and techniques used in deriving these models,
d) Description of the simulation as an implementation of the models as parts of a computable program,
e) Validation of the simulation, involving experiments to calibrate,
f) Since this is a short paper about a long-term project, we will focus on a) and b) which are most relevant to the theme of this volume.

3. MODELING DISCOVERY

It is helpful to think of discovery as a process in which something experienced or believed by one or a few persons is moved from the personal domain of subjective experience into the public domain of objective fact. Initially, different observers using different methods and with different preconceptions about the world will have different experiences. Moreover, the descriptive language of an emerging field is not unlike the state of a natural language prior to the standardization of spelling and grammar: there is considerable variety in the way that things can be described. The transition from the personal to the public domain requires clarification and standardization of the language of observation. For example, an hypothesis can neither be confirmed nor falsified if it is not expressed clearly. This clarification is achieved by operationalising descriptors in terms of the procedures that make it possible to observe and communicate them. In other words, formulation, clarification, communication and validation of hypotheses is inextricably linked to the activities that observers engage in. Actions require situations as well as actors (Suchman, 1987; Gooding, 1992b).

To develop a model of this process which recognizes the complexity of human agency we approach key features of observation and experimentation *via* the much simpler designs of controlled experiments. Since the interaction of human beings is integral to the discovery process, we have selected experiments by psychologists on human subjects.

3.1 Defining an experiment

An investigation typically involves considering a range of hypotheses, constructing experiments that are expected to bear directly on one or more of the hypotheses (to provide confirmatory or disconfirmatory evidence) or obtaining such information by consulting others (either directly or through publications). The model must therefore represent the dynamics of the interaction of hypotheses and new information resulting from experiments and communication.

Historical studies of the initial response to discoveries shows that usually several hypotheses are in play before a situation is reached in which experiments can be designed to discriminate between two contending hypotheses. We can emulate the generation of new hypotheses by manipulating measures of confidence in each of a range of hypotheses. We use *confidence*, a numerical measure of belief, as the primary means of differentiating hypotheses that are actively considered (or "in play") from those that are not being considered, either because they have not yet been invoked or because they are no longer regarded as plausible. We assume that all historically documented experiments (including thought-experiments) can be known by agents (subject to access restrictions, of which more later) and that the differences between those in view and those not in view is a question of assessed usefulness.

3.1.1 The Wason 4-card test

To determine what might be meant by "useful" we consider the well-known four card test devised by Wason and Johnson-Laird (1970, 1977). This explores the ability of subjects to design a very simple experiment in order to prove or disprove a rule about a specific situation. In the initial version the subject is presented with four cards showing a set consisting of two letters and two numbers (e.g., "A", "D", "4", "7"). Subjects know that every card, of which these are a subset, has a letter on one side and a number on the other side. They are given a rule about the four cards, such as: "*If a card has a vowel on one side, then it has an even number on the other side*." The task is to say which of the cards need to be turn over in order to find out whether the rule is true or false.

The most frequent answers in Wason's original experiments were "A and 4" or "only A". Both answers are incorrect with respect to a pure logic view which assumes that subjects will select the minimum set of observations needed to *falsify* the rule. The correct, or logical answer would be "A and 7". These results suggested that people have difficulty in dealing with tests that involve disproof rather than proof. Few people approach the problem by setting out the logical possibilities as a matrix that relates observations to the rule (see Table 1 below). When the task is reformulated into more concrete, meaningful terms, however, a larger proportion of subjects apply a falsificationist strategy. This suggests an alternative to Wason's initial conclusion that humans don't reason in a strict, logical way about the empirical consequences of a rule. Instead, the original design exposes the difficulty most people have reasoning about problems that are expressed abstractly, without reference to a context of *meaningful actions*.

The problem presented in Wason's 4-card test this is not to discover a rule. How is this relevant to discovery? The experiment creates a world constrained by a particular rule in order to find out how human subjects reason about deciding its truth. Natural scientists cannot define the way the world is. To discover constraints they must entertain a wider range of possible hypotheses in order to discover which one(s) may be true. We therefore extend the four-card experiment to incorporate the evaluation of a range of possible hypotheses. Consider a scenario in which subjects are shown a large set of cards spread on a table, with each card showing only one face. Subjects are told that each card has a numeral on one side and a character on the other and that for all cards there is a simple relationship between the number type (odd or even) and the character type (vowel or consonant) on each card. The task is to determine this relationship by turning over as few cards as possible.

The model of experimentation is based on a pragmatic notion of belief as something which governs the actions of the subject. To illustrate, the experiment can be tied to some sort of gamble. The subject starts with a certain amount of money (say £100). Each turn of a card has a cost (say £1) and only if the subject finally reports the answer (or its logical equivalent) correctly does he/she keep what remains of the money. Otherwise all the money is lost. We can ask "how *confident* is the subject of a particular view before they dare answer?" Moreover, each turn of a card is an *action* that must count towards increasing confidence. Our model treats making an experiment and consulting another scientist as *actions* in this sense (see section 3).

Another modification is to state the relationships between possible observations and a rule (or hypothesis) in probabilistic rather than logical terms. The specificity of an hypothesis is an important factor in experimental design. Given a result (e.g., a card with face A turned to reveal a 4), confidence in a very general hypothesis (e.g., Vowel OR Even) is affected less than confidence in a very specific one (e.g., Vowel implies Even). Table 1 shows the a priori probabilities of a result for each act of turning a card, given an hypothesis.

Hypothesis: Vowel implies Even	**Probability of each Result (displayed on the other side of card)**			
THE EXPERIMENT:	**VOWEL**	**EVEN**	**ODD**	**CONSONANT**
Turn over a Vowel		1.0		
Turn over an Even	0.5			0.5
Turn over an Odd				1.0
Turn over a Consonant		0.5	0.5	

Table 1. An example of a set of *a priori* probabilities of the outcome of an action, given the constraint "Vowel implies Even".

Given the known constraints (cards carry only a number and a character) and the limitation of only four logical connectives (implies, or, and, exclusive-or) we can identify 4*4*2 = 32 possible logical statements of the form exemplified by "even implies vowel". After removing the logical equivalencies, this number reduces to 10 logically distinct hypotheses that describe all the constraints. This minimum set can be laid out in the form of ten tables each having the same structure as the example in table 1. Analysis of the a priori probability of results for each action (turning a card) given each of the ten hypotheses shows that hypotheses can be ranked according to their generality.

We show elsewhere that the logical relationships fall into two distinct series, where each series forms three groups of decreasing generality (Addis and Gooding, 1999). The series and groups are independent of the particular set of statements since membership depends upon the relative form of the truth tables. We obtain two series of three distinct groups where group 1 "contains" the results of group 2 and so on. Further, the intersection of group 1 in each series is equivalent to the conjunction of group 3 in the same series.

Our interest in the four card experiment is twofold: first, the refinement of Wason's test illustrates the importance of context to the ability to make decisions about *actions*. Second, as in real science, development of the four-card test shows that both confirmatory and disconfirmatory experimental strategies are needed, depending on how much is understood about the context of action. This is therefore an appropriate first approximation to the scientific situation. It is only a first approximation, however, because these experiments are unlike science in at least three important respects: first, the 4-card problem can be solved by an individual without reference to other persons; second, making observations of letters or numbers is wholly unproblematic, and third, the nature of the task requires reasoning that is categorical and deductive. As we indicated earlier, scientific investigators have to invent (or learn) new categories. This calls for a different kind of reasoning, usually labeled abduction.

3.1.2 The Gruber shadow-box experiment

A better approximation is achieved by including communication about observations and hypotheses, while recognizing that communication is inherently variable and ambiguous. Historical and sociological studies provide many detailed examples (Gooding, 1985; Rudwick, 1985). We amplify our model of experimentation using Gruber's design for experiments that make the communication of perceptual experience central to the task of discovering something about the world (Gruber, 1985, 1990). Gruber's experiments ensure that each subject perceives a different image, created as a shadow projection (or profile) of an object enclosed in the shadow box. Subjects are told that there is only one object in the box. They can see only the image facing them and are instructed to construct objects that could produce the observed shadow. Variations in the design involve giving subjects no instructions about collaboration, instructing them to collaborate or preventing them from doing so. Collaboration involves communicating one's own perceptual experience verbally or with sketches.

The design of these experiments is important because it highlights the roles of *communication* in increasing one's information about the world and of *creative construction* as a response to new information. Gruber has argued that major scientific advances such as those achieved by Darwin, involve the synthesis of disparate points of view (Gruber, 1990). The example from electromagnetism illustrates that although each viewpoint or theoretical approach could be developed independently over a long period of time, a major advance is possible only when resources are developed that deal synthetically with all the evidence (see section 1). Gruber's experimental results largely confirm the expectation that collaborating subjects achieve a larger number of acceptable solutions than non-collaborating subjects. The design can also ensure that additional information, if sought, will confound initial constructions based on a single point of view. Finally, communicating subjects are often at pains to convey the content of their own perceptual experience, especially when this appears to conflict with the interpretation each subject has placed on his/her own experience.

This design is more realistic as a model of empirical science than the four-card test. All three of the features highlighted by the design are found in science. It often happens that an interpretation based on evidence available to one person or group is shown to be inadequate in the light of new evidence from research by another group. In Gruber's experiments, a subject that interprets a triangular shadow as being compatible with (say) the existence of a solid cone in the box, has this hypothesis ruled out by a second subject's report of a circular shadow. Table 2 illustrates this progression, in which the sharing of information at first defeats observer constructs based on one or two observations but eventually supports a model that is compatible with, and explains, the subjective perceptions of three observers.

The compatibility of observations and hypotheses is defined by reference to rules governing mental transformations between 2-D objects (the shadow images) and 3-D constructs (hypotheses about what is in the box). These rules are analogous to the logical rules used to derive the compatibility of experimental outcomes and hypotheses. (The probabilistic version of a truth table in Table 1 is only one of ten examples).

Observer	What Each Observer Sees	Possible objects: No communication	Possible objects: Information one way from	Possible objects: Information one way from:	Possible objects: Information: two-way between:
A	□ Filled triangle	1. flat triangular card, 2. opaque cone, 3. triangular bar	B to A # 2 only	B and C to A None of 1-8, New construct needed	A, B # 2 only
B	□ Filled circle	4. flat circular disc 5. opaque sphere 6. circular rod	A to B: # 2 only	From A and C to B None of 1-8, New construct needed	B, C None of 1-8
C	□ Cross	7. flat cross 8. flat sheets intersecting at 90°	C to B:	From A and B to C None of 1-8, New construct needed	C and either A or B: None of 1-7, but A and B's observations suggest modification of 8: vary the shape of intersecting sheets
A + B + C	Collective view: □□□	N/a	N/a		A, B, C: None of 1 – 8. *New construct combining Objects 1, 4 and 8.*

Table 2. How the range of interpretations of a perception is constrained by new information gained via consultation with other observers, generating the need for new constructs.

Of course the transformation rules are not as formal as the logical rules. This is one reason why it is appropriate to represent them probabilistically. But this difference between the Wason and the Gruber tasks pales into insignificance compared to the fact that, just as Wason found in the case of deductive (falsificationist) reasoning, humans vary considerably in their ability to make mental transformations based on 2-D and 3-D images. Nevertheless, as Gooding (1992, 1996), Tweney (1992) and others have shown, reasoning with and about images is crucial to the development of new scientific concepts. It is therefore worth pursuing the attempt to model such complexities of experimental reasoning.

The Gruber experiment is a more realistic model because there is more scope for ambiguity of description both of observations and of constructs and because this is resolved only through the acquisition of additional information through two way communication (see Table 2). Taken together, the design of the shadow box and the four card experiments suggest an approach to modeling experimentation. In particular, they highlight the effect of constraining observation to predefined outcomes whose relevance to a rule can be specified either logically (as in the Wason experiments) or conceptually (through the application of rules of spatial construction and transformation, as in the Gruber experiment). The first two stages of this process are set out in Table 3.

3.2 Agents, beliefs and actions

Agents

We model a scientist as an *agent* which holds a range of *beliefs* about the world. The state of the agent depends upon how each of the beliefs are modified by the experience of that individual. These beliefs govern the actions of the agent, determining:

- how he/she decides to consult another scientists or to carry out an experiment,
- which fellow scientist is to be consulted or which experiment is to be performed, and
- what interpretations to make about the outcome, whether observed directly or via the state of beliefs of another scientist.

Thus we define an agent in terms of:

- a set of beliefs,
- a mechanism for acting according to these beliefs, and

- a means of adjusting the set of beliefs in the light of the results of its actions.

Experimental Design	Observations are:	Descriptors are:	Scope For Interpretation and Construction	Communication: needed to solve task	Strictness of rules defining compatibility of observation and hypotheses
1. Wason (1977) 4-card (abstract task)	unambiguous	fixed	none, not needed	none, not needed	deductive logic, modeled as truth table
2. Modified Wason 4-card (concrete tasks)	unambiguous	fixed but meaningful to subjects	slight, culturally given	none, not needed	deductive logic, modeled as truth table
3. Gruber (1985) Shadow box – No communication	unambiguous	unambiguous	considerable	n/a	learned geometrical rules of spatial transformation, no consensual constraint
4. Gruber (1985) Shadow box with communication; Gruber version of Wason 4-card	unambiguous	made ambiguous during communication	considerable, necessary	essential	learned geometrical rules of spatial transformation, but with consensual constraints
5. Historical cases involving response to novel phenomena or anomalous data	ambiguous	ill-defined, ambiguous	considerable, necessary	essential	learned and invented rules of classification, evidential relevance, etc., with consensual constraints

Table 3. A progression from experiments involving predefined observational possibilities and logically defined relationships between observations and hypotheses (Wason, 1977) to less formally defined observations and hypotheses and the exchange of data and constructs about disparate observations (Gruber, 1985, 1990), to the situation found in exploratory, experimental science, where neither the observations nor the hypotheses are well-defined (Gooding, 1986, 1990, 1996b).

Beliefs

A *belief* is represented as the pairing of an hypothesis about the world and a number between zero and one that reflects the subjective probability that the hypothesis is "true". Where a set of hypotheses are mutually exclusive the sum of the subjective probabilities will be taken as 1.0 even though the set may in principle not be complete or exhaustive. Each agent is characterized at any time by a set of values or *confidence profile* which represents that agent's view of the world. The set represents only those hypotheses being considered as possible by the agent.

For our simulation of the response to the discovery of electromagnetism, at least six documented hypotheses need to be included (Gooding, 1990, chapters 1-2). Two of these eventually emerged as core hypotheses of the two types of electromagnetic theory. Our model therefore assumes all historically documented hypotheses are available to the agent at all times. This reflects the fact that, contrary to the suppositions of deductive models of scientific inference, hypotheses are never tested in isolation from other possible hypotheses. An important and realistic feature of the model is that an agent acts on the basis of the whole set of hypotheses. (A further virtue is that we do not need to incorporate a model of the process of hypothesis generation).

Additional properties of the agents in an experimental situations include:

- *ability to discriminate between observed results* that are significant in an experimental set-up. This recognizes that observers are sometimes unsure about which interpretation to place on an observation and that experimental methods and observation reports often conflate or confuse one phenomenon with another;
- *ability to make judgements about the likelihood of the occurrence of a result* given the range of *a priori* results for a given experimental set-up if a particular hypothesis were to be "True". This *a priori* judgement is a prediction that assumes a perfect experimental procedure and a known state of the world with no problems due to errors in the performance or interpretation of the experiment. This degree of perfection is normally achieved only in thought experiments (Gooding, 1992c);
- *flexibility* of the agent's change in perception of the world, i.e. the impact of new results on the agent's confidence in each of the hypotheses in play. Flexibility reflects the variable capacity of an agent to respond to new information obtained by observation. Since new views rarely if ever become accepted knowledge without considerable consultation, discussion and negotiation, we denote flexibility of response to information gained by consultation by a separate parameter labeled *receptivity*.
- *competence* to perform an experiment and observe the results. This is an important variable which represents that fact that scientists undergo a passage from being students and novice researchers, through being competent performers to becoming experts. With the increased technological

complexity and scale of research in the 20th Century, expertise is usually limited to a particular aspect of the investigation (Galison, 1987, 1997).

An agent's confidence in each hypothesis is modified in the light of the result of experiments and consultations. For experiments we use a modified version of Bayes' Rule to convert probabilities (of specified outcomes for each hypothesis and situation) into subjective probabilities (the modified belief profile). We do not use Bayes' Rule to calculate the accumulation of evidence for an hypothesis. Consider the equation from the point of view of an agent who performs experiments in order to determine what hypothesis (model) is most likely. One form of Bayes Rule is given by the equation:

$$P_{n-1}(H/R) = \frac{P_{n-1}(H).P_{n-1}(R_e/H)}{P_{n-1}(R_e)}$$

Here R_e is the result of an experiment e and H is a hypothesis. We take $P_{n-1}(H)$ to mean the confidence, as a degree of "belief" that an agent has in a hypothesis H *before* an experiment (denoted hereafer as "belief"). Then the top component of Bayes Rule describes the bias that a particular agent has in its expectation of a result in performing an experiment e for a *particular* hypothesis (H).

In the extended Wason card task above an *experiment* would be the turning of a chosen card (say, a vowel). A *result* will be what is observed on the other side of the card (say, an even number). The expected result $E_{n-1}(R_e/H)$ for any experiment will depend upon the perceived probability of each hypothesis and an a priori understanding of the probability of a result for the hypothesis as though it were the case. This is represented as follows:

$$E_{n-1}(R_e/H) = P_{n-1}(H).P_{n-1}(R_e/H)$$

The confidence profile is retrospective in that it reflects only past experience. Decisions about what to do next involve the application of expectations about the outcomes of future experiments or consultations. In order to represent this overall view of the world we need to calculate a unified value that represents these expectations. We determine an agent's view of the expected result for *all* hypotheses $E_{n-1}(R_e)$ as follows:

$$E_{n-1}(R_e) = \sum_H E_{n-1}(R_e/H)$$

We can now replace the *probability of a result* by this *expectation*, in order to reflect the agent's larger view of the world:

$$P_{n-1}(R_e) \rightarrow E_{n-1}(R_e)$$

This expectation is then substituted into Bayes' Theorem to give the *expected hypothesis given a result from a particular experiment e*:

$$E_{n-1}(H / R_e) = \frac{E_{n-1}(R_e / H)}{E_{n-1}(R_e)}$$

This generic equation represents a personalised view of the world for a specific agent.

4. SIMULATING EXPERIMENTAL SITUATIONS

We specify experimental situations in terms of those features that bear most directly on the probability of obtaining a particular outcome. These *occurrence probabilities* reflect the influence of variables such as the material resources available, the design of the apparatus and, of course the person(s) performing the experiment. These probabilities are objective with respect to the agents. Just as the rule stipulated by Wason and the object placed by Gruber in his shadow box define the way the world is, so the occurrence probabilities represent an independent, informed, historian's view of how things should turn out. Thus, the model incorporates a further, historically valid feature: *an event-result linkage* which specifies an experimental result for a given event.

A *result* is a time and sequence independent observation made with an experimental set-up (a set of cards, a shadow-box, a current-balance – plus the protocols needed to experiment with them); it can occur many times. An *event* is a situated, context dependent result (e.g., Faraday's fourth attempt of 5 September 1821, to elicit circular motion from a particular configuration of wires, currents and magnets). An event only occurs once. A particular performance of an experiment by a particular individual (or team) is a specific event which produces a specific result. This represents the historical fact that only a small subset of the many observations and measurements made by individuals or research teams emerge from the laboratory to acquire the status of real phenomena or natural facts (Latour and Woolgar, 1979).

Access to other investigators and access to experimental set-ups are the other main features of an experimental situation designed into the simulation. For each agent, the accessibility of each experimental set-up within the current historical epoch is specified. Similarly, the accessibility of other investigations and the possible directions of information flow are also set for every agent-agent pair. The logic of this arrangement will be apparent from consideration of the Gruber experiments, if not from a consideration of historical examples. The likelihood of access and communication are specified as probabilities which, like the occurrence probabilities, reflect an historian's judgement. However they could as readily represent probabilities based on relative frequencies with which, say, subjects in a shadow-box experiment actually do consult.

4.1 Revising beliefs

Our model of belief-revision is not Bayesian. We use Bayes Rule to convert the objective probabilities of getting a certain result (given an hypothesis and a set-up) into subjective probabilities (confidence in that hypothesis, given that outcome, if it occurs). It has been argued that a Bayesian model of inference can account for the process of theory choice in science, providing a rational basis for major, revolutionary changes in what scientists believe about nature (Salmon, 1990). Despite its virtues, however, we do not claim that a Bayesian model of inference is a true representation of how experimental scientists learn.

The usual application of Bayes Rule to accumulating evidence for a hypothesis is inappropriate. This is because Bayes Rule assumes a constant and unchanging world *and* that the order of the events is irrelevant. The extension of Bayes Rule that deals with accumulating evidence requires the further assumption that the evidence is independent (i.e., that there is no correlation between one result and any other). In particular, note that in the special case of $P(R_{en}/H) = 0$ for any event n then for any one hypothesis $P(H/..R_{en}..) = 0$ what ever occurs before or after this event. Change in the absolute state of the world is not possible in this form of the calculations.

This means that a hypothesis is eliminated forever once there is evidence it cannot be true. In this way, a Bayesian inductive model of inference generates a naive falsificationism. Although it would be logically correct in a stable universe, this does not reflect either the observed behavior of scientists (who tend to mix confirmation and falsification) or the relatively unstable world of real experiments (Galison, 1987; Gooding, et al., 1989). We retain the assumption of the independence of results, but a static world has no place in our model. We assume that changes in the world will be gradual relative to the number of events that can occur during a change. Thus any evidence that seems to suggest that $P(H/..R_{en}..) = 0$ will have some effect, but no *single* occurrence can eliminate the hypothesis altogether.

In a changing world of information-bearing experiments and consultations the changing probability of an hypothesis' being true can be calculated dynamically, using a window that takes in range of N events, during which changes in the probability of the hypotheses is taken to be relatively slight. At each new event the calculation for the expected occurrence of a hypothesis can be recalculated by taking the new ratio of N-1 expected results plus the certain new result and the total number N thus:

$$E_n(H) = \frac{(N-1).E_{n-1}(H) + x}{N}$$

where x is 0 if the hypothesis is considered false or x is 1 if it is considered true.

Hypotheses rarely achieve absolute certainty – or if they do, scientists re-label them as facts, laws or principles. To call something an *hypothesis*, H is to say that there is some support for H given evidence $R_{e.}$. This support is $E_{n\text{-}1}$ (H/R_e) and is a value which lies between 0 and 1 for each hypothesis H. As a new result R_e is obtained from an experiment e then the probability which represents the confidence an agent has concerning a particular hypothesis, can be modified to $E_n(H)$ by adapting the above equation to the following:

$$E_n(H) = \frac{(N-1).E_{n-1}(H) + E_{n-1}(H / R_e)}{N}$$

The size of the event window N can range from 1 to infinity. The larger N the smaller the effect any evidence has on the change in confidence. We can thus define the parameter *flexibility* which ranges from 0 to 1 for each agent such that:

flexibility = *1/N.*

Flexibility reflects an agent's responsiveness to evidence. In our model this represents the historical fact that crucial, falsifying experiments are exceedingly rare (Worrall, 1989). Only with hindsight, and in textbooks does it appear that scientists throw out an hypothesis on the basis of a single negative result.

Similar considerations apply to the influence of other agents. If an agent consults another (by any form of communication) about a particular topic then this is equivalent to having access to the consulted agent's complete confidence profile. This access has the effect of modifying the consulting agent's confidence in each hypothesis in the same way as if it had performed the same experiments as the consultee. Each hypothesis will be modified according to the following equation:

$$P_n(H) = \frac{(M-1).P_{n-1}(H) + P_{n-1}(H_{Consultee})}{M}$$

The larger M the smaller the effect any consultation has on the change in confidence. This defines an agent's *receptivity* to the opinions of other *agents,* analogous to *flexibility* (see section 4.1). This ranges from 0 to 1 for each agent as:

receptivity = *1/M*

and represents how much notice is taken of any Consultee by the agent.

4.2 Actions

According to our pragmatic theory of belief agent's beliefs are expressed through its actions. Agents either make an experiment or consult another agent. The decision procedures and criteria are described in detail in Addis and Gooding (1999). In the case of an experiment, to discriminate between hypotheses involves choosing an experiment such that (i) the hypotheses are clearly separated by a result, and (ii) the results that distinguish the hypotheses are likely. In keeping with the model devised for the Wason and Gruber experiments we represent experimental set-ups as tables of real numbers that indicate the *a priori* probability of a result, *supposing that* a hypothesis (or model) is "True". This set of *occurrence probabilities* defines each possible experiment. Note that if the list of occurrence probabilities is the same for all hypotheses then from the agent's point of view it is the same experiment, whatever the arrangement is in practice. The list sums to unity for each hypothesis showing that at least one of the results will occur if that model were the case. The table of occurrence probabilities for each experiment represents an informed, historical judgement about the likelihood of obtaining a given result in the situation specified.

An agent chooses an experiment according its bias, represented by the confidence profile. This is represented as a list of real numbers. The choice of an experiment is derived from how effectively an experiment as perceived by an agent can discriminate between hypotheses. This choice is exactly governed by the agent's confidence ($E_{n-1}(H)$). The following equation represents the *entropy* of an experiment, for a particular hypothesis, as perceived by a particular agent:

$$Entropy\,(Experiment\; e) = -\sum_R \sum_H E_{n-1}(H / R_e) * Log_2 E_{n-1}(H / R_e)$$

We multiply an agent's belief profile by the matrix representing the result probabilities of all possible experiments to generate a biased perception for each agent of each of the experiments. As in real life, each agent perceives the potential outcome of an experiment differently. The result of the multiplication is then used to modify the agent's *a priori* confidence value for each hypothesis. The degree of modification to the confidence profile depends upon the agent's flexibility.

The experiment with the *minimum* entropy value fulfils both the above conditions. This experiment is the *most likely* to be chosen by the agent in our simulation. ("Most likely" because actions are governed by a probability distribution based upon an agent's confidence profile). In the case of consultation what is communicated through consultation is an overall influence for all hypotheses *from* the transmitting agent (consultee) *to* the recipient (consulter). An agent selects a consultee from those agents whose *hypothesis entropy* (calculated over the product of the range of hypotheses over all those experiments available to each agent), when combined with an agent's own, will be a minimum. Learning through consultation is similar to that for experiment, except that there is no need for Baye's Rule. This is because values in the confidence profiles of consulters and consultees are already expressed in the same terms as subjective probabilities.

4.3 Experiments with the simulation

We have combined the Wason and Gruber designs to create a simulation of the extended 4-card experiments. This combines the properties of experiments numbered 2 and 4 in Table 3. We are using this to experiment with properties of the simulation, in particular, the effect of the initial configuration of belief profiles and the formation of groups of agents having various patterns of access to experiment. We have also created a detailed representation from historical sources of the early response to Oersted's discovery between 1820 and 1822, with particular reference to the activities of Davy, Faraday and Ampère. Points for the evaluation of the behaviors of the Gruber-Wason and the Faraday-Ampère simulations are summarized in Table 4.

5. CONCLUSION

The simulation described here is an approach to a theory of scientific discovery. While all of the models incorporated into it can be justified on historical, psychological or other grounds (such as simplicity or computational efficiency), some of the constituent models are designed only to reproduce the functional equivalence of processes which we can observe but which we do not yet have enough information about to model. At this stage in our investigation, therefore, validation of the simulation is a matter of evaluating its ability to reproduce known processes and historical scenarios, prior to developing it as an explanation of scientific change.

SCENARIOS

The following should correspond to an historical consensus, for each important event or episode in a case study:

1. **Confidence profiles** of each agent, given realistic, justifiable values for fixed parameters (such as flexibility, receptivity etc.)
2. **Selection of set-ups** by agents, given realistic access setting and agreement about representation of experimental setups (occurrence probability matrices)
3. **Objective view** of world
   (e.g., for Wason: the correct (falsifying) strategy, for Gruber: what's in the shadow box; for the historian: a later consensus (hindsight))

Departures from this can be evaluated counterfactually by simulation, but only after the simulation has been evaluated

AGENTS

1. **Acceptance and rejection of hypotheses:**
   Confidence in particular hypotheses at particular points in the sequence should resemble that of the historical actors represented.
   Hypotheses that fall out of play
   Introduction of new hypotheses
2. **Patterns of activity:**
   experimentation / consultation; experimental set-ups chosen, consultees selected.

GROUPS

1. Formation or persistence of groups should resemble those identified in the historical epochs represented
2. Consensual positions regarding hypotheses should emerge or persist in line with the historical epoch represented
3. Consensual positions regarding best experiments should emerge or persist in line with the historical epoch represented

Table 4. Criteria for the evaluation of the simulation.

ACKNOWLEDGEMENTS

This paper is based on work conducted with the support of research grants from the following UK research councils: MRC-ESRC-SERC joint initiative on HCI and Cognitive Science (SPG # 9107137, Knowledge Design from Natural Intelligence, 1991-93) and the Economic and Social Sciences Research Council (R000 #23 5286, Graphic Programming Methods for Modelling Cognitive and Social Processes of Science, 1994-96).

REFERENCES

Addis, T.R., 1985, *Designing Knowledge-Based Systems*, Kogan Page, London.

Addis, T.R., and Gooding, D.C., 1999, Interacting agents: simulating a community of scientists, in preparation.

Addis, T.R., Gooding, D.C., and Townsend, J., 1993, Knowledge acquisition with visual functional programming, in: *Knowledge Acquisition for Knowledge-Based Systems: 7th European Knowledge Acquisition Workshop (EKAW '93)*, N. Aussenac, G. Boy, B. Gaines, M. Linster, J.-G. Ganascia, and Y. Kodratoff, eds., Springer, Berlin.

Galison, P., 1987, *How Experiments End*, Chicago University Press, Chicago.

Galison, P., 1997, *Image and Logic: A Material Culture of Microphysics*, Chicago University Press, Chicago.

Gillies, D.A., 1992, Comment on Herbert Simon, "Scientific discovery as problem solving", *Int. Stud. Phil. Sci.* 6:29-31.

Gooding, D.C., 1985, "He who proves, discovers": John Herschel, William Pepys and the Faraday effect, *Notes and Rec. Roy. Soc. Lond.* 39:229-44.

Gooding, D.C., 1986, How do scientists reach agreement about novel observations?, *Studies in History and Philosophy of Science* 17:205-30.

Gooding, D.C., 1989, "Magnetic curves" and the magnetic field, in: *The Uses of Experiment*, D. Gooding, T.J. Pinch, and S. Schaffer, eds., Cambridge University Press, Cambridge.

Gooding, D.C., 1990a, Mapping experiment as a learning process, *Sci., Technol.. and Human Values* 15:165-201.

Gooding, D.C., 1990b, 1994, *Experiment and the Making of Meaning*, Kluwer, Dordrecht.

Gooding, D.C., 1992a, Mathematics and method in Faraday's experiments, *Physis* 20:121-147.

Gooding, D.C., 1992b, Putting agency back into experiment, in: *Science as Practice and Culture*, A. Pickering, ed., Chicago University Press, Chicago.

Gooding, D.C., 1992c, The procedural turn, or, Why do thought experiments work?, in: *Cognitive Models of Science*, R. Giere, ed., University of Minnesota Press, Minneapolis.

Gooding, D.C., 1996a, Discovery as creative exploration, *Creativity Research Journal* 9:189-205.

Gooding, D.C., 1996b, Creative rationality: towards an abductive model of scientific change, *Philosophica* 58:73-102.

Gorman, M.E., 1987, Will the next Kepler be a computer?, *Sci. and Technol. Stud.* 5:63-65.

Gorman, M.E., 1992, *Simulating Science: Heuristics, Mental Models and Technoscientific Thinking*, Indiana University Press, Bloomington.
Gruber, H., 1985, From epistemic subject to unique creative person at work, *Archives de Psychologie* 53:167-185.
Gruber, H., 1990, The cooperative synthesis of disparate points of view, in: *The Legacy of Solomon Asch: Essays in Cognition and Social Psychology*, I. Rock, ed., pp. 143-158.
Kulkarni, D., and Simon, H.A., 1988, The processes of scientific research: the strategy of experimentation, *Cognitive Science* 12:139-175.
Kulkarni, D., and Simon, H.A., 1990, Experimentation in machine discovery, in: *Computational Models of Scientific Discovery and Theory Formation,* J. Shrager and P. Langley, eds., Morgan Kaufmann, Palo Alto.
Hesse, M., 1992, Comment on Herbert Simon, "Scientific discovery as problem solving", *Int. Stud. Phil. Sci.* 6:33-34.
Johnson-Laird, P., and Wason, P.C., 1970, Insight into a logical relation, *Cognitive Psychology* 1:134-48.
Johnson-Laird, P., and Wason, P.C., 1977, A theoretical analysis of insight into a reasoning task, in: *Thinking: Readings in Cognitive Science*, P. Johnson-Laird and P.C. Wason, eds., Cambridge University Press, Cambridge.
Latour, B., and Woolgar, S.,1979, *Laboratory Life*, Sage Publications, London.
Nersessian, N.J., 1984, *Faraday to Einstein: Constructing Meaning in Scientific Theories*, Nijhoff/Kluwer, Dordrecht.
Nersessian, N.J., 1985, Faraday's field concept, in: *Faraday Rediscovered*, D. Gooding and F. James, eds., Macmillan, London/American Institute of Physics, New York.
Nersessian, N.J., 1992, How do scientists think? Capturing the dynamics of conceptual change in science, in: *Cognitive Models of Science*, R. Giere, ed., University of Minnesota Press, Minneapolis.
Newell, A.,1977, On the analysis of human problem solving protocols, in: *Thinking: Readings in Cognitive Science*, P.N. Johnson-Laird and P.C. Wason, eds., Cambridge University Press, Cambridge.
Nickles, T., 1989a, Justification and experiment, in: *The Uses of Experiment*, D. Gooding, T.J. Pinch, and S. Schaffer, eds., Cambridge University Press, Cambridge.
Nickles, T., 1989b, Truth or consequences? Generative versus consequential justification in science, in: *PSA 1988 Volume 2*, Philosophy of Science Association, East Lansing.
Rudwick, M., 1985, *The Great Devonian Controversy*, Chicago University Press, Chicago.
Salmon, W., 1990, Rationality and objectivity in science or Tom Kuhn meets Tom Bayes, in: *Scientific Theories*, C.W. Savage, ed., University of Minnesota Press, Minneapolis.
Simon, H.A., 1992a, Scientific discovery as problem-solving, *Intl. Studs. Phil. Sci.* 6:3-14.
Simon, H.A., 1992b, Reply to Critics, *Intl. Studs. Phil. Sci.* 6:69-88.
Suchman, L.A., 1987, *Plans and Situated Actions: the Problem of Human-Machine Communication*, Cambridge University Press, Cambridge.
Tweney, R.D., 1990, Five questions for computationalists, in: *Computational Models of Scientific Discovery and Theory Formation,* J. Shrager and P. Langley, eds., Morgan Kaufmann, Palo Alto.
Tweney, R.D., 1992, Stopping time, *Physis* 20:149-164.
Worrall, J., 1989, Fresnel, Poisson and the white spot: the role of successful predictions in the acceptance of scientific theories, in: *The Uses of Experiment*, D. Gooding, T.J. Pinch, and S. Schaffer, eds., Cambridge University Press, Cambridge.

Scientific Discovery and Technological Innovation: Ulcers, Dinosaur Extinction, and the Programming Language Java

Paul Thagard[*] and David Croft[**]
[*]*Philosophy Department, University of Waterloo, Waterloo, Ontario, Canada, pthagard@watarts.uwaterloo.ca;* [**] *Department of System Design Engineering, University of Waterloo, Waterloo, Ontario, Canada, dwcroft@ uwaterloo.ca*

Abstract: Whereas scientists formulate laws and theories to account for observations, inventors create new technology to accomplish practical goals. Scientific discovery and technological innovation are among the most important accomplishments of the creative human mind. The aim of this paper is to compare how scientists produce discoveries with how inventors produce new technology. After briefly reviewing an account of the recent discovery of the bacterial theory of ulcers, we show that a similar account applies to the discovery that dinosaurs became extinct because of an asteroid collision. Both these discoveries involved a combination of serendipity, questioning and search. We then describe how these three processes also contributed to a very important recent technological innovation, the development of the programming language Java. The paper concludes with a more general assessment of the similarities and differences between cognitive processes involved in discovery and invention.

1. ULCERS

Until recently, peptic ulcers were thought by medical practitioners to be caused by excess acidity, and were treated using antacids. The popular view that ulcers are caused by psychological stress was never substantiated by medical research. In the past decade, there has been a dramatic shift in medical thinking about ulcers, prompted by the discovery by two Australian physicians, J. Robin Warren and Barry Marshall, that most peptic ulcers are caused by infection by a bacterium, Helicobacter pylori. Thagard (1999) provides an account of the psychological, social, and physical processes involved in the discovery and acceptance of the bacterial theory of ulcers. Part of the account was a discussion of how serendipity, questioning, and search each contributed to the discoveries of Warren and Marshall.

Research on the cognitive processes involved in discovery has tended to ignore the substantial role of serendipity in scientific research. Research is

Model-Based Reasoning in Scientific Discovery, edited by L. Magnani, N.J. Nersessian, and P. Thagard, Kluwer Academic/Plenum Publishers, New York, 1999

serendipitous when scientists make discoveries that they were not seeking. Because the creation of laws and theories requires substantial focussed cognitive effort, their discovery is never purely serendipitous. But things and phenomena are often surprisingly encountered as part of the physical process of interacting with the world, as when Roentgen discovered X-rays and Fleming found penicillium mold capable of killing bacteria. In 1979, J. Robin Warren, a pathologist in Perth, Australia, microscopically noticed spiral bacteria in gastric biopsies. He was not looking for bacteria in the stomach – he just happened to find them in the course of his normal duties as a pathologist. After much investigation and conceptual change, these bacteria were later named Helicobacter pylori and found to be responsible for most peptic ulcers. Because it was never the intention of Dr. Warren to search for stomach bacteria, discovery of H. pylori was clearly a matter of serendipity. Serendipity also contributed to the culturing of the bacteria after many failures, when technicians accidentally kept the bacteria incubating for five days rather than two.

In 1981, Dr. Barry Marshall began working with Dr. Warren to determine whether the bacteria found by Warren had any medical significance. They generated numerous important questions, including:

- Are the bacterial correlated with disease?
- How can the bacteria be eliminated?
- Would elimination of the bacteria affect disease?

Questioning is an important part of the early stages of scientific discovery when problems to be solved are being recognized and constructed. Once a problem is well-formed, it may be possible to use much more constrained searches to generate answers to questions. After Marshall and Warren had raised the question of whether the bacteria might be responsible for disease, it was relatively straightforward to design an experiment to determine whether the bacteria correlate with common gastrointestinal diseases such as gastritis, gastric ulcer, and duodenal ulcer. The correlations found between bacterial presence and ulcers were surprising, but they were not serendipitous because the researchers were looking to see whether they occur. The search for such correlations could be done systematically once questioning had posed a problem to be solved. Similarly, once Warren and Marshall questioned how the bacteria can be eliminated, they could do a systematic search involving a known array of antibiotics to determine which ones might be effective at eliminating the disease. Subsequent experiments determined that a combination of antibiotics was capable of eliminating the bacteria and curing peptic ulcers.

A much less condensed version of this account of discovery can be found in Thagard (1999). Serendipity, questioning, and search each contributed to the discoveries that produced the bacterial theory of peptic ulcers. For convenience, we shall refer to this account of discovery as the SQS model, for serendipity-questioning-search. The task at hand is to determine whether the SQS model applies to other cases of scientific discovery and technological innovation.

2. DINOSAUR EXTINCTION

Walter Alvarez (1998) has written a vivid personal account of one of the most important recent discoveries in geology and paleontology. In the mid-1970s, while investigating the geological origins of the Apennine mountains in Italy, he discovered a layer of clay at the boundary between the Cretaceous and the Tertiary sediments in rocks near Gubbio. The KT boundary marks the time when dinosaurs and many other organisms became extinct, so Alvarez and his collaborators began investigating the clay for clues as to what might have happened. They found that the clay contains an unusually high amount of iridium, more typical of asteroids and comets. This led to the hypothesis that an asteroid might have hit the earth 65 million years ago and caused the mass extinction, a hypothesis that has received considerable empirical support. (For a discussion of the explanatory coherence of the collision hypothesis and its competitors, see Thagard, 1991).

Alvarez's detailed description of the chain of discoveries leading to the new theory of dinosaur extinction shows that it fits the SQS model developed for the ulcer case. Serendipity figured in several of the empirical discoveries. Initially, Alvarez's interest was in deciphering the origin of the Apennine mountains, and he began collaborating with Bill Lowrie, a geophysicist specializing in paleomagnetism. They discovered surprising magnetic reversals in rocks near Gubbio, and were fortunate that the rocks were full of forams (marine protozoans) that enabled them to date the reversals. Discovery of these reversals was serendipitous in that they were not being sought by the researchers. According to Alvarez (1998, p. 38): "In a roundabout, unexpected, very lucky way, we had gotten the break that all young researchers dream of!".

The discovery of the centimeter-thick layer of clay at the KT boundary was also serendipitous. At Gubbio and at each new outcrop in the Apennines, Alvarez and Lowrie found the thin layer of fossil-free clay between the last limestone bed with Cretaceous forams and the first limestone bed containing only new Tertiary forams. In order to determine the duration of the extinction, Alvarez wanted to find out how long it had taken to deposit the clay layer. He consulted his father, the Nobel prize winning physicist Luis Alvarez, concerning how to measure the sedimentation rate of the clay at the KT boundary. After the failure of an initial attempt to measure sedimentation via the abundance of the isotope beryllium-10, Luis Alvarez suggested that they look at the amount of iridium in the clay. Iridium is much more common in meteorites than in the earth's crust, and they reasoned that a relatively large amount of iridium would indicate a slow sedimentation rate that had allowed meteorite dust to accumulate. Then came another big surprise. The researchers had expected almost no iridium if the clay had been deposited rapidly, and about 0.1 parts per billion if it had been deposited slowly, but they discovered that the clay samples actually contained around 9 parts per billion. Although the researchers were definitely looking to find out how much iridium was in the clay, the discovery of an extraordinarily large amount of it, suggestive of an asteroid collision, was unexpected. Thus serendipity struck several times in the process of inquiry, including the discovery of magnetic reversals, the finding of the clay layer, and the detection of large amounts of iridium.

Of course, serendipity is not just a matter of blindly stumbling on important phenomena, but occurs in the context of highly insightful questioning. Alvarez had initially set himself the task of determining how the Apennines had originated, and generated new questions as the research evolved, including:

- Why was there a layer of clay at the KT boundary?
- What happened to the Cretaceous forams?
- What happened to the dinosaurs?
- What was the sedimentation rate of the KT clay?
- What extraterrestrial event could have caused both extinction and iridium in clay?
- How could a giant impact cause extinction?

Alvarez (1998, p. 42) recounts his awareness that he had come across a major area of scientific investigation:

> I remember very clearly walking around the grounds at Lamont one day shortly after Al Fischer's talk there and realizing fully that this was a world-class scientific problem. Much of the work we do as scientists involves filling in the details about matters that are basically understood already, or applying standard techniques to new specific cases. But occasionally there is a question that offers an opportunity for a really major discovery. Choosing what problems and what kind of problems to work on is a critical strategic decision for a scientist. The question of the KT

extinction looked like one that could lead in totally new directions, and by the time I finished my walk, I had decided that I would try to solve it.

Much of Alvarez's success, like that of Marshall and Warren in the ulcers case, was the result of generating acute questions that lead to new hypotheses and empirical research.

The cognitive sciences, including psychology, artificial intelligence, and philosophy, currently lack a good theory of how questions, especially fruitful ones, are generated. In the ulcer case, questions arose as the result of surprise (at finding the novel bacteria), curiosity (about whether they are medically significant), need (to help people with stomach problems), and other subsidiary questions generated to help answer the original questions. In Alvarez' thinking, the main sources of questioning appear to be surprise (at finding the clay and at its high Iridium content), curiosity (about the causes of extinction), and the posing of subsidiary questions aimed at providing answers to ones already generated. We need a general cognitive theory of how such sources lead people to ask particular questions, but no attempt to generate such a theory will be made here. Note that a theory of question generation will need not only to consider linguistic and reasoning processes, but will also have to take into account emotional aspects of cognition, since surprise and curiosity are in part emotional states.

Although question generation is little understood, cognitive theory can tell us a lot more about the processes by which questions are answered. A why-question is answered abductively, by generating a hypothesis that explains the puzzling phenomena that generated the question. Analogies are often useful in this abductive process, when explanations of similar phenomena are used to suggest potential hypotheses for the puzzling phenomena (Holyoak and Thagard, 1995). Alvarez (1998, p. 75) does not remember when the hypothesis of an asteroid impact first came up, but he does recall being very interested in impact craters on the Moon and the planets. It is possible, therefore, that the impact hypothesis was generated analogically:

- Asteroids have hit the moon and left large craters.
- So, similarly, asteroids might have hit the earth and produced the iridium layer.

Unfortunately, no one could explain how an asteroid impact could cause world-wide extinction, until Luis Alvarez remembered reading about the 1883 explosion of the Indonesian volcano, Krakatoa, which had blown so much dust into the air that weather all over the world was affected. Analogically, Alvarez conjectured that the impact of a huge asteroid could send so much dust into the air that it would get dark all around the world and make the whole food chain collapse, producing extinction of the dinosaurs and many other organisms. Hence it appears that analogical reasoning helped to provide answers to some of the important questions that Alvarez was pursuing.

Other problems generated by Walter and Luis Alvarez were so well-defined that their solution can well be described in terms of search through a space of possibilities:

- What is the sequence of magnetic reversals above and below the KT boundary?
- What element could be used to measure the sedimentation rate of the KT boundary clay?
- Where is the impact site?

In each of these cases, there was a determinate set of possible answers, and researchers could proceed by searching for solutions. In contrast, the process of generating questions in the first place is much more open-ended and is not well characterized in terms of search. The space of possible questions is as large as the space of possible sentences, and little is known about how to constrain search of this space to generate questions that are interesting and potentially fruitful.

In sum, the discovery of the asteroid impact theory of dinosaur extinction seems, like the ulcers case, to have involved serendipity, questioning, and search. Now we can consider whether the SQS model also applies to technological innovation.

3. JAVA

Since computer programming was invented in the 1940s, many hundreds of programming languages have been developed. Most have turned out to be of little practical use, but a few languages such as FORTRAN and C have had widespread application. In the 1990s, the most successful new programming language has been Java, constructed by James Gosling in 1991 as part of a project at Sun Microsystems. Its spectacular success is in part the result of its suitability for Internet use. Java applets are programs that can be easily accessed using World Wide Web browsers such as Netscape. When a user opens a web page containing a Java applet, the program is downloaded onto the user's own computer and runs regardless of the kind of computer used. Java was enormously innovative in being platform independent, producing programs that can be run on computers with different CPU chips and operating systems. But Java was not designed with the Web in mind, for the first Web browser Mosaic was only developed in 1993, two years after the creation of Java. Java's development provides an interesting story of serendipity, analogy, and creative problem solving.

In 1990, Patrick Naughton, a top programmer at Sun Microsystems told Sun's chief Scott McNealy that he was quitting to join NeXT Computer where he could work on more interesting projects (Bank, 1995). Convinced by Naughton's contention that Sun was becoming insufficiently innovative, McNealy told Naughton he could have a million-dollar budget to put together a small team of outstanding programmers and engineers that would work without corporate interference from Sun. With carte blanche to pursue new projects, Naughton recruited Gosling and a few other top people, and in a hot tub in Lake Tahoe in 1991 they decided to build a prototype of a small device that could control everyday consumer appliances.

To control this device, they originally decided to program the device in C++, a popular computer language that combined the language C, originally build for applications using the operating system Unix, with ideas about object oriented programming borrowed from another language, Simula. However, for use in consumer appliances, the program needed to be more reliable and simpler than was possible with C++, so Gosling decided to develop a new computer language. The result was initially called "Oak" after a tree outside his window, but was later renamed "Java" for marketing reasons.

Oak was basically a simplified, more reliable adaptation of C++, but its major innovation was in the way it could be used by different kinds of computers. Normally, a program written in a particular programming language is compiled into machine language on a specific kind of computer. For example, a C++ compiler running on a Macintosh will produce very different machine code than will be produced by a C++ compiler running on a Windows computer. Because Oak was meant to be used on a wide variety of different devices, Gosling designed it to be compiled, not into machine language, but into bytecodes, an intermediate language that can then be interpreted for use on particular machines. Any computer that has a bytecode interpreter can run a compiled Java program using its own machine language. The resulting process may be slower than running a program written in C++, but much more flexible in that it does not require a different compiler for every computer platform.

By August, 1992, Naughton, Gosling and their seven-person team had produced a prototype personal digital assistant with a small, touch-operated screen that could control TVs and VCRs, but attempts to sell the device for use in interactive television and computer games failed. Then in June, 1993, the first Mosaic browser was released and the World Wide Web began to take off. Bill Joy, Sun Microsystems' co-founder, realized that Oak could naturally be adapted for use on the Internet. By 1995, Gosling had produced a Web-suitable version of Oak, now renamed Java, and Naughton had written HotJava, an interpreter for Web browsers. Since then, many thousands of Java applications have been produced and are available on the World Wide Web. For example, Thagard has made available on his Web site a Java version of his program ECHO, which evaluates hypotheses based on their explanatory coherence (http://cogsci.uwaterloo.ca/JavaECHO/jecho.html). Previously, researchers wanting to run ECHO had to be sent special versions for their own particular computer, but the Java version of ECHO is usable by anyone with an up-to-date Web browser.

Let us now consider whether the SQS model of scientific discovery applies to the invention of Java. At first glance, serendipity was not involved: Gosling designed the language to solve the problem of programming a variety of consumer appliances. But the application to the Internet was clearly serendipitous, because none of the Sun Microsystems team had any idea of eventual Web applications when they were developing their prototype product. By luck, a programming language designed for one purpose turned out to be incredibly useful for another purpose that arose only later. We can distinguish at least three kinds of serendipity:

- Finding something not sought for, as in the discovery of H. pylori and the clay layer at the KT boundary.
- Finding something sought for, but by an unplanned method, as in the accidental culturing of H. pylori through longer incubation (Roberts, 1989, calls this pseudoserendipity).
- Finding something sought for that turns out to useful for a purpose unconnected with the reason for which it was sought, as in the creation of Java.

Another example of the last sort of serendipity is the invention of Post-It Notes, which came about because a of much delayed realization that a weak glue could actually be useful.

Much more important than serendipity to the development of Java was the generation of interesting and fruitful questions. Naughton's team started with very general questions that they gradually refined:

- What will be the next wave of computing?
- How can computers be used more effectively in consumer electronics?
- How can a sophisticated personal digital assistant (PDA) be built for consumer appliances?
- What programming language could be used for the new PDA?
- What features should a new programming language have?

- How could programs be made to run on a wide range of electronic devices?
- What could the new PDA be profitably use for?
- How could the newly developed programming language be used on the Web?

Producing the PDA and the new programming language were prodigious technological feats, but initiation of these projects required the generation of questions that provided a focus for ongoing work.

Once questions were formed, then more constrained processes characterizable as search could be used to generate answers. The question of what programming language to use at first seems amenable to answer by search, since there were only a small number of reasonably appropriate languages to consider. But Gosling's decision that C++ was inadequate provoked a less constrained effort to produce a new programming language. Here analogy was a crucial ingredient, since Gosling based much of Oak on C++, although he also incorporated features of other programming languages that he liked. Use of multiple analogies is common in the development of programming languages, as we described in the creation of C++ from C and Simula.

An analogy of a very different sort seems to have brought Gosling the key insights about how a programming language could be used to control many kinds of devices. At a Doobie Brothers concert in 1991, Gosling reacted to the wiring, speakers, and lights: "I kept seeing imaginary packets flowing down the wires making everything happen. I'd been thinking a lot about making behavior flow through networks in a fairly narrow way. During the concert, I broke through on a pile of technical issues" (quoted by Bank, 1995). The flow of sound and light at the concert suggested to him ideas about the flow of information in computer networks. A much more prosaic analogy may have contributed to the application of the crucial bytecode concept. Gosling (1995) wrote: "The solution we settled on was to compile to a byte coded machine independent instruction set that bears a certain resemblance to things like the UCSD Pascal P-codes." More constrained search processes must have been involved in solution of more low-level problems concerning the details of the construction of Java.

As the next section will discuss, there are differences as well as similarities in the cognitive processes involved in scientific discovery and technological innovation. But at a higher level, it is clear that technological innovation in the case of the Java programming language involved all three ingredients of the SQS model presented earlier: serendipity, questioning, and search. Let us now do a more fine-grained cognitive comparison.

4. COGNITIVE PROCESSES IN DISCOVERY AND INNOVATION

Scientists and inventors ask different kinds of questions. Because scientists are largely concerned with identifying and explaining phenomena, they generate questions such as:

Why did X happen? What is Y? How could W cause Z?

In contrast, inventors have more practical goals that lead them to ask questions of the form:

How can X be accomplished? What can Y be used to do? How can W be used to do Z?

Despite the differences in the form of the questions asked by scientists from the form of the questions asked by inventors, there is no reason to believe that the cognitive processes underlying questioning in the two contexts are fundamentally different. Scientists encounter a puzzling X and try to explain it; inventors identify a desirable X and try to produce it. As mentioned above, there is no currently available theory of question generation, but roughly the same cognitive theory should be able handle both scientific and technological questions. Both kinds of question are produced in response to surprise, curiosity, need, and other questions, and both kinds involve emotions such as puzzlement and excitement.

In scientific discovery, why-questions are answered abductively, by generating hypotheses that provide explanations. As identified by C.S. Peirce and subsequent students of scientific reasoning, the basic pattern is:

Why X? Y would explain X. So maybe Y.

By definition, abduction concerns the generation of explanatory hypotheses, so abduction is not central to invention, which primarily involves the generation of answers to questions that are practical rather than explanatory. But there is a reasoning process in technological innovation that has a very similar structure to abductive inference (cf. Dasgupta, 1996):

How to do X? Y might do X. So maybe try Y.

For both invention and discovery, the difficult task is to generate the answer Y that provides a possible solution for the problem X. Abductive generation of hypotheses can be performed using rules such as "If Y then X", by causal schemas that connect Y and X, by visual representations of causal connections (Shelley, 1996), and by analogical inference described in the next paragraph. Invention, generation of inventive solutions can similarly be performed using rules, causal schemas, visual representations such as diagrams, and analogical inference. Abduction may also be a subsidiary process in technological innovation, when innovation depends on generating explanations of physical processes relevant to the new technology.

Analogical inference in discovery and innovation has a difference focus, but a common structure. Analogical discovery proceeds roughly as follows:

> Why X? X is like Y, which is explained by Z. So maybe X is explained by something like Z.

Analogical invention is analogous:

> How to do X? X is like Y, which is accomplished by Z. So maybe X can be accomplished by something like Z.

Although the purposes of analogical discovery and analogical invention are different, they both involve solving a target problem by finding a source problem that can be mapped to the target problem in order to suggest a solution to it. Analogies have often contributed to technological innovation:

- Laennec invented the stethoscope by analogy to a child's game involving listening to the end of piece of wood (Thagard, 1999);
- Alexander Graham Bell modeled the telephone in part on the human ear (Gorman, 1992);
- Thomas Edison used his earlier work on telegraphy to suggest solutions to problems involved in designing electric lights (Israel, 1998, p. 168).

An important part of scientific discovery that this paper has not yet addressed is the formation of new concepts such as Helicobacter pylori. Whereas new scientific concepts are generated to refer to entities and processes that are discovered or postulated as part of explanatory hypotheses, new technological concepts are formed to describe newly created entities and processes. In the Java case, the most important new concept was applet, introduced to refer to programs that could be sent over the Internet as Java bytecodes to be used by Web browsers. In science, theoretical concepts are largely generated by combining existing concepts, and the same seems to be true for technological innovation. Applet is roughly the result of combining "program application" and "Internet transfer". Bytecode was not a completely new idea generated for Java, but was rather modified from use in earlier programming languages such as Smalltalk; the key extension was that bytecodes could be sent over the Internet to run on many different kinds of computers. New technologies have generated numerous new concepts: automobile, VCR, computer chip, graphical user interface, and so on. Producing new concepts can be a crucial part of generating novel questions, as when Marshall and Warren named the newly discovered bacteria and used the new name to ask questions about it. In both scientific discovery and technological innovation, creative thinkers build on their familiarity with the knowledge available in a domain to produce new concepts, problems, and solutions.

It is important to recognize that technological innovation is not a purely cognitive process operating in the mind of individuals. Even though Gosling was the major programmer for Java, he worked extremely closely with a team of programmers and engineers at Sun Microsystems, and other people were responsible for key ideas such as adapting Java for the Internet. Scientific discovery is similarly a social as well as a cognitive process (Thagard, 1997, 1999). Barry Marshall and Robin Warren worked with many collaborators on the testing of the bacterial theory of ulcers, and Walter Alvarez worked with his father and others on the development and testing of the collision theory of dinosaur extinction. Like scientific research, technological innovation is a physical as well as a cognitive and social process, involving interactions with the world and multiple refinements as devices are improved. In science, a theory is often refined after experiments based on it display problems. Even more frequently, new technology requires many rounds of alteration and adjustment before it can accomplish the practical goals that it was designed to meet. Java went through numerous changes as it was improved and adapted for Internet and other uses.

5. CONCLUSION

Two cases of scientific discovery and one of technological innovation are hardly sufficient to generalize conclusively about the relation between discovery and invention. But the ulcers, dinosaur, and Java cases are rich examples of the creative mind at work, and our analysis of them has suggested some major similarities and minor differences in the cognitive and other processes involved in discovery and innovation. Discovery and innovation both involve serendipity, when physical and social processes introduce unexpected elements into the cognitive process of problem solving. Both involve questioning, although the questions generated differ in whether they are directed at explanatory or practical goals. Both often involve the use of analogy, with fundamentally similar processes of analogical inference. Both often involve the generation of new concepts, although technological concepts usually apply to newly created devices rather than to discovered or postulated entities. Although abductive inference to explanatory hypotheses is much more central to scientific discovery than to technological innovation, inference to possible solutions to technological problems seems to involve very similar representations and processes. We can therefore conclude that scientific discovery and technological innovation are cognitively very much alike. Scientists and inventors are among our most creative thinkers, but there is little fundamental difference in the mental origins of their creativity.

REFERENCES

Alvarez, W., 1998, *T. Rex and the Crater of Doom*, Vintage, New York.

Bank, D., 1995, The Java saga, *Wired* 3(12).

Dasgupta, S., 1996, *Technology and Creativity*, Oxford University Press, New York.

Gorman, M.E., 1992, *Simulating Science*, Indiana University Press, Bloomington, IN.

Gosling, J., 1995, Oak intermediate bytecodes, available at *http://www.javasoft.com/people/jag/.*

Gosling, J., n.d., A brief history of the green project, available at *http://www.javasoft.com/people/jag/green/index.html.*

Holyoak, K.J., and Thagard, P., 1995, *Mental Leaps: Analogy in Creative Thought*, MIT Press/Bradford Books, Cambridge, MA.

Israel, P., 1998, *Edison: A Life of Invention.* Wiley, New York.

Roberts, R.M., 1989, *Serendipity: Accidental Discoveries in Science*, Wiley, New York.

Shelley, C.P., 1996, Visual abductive reasoning in archaeology, *Philosophy of Science* 63:278-301.

Thagard, P., 1991, The dinosaur debate: explanatory coherence and the problem of competing hypotheses, in: *Philosophy and AI: Essays at the Interface*, J. Pollock and R. Cummins, eds., MIT Press/Bradford Books, Cambridge, MA, pp. 279-300.

Thagard, P., 1997, Collaborative knowledge, *Noûs* 31:242-261.

Thagard, P., 1999, *How Scientists Explain Disease*, Princeton University Press, Princeton, NJ.

A Hierarchy of Models and Electron Microscopy

Todd Harris
Department of Philosophy, University of California, Davis, CA, USA, trharris@ucdavis.ed

Abstract: This paper examines a hierarchical system of models connecting data and theory that was proposed by Patrick Suppes, and later developed by Deborah Mayo. After it summarizes examples found in Suppes and Mayo it takes a closer look at the experimental model and concludes that the experimental model has an evaluative role in the hierarchy, as well as a representational role. The last section of the paper illustrates with the example of electron microscopy and presents some problems that require further study.

This paper examines the hierarchy of models presented by Patrick Suppes, and later elaborated by Deborah Mayo. I will first present examples found in both Suppes and Mayo. I will then generalize from these examples to produce a notion of an experimental model that can be applied more generally, as I will illustrate with the example of electron microscopy. It will be shown that experimental models, which lie in the middle of the hierarchy, differ from both data models and theoretical models in that they include criteria for judging other models. I will then apply this general description to electron microscopy and present some problems that should be examined in much more detail.

Before discussing experimental and data models, it might be helpful to quote a general description of the model view of science from Paul Teller. This will help put my paper in the context of a larger debate occurring in the philosophy of science.

> An outline of Giere's 1988 formulation will provide a kind of "canonical" presentation of what I will refer to as the "model view". In the first instance science produces models, which are sometimes concrete physical objects, but which in most cases of interest are abstract objects. Models are connected to the world by theoretical hypotheses, sometimes more, sometimes less explicitly stated. Crucially, Giere takes each theoretical hypothesis to characterize some specific system, or some kind of system, as similar to a model, but only in limited respects, and even in those respects only to a limited degree of accuracy. Laws function as central, but not exclusive, means of defining models, so that laws are true by definition of the models they serve to characterize. Laws connect to the world only indirectly, through the theoretical hypothesis, in which all

Model-Based Reasoning in Scientific Discovery, edited by L. Magnani,
N.J. Nersessian, and P. Thagard, Kluwer Academic/Plenum Publishers, New York, 1999

the relevant empirical content resides. Theories are loose collections of models, organized inexactly in terms of the laws, methods, and other aspects which one uses to characterize models (Teller, forthcoming, p. 1).

1. SOME PROPERTIES OF MODELS

Let me develop this general statement further by discussing in more detail three points which will become important later in this paper.

First of all, models are often used to represent real world objects or processes. For example, a ball-and-stick chemistry set can be used to construct models that represent chemical compounds. Notice that the chemical model is said to represent a particular compound because it is similar to that compound in a relevant respect. Inaccuracies are allowed. A chemical compound has many properties lacking in our ball-and-stick model, and the model has many properties, such as color, that the real world object lacks. The point here is that the similarity relation is not the same as exact correspondence. We cannot say that the model is accurate to the real world object in every respect, we can only say that it is similar in relevant respects.

Second, a model is often constructed to serve a purpose. For example, model planes can be built to test aerodynamics in wind tunnels. Other models could be built to test the stress on particular joints in an airplane during take off. When an engineer constructs the model, this purpose is kept in mind. Notice that this is not always the case, a spiral staircase can be considered a scale model of a DNA strand, yet it was not built for this purpose. If you ask the model builder why the model is inaccurate in a certain respect, the model builder might bring up the purpose of the model. For example, one might ask an aeronautical engineer why a model airplane wing is smaller than the real world object. The engineer might answer that because the model was built to test the air flow around the wing, the size of the wing does not matter.

Third, models are most often, explicitly or at least implicitly, accompanied by a key. Without the key, it would be impossible to identify the real world object the model was intended to represent. Using the ball-and-stick chemistry set example again, we must know which colors stand for which elements before we know what compound is being represented by a given structure. Theoretical hypotheses can be thought of as providing such keys, as they connect the model to the object, phenomena, or process that the model is intended to represent.

2. A HIERARCHY OF MODELS

Patrick Suppes (1969) introduced the idea of a hierarchy of models connecting raw data with theory in his article, "Models of Data". Deborah Mayo later developed this hierarchy in her book, *Error and the Growth of Experimental Knowledge*. Before discussing their examples, we should discuss the concept of a hierarchy of models in general.

At the top of the hierarchy, theoretical models are usually presented as being an abstract object, such as the linear oscillator. There are also laws, such Hooke's law, which, as noted above, are true by definition of the theoretical model.

We create experimental models to test theoretical models. There are at least three differences between experimental models and theoretical models that are relevant to this paper. First of all, they are more specific than theoretical models. Although they use some idealizations of the theoretical model, they also refer to a specific experimental procedure. They are also not purely idealized, in that they refer to concrete objects such as experimental apparatuses. Second, they include directives for action, such as "check and record the needle position every ten minutes". And lastly, they include norms of evaluation. We will see this illustrated later.

At the bottom of this hierarchy are data models, which are distinct from the raw data. Much more will be said about data models later in the paper.

To understand this hierarchy it will be helpful to examine the example presented in Suppes' paper. Suppes takes his example from an experiment on learning theory. He chooses a particular experiment designed to test a linear response learning theory.

In Suppes' example, when specifying an experimental model, one must decide how many responses and how many reinforcements one will have in the experiment. One might consider a simple contingent reinforcement schedule in which there are two responses and two reinforcements, then add the condition that the probability of reinforcement should be independent of the trial number and any events that precede it. Then you can derive two conditions that must be met for such a contingent reinforcement schedule. These conditions give the arbitrarily set probability of a reinforcement, given a subject's response. For example, if the subject gives response #1, the probability of reinforcement #1 is 40%, while the probability of reinforcement #2 is 60%. An experimental model specifies a range of values which satisfy these two conditions. That is, the experimental model will contain all possible outcomes which satisfy the contingent reinforcement schedule.

Data models are at the bottom at the hierarchy. These are sets of data points that have been manipulated by the scientist, and so differ from the original raw data. For example, certain data points may have been thrown out because of a glitch in an instrument, such as a burned out indicator light. How do we decide if a given data model is a good model of the raw data? To be a good data model, it must satisfy constraints stated in the experimental model. For example, we might perform a goodness of fit test specified by the experimental model in order to see if the constraints are satisfied. In Suppes' example, we check to see if the relative frequency of reinforcements in the data model are close to the probabilities of reinforcement in the experimental model. To do this we can employ a variety of statistical tests, and Suppes has examples of such tests in his paper. I will not list them here, as I am trying to find a more general notion of experimental model.

Turning to Mayo's examples, she presents a study that examined the link between birth control pills and a certain blood clotting disease. The theoretical model is represented by a null hypothesis, H_0: "H_0: the incidence of clotting disorders in women treated (with the pill) *does not* exceed the incidence among the controls" (Mayo, 1996, p. 141). The hypothesis is stated in terms of idealizations of the theory.

In order to test this hypothesis, one must create an experimental model. This experimental model will (among other things): 1. Specify the size of the sample population, as well as the number of trials to be conducted. 2. Define experimental variables, which in this case are the observed proportion of blood clotted women in the treated group, X_T, and the observed proportion of blood clotted women in the control group, X_C. 3. Define a test statistic, in this case the risk increase, *RI*, which is the difference between X_T and X_C. 4. Provide the form of the experimental distribution. An experimental distribution is the distribution of values expected for the test statistic if the theoretical hypothesis is true. 5. Provide a measure of fit or distance between values of test statistic predicted by the hypothesis and observed values of the test statistic. In this example, it is the difference between the predicted value of *RI* (assuming H_0 is true) and the observed value of *RI*.

Mayo believes an experimental model has two main purposes. The first is to predict, or specify the *kind* of data we are looking for. "If the primary question is to test some hypothesis *H*, the job of the experimental model is to say, possibly with the aid of other auxiliary hypotheses, what is expected or entailed by the truth of *H*, with respect to the kind of experiment of interest" (Mayo, 1996, p.133). This information is contained in the experimental distribution. The second function of experimental model is to "specify analytical techniques for linking experimental data to the questions of the experimental model" (Mayo, 1996, p. 134). This is the measure of fit or distance between observed and predicted values of the test statistic.

Most experiments consist of many trials. The researchers do not compare the result of each individual trial with the value of the test statistic predicted by the theory, but instead an average is taken over a number of trials. In the birth control example, the difference between the average incidence of blood clotting in the treated and control groups are used to compute the observed risk increase, which is then compared to the value predicted by the hypothesis. In calculating these averages, some data points may be thrown out. For example, a particular subject may have a family history of blood clotting disease, or take a larger number of birth control pills than the other test subjects, and so be excluded from the final averaging. Because some data points are ignored, and then an average is calculated using a subset of the raw data points, we may say that the researchers have constructed a model of the raw data.

In order for a data model to be a good model of the raw data, we would like the assumptions of the experimental model to hold in the actual experiment. Mayo discusses two ways to address this issue. The first has to do with the design of the experiment. The scientist tries to create an experiment in which none of the experimental assumptions are violated. But this is rarely, if ever, possible, for most idealizations in experimental models cannot be created in the laboratory. The second way is to correct the raw data. This is accomplished through error analysis and data reduction.

3. EXPERIMENTAL MODELS

Suppes' and Mayo's examples are confined to cases in which statistical goodness of fit tests can be applied. I will now generalize from the examples found in Mayo. I believe that in Suppes' and Mayo's examples, the experimental model performs three functions, which I will list with their natural generalizations. 1. The experimental model specifies the sample size, and the procedure by which the sample should be collected. In general, this is specifying a procedure for collecting data. 2. The model also gives acceptable methods of error reduction and data analysis, where data points may be thrown out, or averages taken to eliminate aberrations. In general, this is specifying a procedure for sanitizing data. 3. Mayo's experimental model also specifies a measure of fit between a distribution calculated using theoretical assumptions, and the distribution that results from manipulating the raw data. This is specifying a procedure for judging the bearing of the data, as presented in the data model, on the hypothesis being tested.

Notice that the experimental models are used to evaluate other models, namely, data models. Because of this they differ from both theoretical models and data models. In addition to its representational role, an experimental model must also include prescriptions, and norms of evaluation. Following the above distinctions, these norms of evaluation can be divided into three groups. First, when the procedure for collecting data is specified, the experimental model is giving criteria for evaluating the actions of the scientists. Second, when the experimental model specifies a procedure for sanitizing data, it is evaluating another model, a data model. Third, when the experimental model outlines how the hypothesis or theory should be judged in light of the resulting data model, it is specifying norms of evaluation of a theory or hypothesis. Before turning to electron microscopy I will expand on the evaluative elements in each of the three features stated at the beginning of this section.

First of all, the experimental model will instruct the scientist about how to collect the raw data. This may include, among other things, how to operate the instruments, and how to isolate the specimen or phenomena that is being studied. An example might be "place the scale on a level surface" or "divide the sample population into experimental and control groups using a random method". Another example is the specimen preparation procedure in electron microscopy.

Secondly, an experimental model will tell the scientist what counts as an acceptable data model. This will include telling the scientist what to do when the scientist believes there is an error in the data model, or an outline of procedures for eliminating assumed errors in the data model. An example might be "if no particles are reaching the detector, increase the intensity of the electron beam".

The third way in which experimental models may have an evaluative element is in adjudicating the bearing of the data model on the hypothesis being tested. The model must indicate what is considered to be an acceptable deviance from the predicted outcome. One example is "the average of the data points must be within 5% of the predicted value".

At this point it might be of interest to look at another problem in the philosophy of science, the theory ladenness of observation. This problem begins by noting that observational judgments are affected by beliefs about the world we have. One well known example involves a simple line drawing which looks like a bird when placed in a field of birds, but looks like an antelope when placed in a field of antelopes. This dependency of observation on our beliefs is troublesome because we make observation statements in our test of theoretical hypotheses. We would therefore like to know to what extent the observation statements made during the experiment are dependent on theory.

Experimental models are also theory laden. Our decisions about how to set up an experimental model are heavily guided by a wide range of background theories. Because of this, we should keep in mind the above discussion of the theory ladenness of observation. Initially, one might hope to avoid a situation in which the theory being tested dictates how the data should be modeled, and how the data model should be used as evidence for or against a particular hypothesis. This injunction has been questioned by some philosophers, an example being Glymour's bootstrapping methodology (Glymour, 1980). I will not go into more detail in this paper.

4. A HIERARCHY OF MODELS AND ELECTRON MICROSCOPY

Before applying the concept of a hierarchy of models to electron microscopy, we must see if electron micrographs can be considered models. I will first see if they satisfy the criteria stated at the beginning of this paper.

Electron micrographs represent real world objects or processes. Just as with a ball-and-stick chemical model, micrographs lack many of the features of the specimen, including color, and three dimensional structure. Also, electron micrographs contain artifacts, such as Fresnel fringes or "holes", which are not present in the specimen, but are products of specimen preparation or of the electron microscope itself. This means that we should say that an electron micrograph is similar to the real world object it represents. This is the first property of models discussed at the beginning of this paper.

Second, if one asks a biologist why it is acceptable for a micrograph to lack certain features of the intended real world object, the reply would often be that those features are not important to the project. For example, scientists often underfocus the electron microscope. This reduces the resolution, but enhances contrast. It makes the edges of larger structures appear distinctly, while making finer structure blurry. The reduced resolution is acceptable when the scientist is only interested in representing the larger structures of the cell.

Finally, electron micrographs are constructed with a purpose in mind. In varying the fixation, drying, or embedding procedures, a scientist changes the appearance of the electron micrograph. This might be done to reveal cell structure that would ordinarily be transparent to the electron beam. Because of this, interpretation of some electron micrographs requires knowledge of the specimen preparation procedure. One might think of this as a key (theoretical hypothesis). A scientist must know the details of the specimen preparation procedure in order to know what structures are supposed to be revealed in the micrograph. This is the third property of models stated at the beginning of this paper.

Because electron micrographs share these features of models, I think it is best to think of the micrograph as a model. More specifically, I believe electron micrographs can be considered data models.

If the electron micrographs are data models, what is the corresponding experimental model? We can answer this question by returning to the three features presented in section 3.

First, the experimental model tells the scientist how to collect data. In transmission electron microscopy as performed in cell biology, this will involve, among other things, stating the specimen preparation procedure. This procedure can be divided into stages. First, there is fixation, in which the specimen is placed in a chemical that will bind to the cellular contents. Second, there is drying, in which the specimen is placed in a desiccant. Third there is embedding, in which the specimen is placed in a waxy material that will make it easier to slice. Fourth, there is sectioning, in which the block containing the specimen is sliced into thin sheets, called microtomes. Then there may or may not be additional steps, such as staining. This procedure can be varied during an experiment, and is often fine tuned after seeing what results. In effect, this is specifying a procedure for creating the electron micrograph, or data model.

Second, the experimental model specifies a procedure for sanitizing data. There are criteria for judging a good micrograph, such as a balance between contrast and focus, or the presence of unbroken, bilaminar membranes. There are also directives for eliminating artifacts. An example of these directives might be "if there is a spot in the micrograph, try a different sectioning angle to see if the spot changes its shape, or disappears", or more complicated prescriptions, such as, "to prevent excessive loss of cellular material, reduce the fixation time".

An experimental model also contains experimental assumptions which must hold in the experiment or which must be reflected in the way the data model departs from a literal description of the raw data in order for the data to be considered a data model. There are many assumptions which must hold in an electron microscope experiment. The beam energies must be of a certain strength, the magnetic lenses must produce close to homogenous magnetic fields, the microtomes must be of a certain thickness.

One must keep in mind that when the scientist tries to determine if a given spot on the micrograph is an artifact, she need not go through a detailed examination of the specimen preparation procedure to make sure experimental assumptions are not violated. Instead she changes the procedure and sees what results. For example, changing the embedding material or the fixative until a dark spot disappears. This method of identifying artifacts is discussed by Ian Hacking in *Representing and Intervening*: "Practice - and I mean in general doing, not looking - creates the ability to distinguish between visible artifacts of the preparation or the instrument, and the real structure that is seen with the microscope" (1983, p. 191). The decisions that scientists make to change the specimen preparation procedure might not be based on any scientific theory at all. Often scientists do not try to determine the source of an artifact.

Third, the experimental model is supposed to provide a measure of fit or distance between the data model and the result predicted in the experimental model. This is much easier to identify in statistical experiments, such as the ones discussed by Mayo, because one can state an explicit measure of fit or distance. In electron microscopy, there is no such explicit, quantitative measure. This means that understanding this feature of experimental models in electron microscopy will involve a detailed analysis of the similarity relation that is supposed to hold between a theoretical model, such as a wax model of a mitochondria, and a series of electron micrographs. For example, there is an averaging process in which the scientist looks at a number of micrographs, which might be from different microscopes, or from sectioning the specimen at different angles, and comes to a conclusion about the three dimensional structure of the specimen. Different micrographs will be similar to the theoretical model in different ways.

There is the further question of whether or not the experimental model, as outlined above, is really a model in the same sense that an electron micrograph is called model. If so, the experimental model, like the data model, should satisfy the three criteria stated at the beginning of this paper.

First, it should represent a real world object or process. The experimental model combines three elements, a procedure for collecting data, a procedure for sanitizing data, and a measure of fit or distance between the data model and the result predicted in the experimental model. I believe the first two elements can be seen as representing a model building process. In other words, the experimental model is a protocol that guides the scientist in the construction of data models. The experimental model would function like a road map. Just as a map presents possible routes given a traveler's destination, the experimental model would outline possible courses of action given the scientists goals. This leaves the third element, the measure of fit or distance, to be incorporated in a second model distinct from the experimental model, maybe one similar to the "mediating models" mentioned in Morgan and Morrison (1999). This needs to be worked out in detail in a larger project.

Second, a model is constructed to serve a purpose. I have stated many reasons for constructing experimental models.

Third, the model should be accompanied, explicitly or at least implicitly, with a key. I believe the assumptions that must hold in order for the data model to be considered a model of the data can be considered a key that ties the experimental model to the actual procedure followed in the lab. If these assumptions hold, then we may say that our experimental model represents a real world process, namely, actual process the scientist go through to construct the data model. Again, there are details to be worked out here.

We have seen that the notion of a hierarchy of models, as first presented by Suppes and later expanded on by Mayo, is confined to a special class of experiments, namely, those in which statistical goodness of fit tests can be applied. However, one can develop a general notion of a hierarchy of models that can be useful in understanding a wider range of experiments. Specifically, this generalized hierarchy of models can be applied to electron microscopy.

REFERENCES

Giere, R., 1988, *Explaining Science: A Cognitive Approach*, University of Chicago Press, Chicago.

Glymour, C., 1980, *Theory and Evidence*, Princeton University Press, Princeton, NJ.

Hacking, I., 1983, *Representing and Intervening: Introductory Topics in the Philosophy of Natural Science*, Cambridge University Press, Cambridge.

Hayat, M.A., 1989, *Principles and Techniques of Electron Microscopy: Biological Applications*, Macmillan Press, New York.

Lynch, M., 1985, *Art and Artifact in Laboratory Science*, Routledge and Kegan Paul, London.

Mayo, D., 1996, *Error and the Growth of Experimental Knowledge*, University of Chicago Press, Chicago.

Meek, G., 1976, *Practical Electron Microscopy for Biologists*, John Wiley and Sons, New York.

Morgan, M.S., and Morrison, M.C., 1999, Models as mediating instruments, in: *Models as Mediators*, M.S. Morgan and M.C. Morrison, eds., Cambridge University Press, Cambridge, forthcoming.

Suppes, P., 1969, Models of data, in: *Studies in the Methodology and Foundations of Science*, Reidel, Dordecht.

Teller, P., 1999, Models vs. TOEs, forthcoming.

Expansion and Justification of Models: the Exemplary Case of Galileo Galilei

Vasilis Raisis
University of Athens and National Technical University of Athens, Athens, Greece, vraisis@central.ntua.gr

Abstract: For T.S. Kuhn in his Postscript to *Scientific Revolutions* and R. Giere recently science is as a cluster of models that is produced and expands from a few prototypical cases. Both, they are against any seeking of common characteristics that could function as "necessary and sufficient conditions" of membership in the cluster, leaving the scientific community as the final judge of what constitutes a problem for a specific science. With a sketchy examination of Galileo's work on motion I try to show that two models could belong to same cluster, not because they share common characteristics, but because the one is a limiting case of the other. In this case, we can have rules of transformation that although are not logical are nevertheless "objective" and rigorous. Moreover, such a context reframe the problem of when a model is a good representation of a physical system.

1. SCIENCE AS A CLUSTER OF MODELS

1.1 Kuhn's exemplars

From a commonsensical point of view, and for most philosophers also, examples carry no special or unique information. In his "Postscript" to *Scientific Revolutions* (Kuhn, 1970a) Thomas Kuhn comes to discard this point of view. He argues that the pupil of science has learned nothing before the teacher gives her some examples. She has learned nothing, not because it is difficult to understand concepts and laws directly from definitions, but because concepts and laws mean nothing before they are applied. The application is not just to an example, but to an *exemplar,* a prototypical situation that demonstrates the functions and uses of concepts and laws. Exemplars, in this way, are situations such that words/symbols acquire meaning only within them. This is the first step, where, in Kuhn's words, "nature and words are learned together" (Kuhn, 1970a, p. 191). The problem now, is how the scientist can apply the concepts and the laws in situations that are differ-

Model-Based Reasoning in Scientific Discovery, edited by L. Magnani, N.J. Nersessian, and P. Thagard, Kluwer Academic/Plenum Publishers, New York, 1999

ent from the idealized or specific conditions of exemplars. Here is the second function of exemplars, that is, substituting the rules of how to do science. Instead of applying a general method, a scientist has to "see his problem as like a problem he has already encountered" (p. 189). In this way scientists "solve puzzles by modeling them on previous puzzle solutions" (p. 189). New situations are not confronted directly through conceptual apparatus but they are first seen as instances of the prototypical situation, the exemplar.

1.2 The central character of model-based reasoning

The dispute about the role of models in science started long before Kuhn. Kuhn's exemplars are not the only notion of what we mean when we say that scientists employ models and model-based reasoning in their practice. I started by referring to Kuhn because he is one of the first[1] to explicitly claim that models – or his particular kind of models – are not just irreplaceable in scientific discovery, but they are the only way to understand what scientists are doing. Let me explain.

Before Kuhn there were two conflicting positions about models. According to the first (Duhem, logical positivists, rational reconstructivists), models are redundant from a philosophical point of view. Scientists sometimes use models as heuristic means, but this forms part of the process of discovery. According to the "redundant" point of view, what is interesting for philosophy is the reasoning and the experiments that justify a scientific claim and not the psychological conditions of discovery of this claim. This conception is manifested in the famous distinction between context of discovery and context of justification (Reichenbach, 1938). According to this view, philosophical analysis should confine its attention to theories as finished products.

The opposing view (Campbell, 1920; Black, 1962; Hesse, 1966, and a recent application to historiography of science, in Nersessian, 1992) claims that models are necessary for scientific discovery and, in addition, discovery is not just a psychological but a philosophical problem as well. In Max Black's words, "the employment of models is [...] a rational method having its own canons and principles" (Black, 1962, p. 236).

Kuhn has made a third, stronger claim about models. He claims that the same process – "seeing as" – holds not only for discovery but must be part of the textbook tradition as well. A pupil, for Kuhn, learns science by learning how to move from one "solved problem" to another. In this way, "seeing as" is necessary not only for the exploration of new areas, but for the compre-

[1] For similar approaches see Suppes (1962), Beth (1961), van Fraasen (1970) and Suppe's introduction (pp. 1- 241) in Suppe (1977).

hension of science and the communication of scientific results, as well. For Kuhn, model-based reasoning is not just a legitimate form of reasoning but the core of scientific thought and scientific practice. From his point of view the science is a cluster of "solved problems". I will call Kuhn's account "strong claim about models".

1.3 Two aspects of modeling

So far I have used Kuhn's "exemplars" or "solved problems" as kind of models. This may be in part misleading because Kuhn's account misses one useful aspect of modeling: that between a theoretical idealized construction and a real system. Let me show this using one of his examples. In his words, "Daniel Bernoulli discovered how to make the flow of water from an orifice resemble Huyghens' pendulum [...] from that view of problem the long-sought speed of efflux followed at once" (Kuhn, 1970a, p. 190). But something is missing here. In order to solve his problem, Bernoulli measured actual speeds of water and found them to be in accordance with the theoretical ones. Only after such a measurament can one speak about a "solved problem" in physics. This is to say that a problem is solved, if it is seen as a previously solved problem *and* the predictions from this new applications are conformed by experiment. In order to use exemplars as kind of model-based reasoning we have to distinguish between two different kinds of modeling. Huyghens' pendulum, for example, from one side is a theoretical model, an idealized construction depicting real pendulums. This modeling_1 is to be checked by experiment. On the other side, it is a model on which other theoretical constructions, like the flow of water from an orifice, can be modeled. This modeling_2 is to be done by similarity judgments whatever these latter are based on.

Ronald Giere also makes the "strong claim" about models, but he explicitly differentiates the two kinds of modeling. In Giere's account, it is clear that the problem of theoretical reduction is one thing, and the verification of models via situations of the real world, is another. For Giere a model is a theoretical, idealized construction, that has always a real counterpart, or a whole cluster of counterparts in the real world: "Theoretical models are the means by which scientists [...] represent the diverse systems found in the real world: springs and pendulums, projectiles and planets [...]" (Giere, 1988, p. 80). In this way, if a model is to belong to the cluster it does not only have to be modeled on some (theoretical) exemplars, but also to be a reliable model of a real system. So, for Giere science is

1. A population of models.

2. Various hypotheses linking those systems with systems in the real world (Giere, 1988, p. 85).

In what follows, I will restrict myself to this view of models and to the "strong claim". I am not going to offer explicit reasons why somebody should accept something like this. Instead, I am going to focus on two questions internal to this approach.

2. TWO QUESTIONS ABOUT MODEL-STRUCTURED SCIENCE

Let me take a common example of physics textbooks in order to more carefully frame my questions. Physics textbooks sometimes start with the problem of linear oscillator with a central force and then present the pendulum with a point-mass for a bob. The latter is solved by reducing it to the linear oscillator problem.

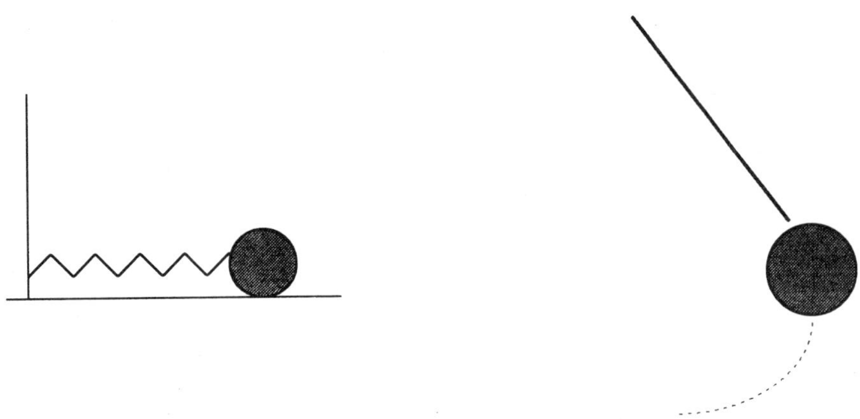

Figure 1.

So we have two different idealized situations, and the second (pendulum) is theoretically manipulated so it can be seen as the first (oscillator). In order to properly use the concepts we have learned in the first situation in the second situation, we must already have seen the second situation as similar to the first. The question that naturally arises is this: If there are no rules for recognizing the situations as similar, what puts them in the same cluster? Why are they both mechanics? An appeal to the success of the reduction could not solve the problem. If, for example, somebody solves the problem of variation of prices in stock market, seeing it on the basis of linear oscillator, this does not mean that the two models belong to the same cluster. So, the first question is: If a science is a cluster of models, what makes this cluster hold together if not similarity rules? This is the problem of model expansion or the question of theoretical similarity.

The second question concerns the verification of models. Let us suppose that we have reduced the pendulum problem to the linear oscillator problem. From this reduction we expect specific rates for the period related with the length. We get into a laboratory, we experiment with pendulums, and we are happy to see that we have only small deviations from the theoretically calculated rates, deviations that we explain by the friction, errors of measuring instruments and so on. But is that so? How do we know that deviations from expected rates are not due to the inappropriateness of our model? Can it be due to leaving out important aspects, of making the wrong idealizations or isolations, or finally, of having chosen the wrong source-model (linear oscillator) for the reduction? In other words, in what way may we judge the permissibility of approximations?

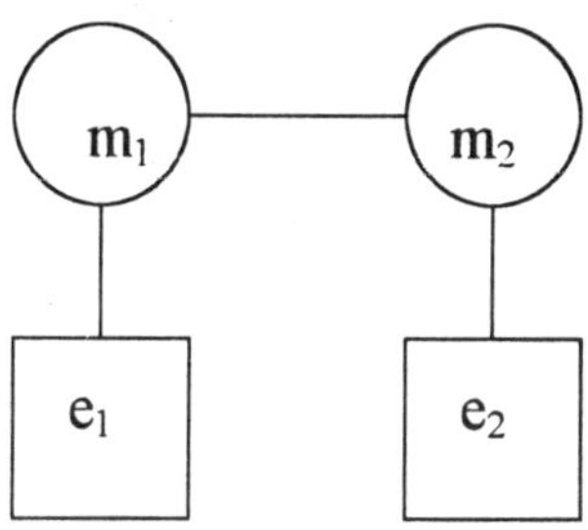

Figure 2.

Let me summarize the two questions. Suppose we have a theoretical model m_1 that is a model of the real system e_1 and a theoretical model m_2 that is a model of the real system e_2. The first question concerns the link between m_1 and m_2. The second question concerns the links between m_2 and e_2 and between m_1 and e_1.

3. KUHN AND GIERE ON THE TWO QUESTIONS

3.1 The problem of expansion

According to Kuhn, the practice of *normal science* depends on the ability "[...] to group objects and situations into similarity sets which are primitive in the sense that the grouping is done without an answer to the question 'Similar with respect to what?'" (Kuhn, 1970a, p. 200), The grouping is not to be done by criteria of similarity but through the "seeing as" process. "Seeing as", for Kuhn, is misconstrued "if reconstructed in terms of rules that are first abstracted from exemplars and thereafter function in their stand" (p. 192). The reason for this is that "seeing as" belongs to perception, it is an automatic and direct "process over which we have no control" (p. 194). The "Gestalt shift" was the central metaphor in *Scientific Revolutions* denoting how scientists move to a new paradigm. In the "Postscript", Kuhn uses the "Gestalt shift" literally. Physicists, literally *see* situations *as* situations they have encountered before.

Kuhn's idea can be clarified with an example. All of us can see two letter-samples as A's without have being given explicit rules of how to do it, and we can only afterwards seek for some criteria of similarity that are always incomplete. This is a quite comprehensible situation but is it really a good example of model-based reasoning in science?

Turning back to Giere's account, we can see that it is quite similar to Kuhn's as he is against any seeking of common characteristics. Giere, uses the Wittgenstein's notion of family resemblances. That is, although some models are central in the cluster, the other models do not "inherit" a definite set of common characteristics. The lack of common characteristics has the result that "no necessary and sufficient conditions determine which models or which hypotheses are part of the theory" (Giere, 1988, p. 86). The criterion of grouping must be sought outside the cluster:

> Nothing in the structure of any models themselves could determine that the resemblance is sufficient for membership in the family. The question it seems, is solely a matter to be decided by the judgements of members of the scientific community at the time. This is not to say that there is an objective resemblance to be judged correctly or not. It is to say that the collective judgements of scientists *determine* whether the resemblance is sufficient. This is one respect in which theories are not only constructed, but socially constructed as well (*ibid.*, p. 86).

The argument here seems to go like this: because there are no common characteristics in models, there can be no explicit rules for treating one model as resembling another. So, between models one may detect only family resemblances. As family resemblances can not constitute a sufficient criterion, something outside the model must be found. For Giere, this is the scientific community; so the community is the final judge.

The first part of the argument is crucial for what we called "strong claim". If a model is just a convenient tool to extract some characteristics that function consequently as similarity rules, then we can keep the rules and forget the model. This is exactly what Kuhn and Giere disapprove of. What I want to contest is the second part of the argument, namely, that family resemblances and community decisions are the only alternatives we have.[2]

3.2 The problem of justification

As for the second question, the justification of models, Kuhn spoke about "solved problems" leaving the impression that the "seeing as" process is also sufficient to recognize a solution. But even if this were the case, we would still have left unexplained why the central model, the *exemplar,* is a solved problem. Exemplars, the original prototypes, are not themselves modeled anywhere, so why treat them as solved?

For Giere, models represent systems of the real world because they are similar to them. In this case, similarity is stated by an explicit and verbal rule. Scientists do have explicit instructions concerning how to test the similarity of a model. Giere calls these instructions *theoretical hypotheses.*

The general form of a theoretical hypothesis is: "Such-and-such identifiable real system is similar to a designated model in indicated respects and degrees" (Giere, 1988, p. 81) For example, the earth and the moon is a system similar to the two-particle Newtonian model, as far as, "positions" and "velocities" are concerned (p. 108). Here, with respect to aspects, the similarity is quite clear. There are explicit verbal instructions of how to extract a theoretical speed from the model, how to measure an actual speed and how to compare these two. The problem starts after comparing. Hypotheses for Giere must specify the acceptable degree of approximation (e.g. "deviations over 2% falsify the hypothesis"). As Giere admits (p. 108), depending upon the specification of the degree, the same hypothesis turns to be either true or false. In this case the justification of the theoretical model *vis-a-vis* a real system depends on an arbitrary factor as the decision as to what is an acceptable degree of approximation has nothing to do either with the model or with the world. Consequently, a crucial factor for the justification of a model is the decision about permissible approximation, which depends, in the final analysis, (again) on the scientific community. This result is unwelcome to Giere, who wants to explain science as a product, not only of interactions among people, but as an interaction between people and nature as well. Accordingly, he gives two answers to the issue above.

The first is that the "specifications need not be tacit, and in science they often are not". In addition, "a judgement on the accuracy of the data must

[2] I don't want to exclude social factors from explanations of science. What I want to argue is that a scientific community could not decide on similarity relations, in the same way as that community decides what is politically useful.

precede the judgement whether that degree of accuracy is acceptable" (pp. 108-109). This claim makes the justification of models an honest game, but it is still a game. Even if we make the decision before we measure, this is still irrelevant as to how the world is.

The second argument of Giere appeals to Darwinian theory. The specification of aspects and degrees "does not depend entirely on social agreements which are totally independent of our models and how the world is" (Giere, 1984, p.15). What is left from the "entirely"? That our abilities to see similarities and to represent our environment have natural constraints as they are "built in" in our perceptual systems. Giere is right about the need of a "naturalistic counterpart of philosophical theories of justification or rationality" (Giere, 1984, p. 16). What I want to dispute is that the other part is nothing but social agreements.

I will try now to answer the two questions through an example, a sketchy presentation of some central episodes of Galileo's effort to create a science of motion.

4. THE NEW SCIENCE OF GALILEO: EXPANDING AS TAKING TO THE LIMIT

4.1 Balance period

Galileo in his early unpublished manuscript, *De Motu*, starts his lifelong effort to found a science of motion. He contests the Aristotelian theory of motion seeking a new one. In order to solve the problem of motion he asks for a model, for a well-solved problem, aiming to model other motions on it. The Archimedean balance, is the best candidate for two reasons: first, the balance has a quasi-axiomatic foundation, as Archimedes had proved the relation between weights and distances, starting from "self-evident" principles.[3] Beyond mathematical demonstrations, the law of balance has undoubted empirical corroboration and of course great practical utility. That is, it fulfils every ideal of what constitutes a well solved physical problem. The second reason is that Archimedes has already used the balance to solve hydrostatic problems (Dijkesterhuis, 1987, pp. 373-380). So, it has already been

[3] Archimedes actually uses two axioms, first that equal weights in the same distances from fulcrum are balanced and second that you can substitute two equal weights with the double weight in the middle of distance between them, see Dijkesterhuis (1987, pp. 286-305) and for a simplified presentation Lindberg (1992, p. 153). Galileo in *Le Meccaniche* (Galileo, 1600, pp. 153-155) constructed a new proof.

proven expandable. So, on the one side we have the "logical" production of the law from principles. On the other side, the law seems to be just the inductive generalization of empirical measurements. So, this is a "prototypical" solution, as both theoretical and empirical ideals of what constitutes sound knowledge seem to fulfilled. In *De Motu,* Galileo wants to transform first problems about free fall of bodies into problems about hydrostatics, and then to treat both as balance problems.

His strategy to transform ("see as") kinematics problems into static ones, was to specify the constant cause of fall in virtue of a balance: the body from the one side, and an equal volume of the liquid of the other side. The difference in two (specific) weights was alleged to be the constant cause of the fall.

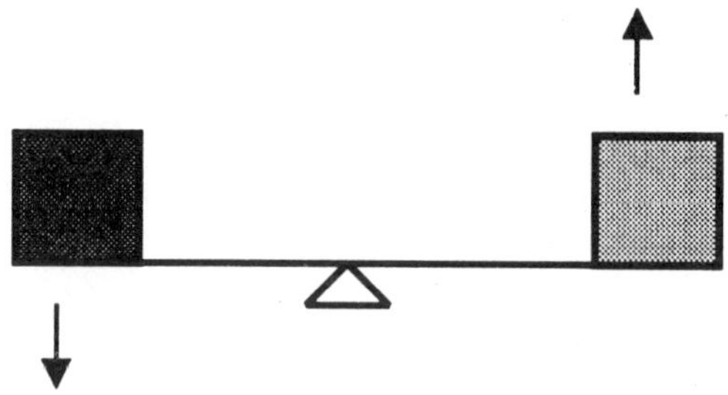

Figure 3.

In case of free fall, the body is falling only by its own weight.

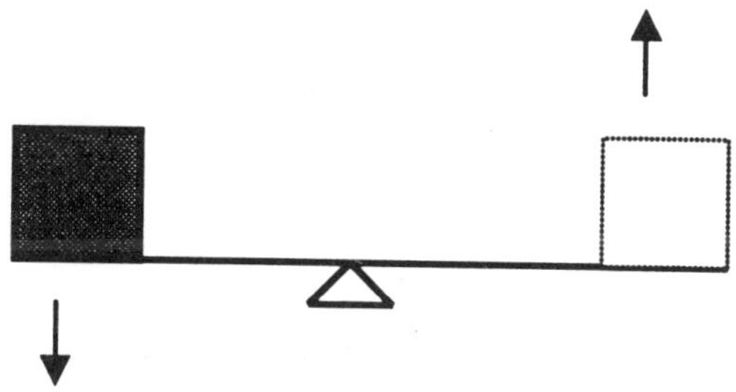

Figure 4.

In this way he specifies the constant cause of fall (specific weight of the body minus specific weight of the liquid) supposing that the magnitude of the final speed[4] of falling is proportional to the magnitude of this constant cause. In the same manuscript he tries to explain the failure of experiment to fit the data admitting that "the ratios we have set down are not observed" (Galileo, 1590, p. 69). He finally comes to understand that not only weight but neither specific weight had anything to do with final or middle velocity of free fall.[5]

In the meanwhile, he had started the inclined plane experiments in order to test his theories. As the reduction to balance seemed not to work, he decided to cope in another way with the inclined plane. He decided to create his own exemplar to solve the problem of inclined plane prototypically.

4.2 Inclined plane period

About 1604, he finally succeed in obtaining the first results for the relations between speeds, times, spaces.[6] Having tested his results in various inclinations he expects them to hold in the limiting case of 90 degrees, namely, free fall. It is not a self-evident transformation. His spheres were not falling but rolling, neither was their motion free, as they were in constant contact with the corridor. The argument here is that, as the results were the same in different inclinations, they should hold in the limiting case of the vertical plane. It should be noticed that this expansion needs to be checked using a different experimental apparatus, falling stones from the Pisa tower, for example.[7]

Another limiting case of his exemplar, is the horizontal plane. But the horizontal plane is a double limit. It is a limit for accelerated motion, and a

[4] Early Galileo believed that acceleration occurs only at the first stage of fall, until the impetus that would be needed to raise the body in such height had consumed, Galileo (1590, ch. 21, pp. 100-105).

[5] I don't mean that it was "nude" experimental results that falsified his early theory. I believe that in order to understand Galileo's achievements neither the generalization of experimental results, nor the "seeing with other way the same things" are appropriate. More fruitful seems to be something like Gooding's account of Faraday's work, namely a presentation that would be characterized by "nonlinearity-reflexiveness and [...] the interaction of concepts, percepts and objects" (Gooding, 1992, p. 47).

[6] In his early stage Galileo uses the speed as proportional to time and proportional to space indifferent. In *Discorsi* (pp. 159-161, [203]-[204]) this view is overthrown, but it is not generally acceptable when Galileo understood that speed must be proportional to the root of space, instead of space.

[7] Galileo refers to experiments of leaving balls of height of 150-200 braccia in Galileo (1638) pp. 225-226 ([276]-[277]). For the famous but debatable experiment of Pisa, see Cooper (1935).

limit for decelerated motion as well. So, the motion here must be neither accelerating, nor decelerating. He supposes that an horizontal motion without resistance would be eternal.[8] Here we also have different experimental apparatus, as Galileo argues for this by referring to ice lakes or to continual smoothing of surfaces.[9]

4.3 Projectiles

Galileo finally reaches what he takes as his greatest achievement. The solution of the problems of the projectiles. He sees the motion of a projectile as a motion that is done on two inclined planes in the same time. That is, the ball is falling in the vertical, following the law of free fall and, in the same time, it travels with constant speed along the horizontal. The justification of model again demands different experimental apparatus from the inclined plane. It needs cannons and correlating the gun powder loads with the heights, ranges and angles of projectiles. Two experimental results were known, before Galileo produced them. The first was that maximum range is attained at elevation of 45°.[10] The second was that the path of projectile is parabolic.[11]

4.4 Pendulums

Let us see one more expansion of the inclined plane model, the pendulum. Here, it is quite sure that Galileo had a first experimental result. Counting many oscillations with different swings he saw that they need the same time. He never stopped to seek proof for this. After solving the inclined plane problem, he had an idea how to show the truth of the claim about the pendulum. One of his inclined-plane theorems was the isochronism of chords. (The time of descent between different chords which end at the same point of the circle is the same.)

[8] According to Galileo, on the large scale it is the circular motion that is "neutral", as only in this case the body does not approach neither recede the centre of the earth (*Discorsi*, p. 197 [243]). Notice that circular inertia is not a universal principle for Galileo, as it pressupposes the presence if acceleration towards the center of the earth.

[9] Moreover, if we believe Stillman Drake's interpretation of unpublished manuscript folio 116^v, Galileo has a more clever idea to verify the inertia principle. See Drake (1973) and for a revised view Drake (1991, pp. 99-115).

[10] Galileo (1638, p. 298).

[11] According to Damerow, et al., (1991, p. 152) Guidobaldo had an experimental demonstration of this that may have been known to Galileo.

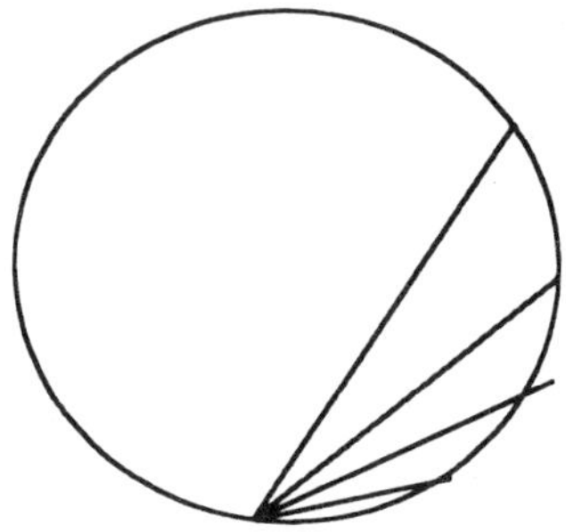

Figure 5.

As the angle becomes smaller and smaller, the difference between the chord and the arc diminishes. So, the limiting case of a pendulum is the motion on inclined plane.

4.5 Summa

Skipping other expansions of inclined plane[12] we can now summarize the whole point. First, we have – after thirty years of effort – a solved problem, the inclined plane. Free fall, and inertia are limiting cases of the inclined plane. Projectile motion is a combination of these two motions. But we do not really need a second notion – that of combination – in order to see projectile motion as motion in inclined plane. As shown in the figure above, in one limit, as the vertical velocity becomes slower and slower, the projectile approximates to free fall.

5. CONCLUSIONS

5.1 The transformation rule

If our sketch is right, Galileo possesses and tries to legitimate a vigorous, visually demonstrative, mathematical manipulatable rule for transforming

[12] The more speculative expansion is to solar system. According to this, the celestial globes were accelerated with the "natural ratio" of acceleration until they reached the "degrees of velocity which originally seemed good to the Divine mind". Measuring the size of orbits and speeds Galileo "proves" that they depart from the same place (Galileo, 1632, p. 29).

the models: taking to a limit. This suggest an answer to our first question: *Two situations are similar/they belong to the same cluster of models, not because they share common characteristics, but because the theoretical model that corresponds to one is a limiting case of the theoretical model of the other.* There are two forms of limiting case analysis. In one case we can take another model as a limiting case of the central model. In the second case the central model is the limiting case of a more complex model that was produced from the superimposition of two or more limiting cases of the central model.

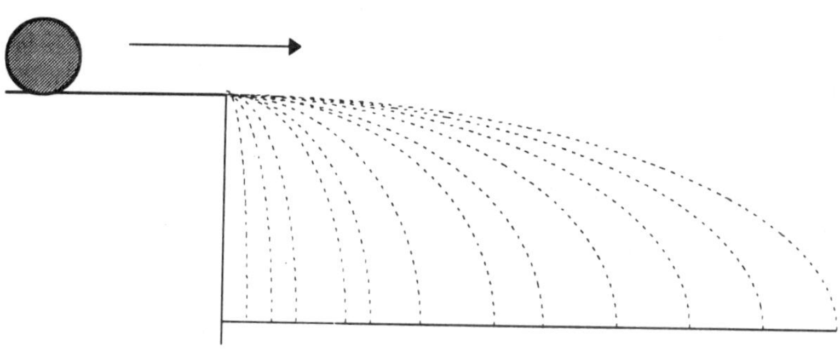

Figure 6.

5.2 The reciprocal character of justification

Let us now turn to the problem of experimental verification. The inclined plane seems to have a good verification. In third day of *Discorsi* Galileo produces a whole cluster of theorems. Many of them could be tested in an easier way than the direct measuring of spaces and times.[13] But in any case, even if we accept Drake's view of Galileo as experimentalist, an experiment could never had given him anything better than, a *very good approximation* that e.g., the spaces traversed are proportional to the squares of times elapsed; or two different stones from the tower of Pisa reach the ground *al-*

[13] Many theorems in the third day of *Discorsi* are formulated in the form "If..., then movements from rest will be made in equal times" (isochronism of chords the most prominent example). We speculate that in this formulation, Galileo could have checked the propositions without an instrument that measures times. In this way he could take a reciprocal justification of more difficult experiments as the one which gives the square-time law. For the reciprocal character of justification see below.

most together; or the motion of a ball soaked in ink leaves a line that *seems to be* parabolic; or the maximum range of a projectile is achieved at around 45 degrees *within the degree of accuracy of the relevant instruments.*

I claim that all these approximations, are mutually justified by each other. Their reliability is reinforced as the respective theoretical models are transformed into each other. For example, the experimental results of parabolic path and 45 degrees maximum range are justified because they are produced by a law (space-time squared) that has independently and carefully been verified through inclined planes. And reciprocally, the reliability of inclined plane experiments is justified by the expanding of this model to different settings in which it gives the successful production of parabolic path, of the angle of maximum range etc.

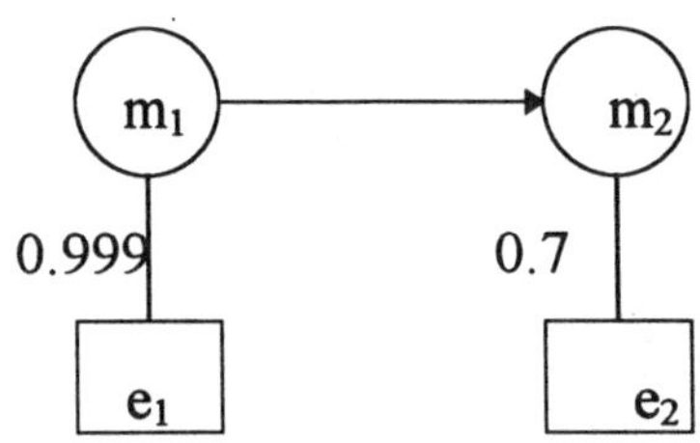

Figure 7.

Assume that m_2 is a limiting case of m_1. Let suppose that e_1 supports m_1 to a 0.999 level. But e_2 supports m_2 only to 0.7 level. In this case the scientist does not conclude that m_1 is adequately supported but m_2 is not. Instead, he searches for something that explains the discrepancy. The result is, that *there are not "good" or "bad" approximations in such a context, unless the links among the models being tested are taken into account.*

If you have a model, that with a quite rigorous rule of transformation works in many *prima facie* different cases you can bet on it. Successors of Galileo actually did.

The latter argument suggests an answer to the second problem, about the justification of models: an experimental test of a theoretical model m_1 is reinforced if there is one at least experimental test of a theoretical model m_2 and m_2 can be reduced to m_1 with specific transformation rules.

6. EPILOGUE: BACK TO MODEL THEORIES

Models not only in philosophy of science but in cognitive science also have been proposed as an alternative to rules. Rules, explicitly or implicitly have been interpreted as the isolation of similar characteristics or properties of objects or situations. As common characteristics seem not to work in some cases, two alternatives have proposed: the first, is "family resemblances". Nevertheless, "family resemblances" is again about common characteristics. The second is variants of "pattern recognition" or "seeing as".

The concept of "seeing as" has many times within philosophical debates been regarded as irrational and relativistic. It seems catastrophic for the philosophical domain, because it may legalize anybody to *see* something *as* anything he/she wants to. I used the example of Galileo to show that, in his case at least, this is not true. The links of the "prototype" with the possible extensions, as we saw, are rigorous and restricted, exactly because the rules of transformation are manifested within the manipulations of the model. These links are both mathematical and visual, and they are not less objective from than the ones based on common characteristics.

The historical case of Galileo shows that beyond family resemblances and pattern recognition there are other alternatives to model-based reasoning. I don't claim that "taking to the limit" is the only method by which models expand. I used it as an example of how we can have rules that are explicit and objective, but are not logical. Furthermore, I tried to show that the possession of transformation rules gives one extra reason for the reliability of experimental results: experiments in different situations corroborate each other, if the corresponding theoretical models are mutually transformed.

ACKNOWLEDGMENTS

This paper would hardly exist without the help of professor P. Machamer. I want also to thank my teacher and supervisor, professor A. Baltas, not only for his help in this paper, but for his general encouragement and guidance and Costas Pagondiotis for reading the draft and make suggestions.

REFERENCES

Beth, E., 1961, Semantics of physical theories, in H. Freudenthal, pp. 48-51.

Black, M., 1962, *Models and Metaphors*, Cornell University Press, New York.

Campbell, N.R., 1920, *Physics: The Elements*, Cambridge University Press, Cambridge. Republished under the title, *Foundations of Science*, Dover, New York, 1957.

Cooper, L., 1935, *Aristotle, Galileo and the Tower of Pisa*, Cornell University Press, Ithaca, New York.

Damerow, P., Freudenthal, G., McLaughlin, P., and Renn, J., 1991, *Exploring the Limits of Preclassical Mechanics*, Springer-Verlag, New York.

Dijksterhuis, E.J., 1987, *Archimedes,* Princeton University Press, Princeton.

Drake, S., 1973, Galileo's experimental confirmation of horizontal inertia: unpublished manuscripts, *Isis* 64:291-305.

Drake, S., 1978, *Galileo at Work,* The University of Chicago Press, Chicago and London.

Drake, S., 1991, *Galileo: Pioneer Scientist,* The University of Toronto Press, Toronto, Buffalo and London.

Freudenthal, H., 1961, *The Concept and the Role of the Model in Mathematics and Natural and Social Sciences*, Reidel, Dordrecht.

Hesse, M.B., 1966, *Models and Analogies in Science,* University of Notre Dame Press, Notre Dame, IN.

Galilei, G., 1590 [1960], *De Motu* [*On Motion*], translated by S. Drake, University of Wisconsin Press, Madison.

Galilei, G., 1600 [1960], *Le Meccaniche* [*On Mechanics*], translated by S. Drake, University of Wisconsin Press, Madison.

Galilei, G., 1632 (1967), *Dialogue Concerning the Two Chief World Systems*, Translated S. Drake, University of California Press, Berkeley, Los Angeles, London.

Galilei, G., 1638 [1974], *Discorsi* [*Discourse on the Two New Sciences*], translated by S. Drake, University of Wisconsin Press, Madison.

Giere, R., 1984, Towards a unified theory of science, in: *Science and Reality*, J. Cushing, C. Delaney, and G. Gutting, eds., University of Notre Dame Press, Notre Dame, IN, pp. 5-31

Giere, R., 1988, *Explaining Science,* The University of Chicago Press, Chicago.

Gooding, D., 1992, The procedural turn; or why do thought experiments work?, in: *Cognitive Models of Science*, R. Giere, ed., Minnesota Studies in the Philosophy of Science, vol. XV, University of Minnesota Press, Minneapolis, pp. 45- 76.

Kuhn, T.S., 1970, *The Structure of Scientific Revolutions,* Second Edition, The University of Chicago Press, Chicago and London.

Kuhn, T.S., 1970a, Postscript, 1969, in: T.S. Kuhn, 1970, pp. 170-210.

Lindberg, D., 1992, *The Beginnings of Western Science*, The University of Chicago Press, Chicago.

Machamer, P., and Woody, A., 1994, A model of intelligibility in science: using Galileo's balance as a model for understanding the motion of bodies, *Science and Education* 3:215-244.

Nersessian, N.J., 1992, How do scientists think? Capturing the dynamics of conceptual change in science, in: *Cognitive Models of Science*, R. Giere, ed., Minnesota Studies in the Philosophy of Science, vol. XV, University of Minnesota Press, Minneapolis, pp. 3-44.

Reichenbach, H., 1938, *Experience and Prediction*, The University of Chicago Press, Chicago.

Suppe, F., ed., 1977, *The Structure of Scientific Theories*, University of Illinois Press, Urbana, Chicago and London.

Suppes, P., 1962, Models of data, in: *Logic, Methodology and Philosophy of Science: Proceedings of the 1960 International Congress*, E. Nagel, P. Suppes, and A. Tarski, eds., Stanford University Press, Stanford, 252-261.

van Fraasen, B., 1970, On the extension of Beth's semantics of physical theories, *Philosophy of Science* 37:325- 339.

Wittgenstein, L., 1953, *Philosophical Investigations*, Blackwell, Oxford.

Simplifying Bayesian Inference: The General Case

Stefan Krauß, Laura Martignon and Ulrich Hoffrage
Center for Adaptive Behavior and Cognition, Max Planck Institute for Human Development, Berlin, Germany, krauss@mpib-berlin.mpg.de, martignon@mpib-berlin.mpg.de, hoffrage@mpib-berlin.mpg.de

Abstract: We present empirical evidence that human reasoning follows the rules of probability theory, if information is presented in "natural formats". Human reasoning has often been evaluated in terms of humans' ability to deal with probabilities. Yet, in nature we do not observe probabilities, we rather count samples and their subsets. Our concept of Markov frequencies generalizes Gigerenzer and Hoffrage's "natural frequencies", which are known to foster insight in Bayesian situations with one cue. Markov frequencies allow to visualize Bayesian inference problems even with an arbitrary number of cues.

1. INTRODUCTION

An essential component of scientific reasoning is postulating hypotheses about the world. Observing the world, though, does not provide proofs that these hypotheses are correct. The world only provides data. In order to check the fit between the data and our hypotheses we usually make use of conditional probabilities. This can happen in two ways:

1. Considering P(D | H)

P(D | H) is the probability of the data (D), given a hypothesis (H). It underlies, for instance, Fisher's null hypothesis testing.[1] Relying on P(D | H) is the path of statistical inference common in the social sciences as, for example, sociology or psychology. In most standard statistical textbooks only this approach is taught and presented as an undisputed methodology.

Yet, if we want to get an idea on the fit of the data and a hypothesis we can assess the conditional probability in the other direction:

2. Considering P(H | D)

Here we compute the probability of a hypothesis (H), given the data (D), that is, P(H | D), by means of *Bayes' formula*:

[1] In this case D is the present or more extreme data and H represents the null hypothesis.

Model-Based Reasoning in Scientific Discovery, edited by L. Magnani, N.J. Nersessian, and P. Thagard, Kluwer Academic/Plenum Publishers, New York, 1999

$$P(H|D)=\frac{P(D|H)\cdot P(D)}{P(D|H)\cdot P(D)+P(D|\neg H)\cdot P(\neg H)}$$

Scientific inference methods derived from this formula ("Bayesian models") were proposed during the Enlightenment. But their intractability for realistic problems was a serious roadblock to their use as statistical models. The advent of fast computers has circumvented intractability at least partially. Bayesian models for inference have recently played a fundamental role in scientific discovery in a variety of applications, such as the following (*Neue Zürcher Zeitung*, May 13, 1998, S. 39):

- The Max Planck Institute for Meteorology in Hamburg is presently applying a Bayesian analysis to hypotheses about changes in climate.
- The Socicty for Mathematics and Data Analysis in St. Augustin is using Bayesian methods for estimating credit risks.
- The "Krebsatlas" (almanac of cancer patients) in Germany is being reviewed at the Ludwig Maximilian University in Munich by means of Bayesian techniques. The task is to detect and eliminate spurious correlations.
- Mathematicians elicit hydrological conductivity of the ground at Sellafield from experts on behalf of the Britsh government. They use the experts' beliefs to estimate the fitness of the area as a permanent disposal site for nuclear.
- Most expert systems are based on Bayes' formula. A famous example is MUNIN (Muscle and Nerve Inference Network) from Lauritzen and Spiegelhalter (1988), used for diagnoses based on measurements of muscular electrical impulses ("electromyography").

Cognitive psychologists debate the question, whether the laws of probability theory, Bayes' formula in particular, are psychological. The Enlightenment thinkers viewed probability theory – an extension of the logical calculus – as the calculus of reason and the cornerstone of rational choice. Thus, the laws of probability were seen not just as normative, but also as descriptive of enlightened human thinking. Laplace saw probability theory as common sense reduced to a calculus. Yet, the identification of the laws of human reasoning with those of probability were not meant to last. The focus slowly shifted, and probability theory became just another mathematical tool of the natural sciences. As a description of human reasoning, probability theory was rejected from the early 19th Century onwards and except for some inveterate rationalists, this rejection lasted until the middle of the 20th Century.

Modern experimental psychology opened up another front in the war against the view of rational man as probabilist, in a flurry of work documenting the ways in which actual human reasoning differs from the probabilistic norm. Kahneman and Tversky (1973, p. 237) wrote: "In making predictions and judgments under uncertainty, people do not appear to follow the calculus of chance or the statistical theory of prediction." Deviations from the probabilistic ideal were regarded by many as proof that unaided human reasoning is riddled with fallacies. These claims were summarized in the famous book *Judgment under Uncertainty* (1982), edited by D. Kahneman, P. Slovic, and A. Tversky. This view has raised again new controversies that persist until today. Many of the so called fallacies can be explained and eliminated (e.g., Hell, Fiedler, and Gigerenzer 1993). In this debate we concentrate on one of the most controversial issues: Bayes' formula.

Bayes' formula allows to "update" the probability of a hypothesis when new data arise. This "updating" occurs in real life reasoning in a variety of situations: a judge has to change his belief in the guilt of a suspect (hypothesis) when new indicators (data) are gathered. A medical doctor has to change his belief on the state of his patient (hypothesis) when provided with new test outcomes (data).

1.1 Does unaided human reasoning follow Bayes' formula?

Ward Edwards, one of the chief proponents of Bayesian decision making, was the first to treat this question empirically. He tested people's reasoning and found that people are good intuitive "conservative" Bayesians, which means: if people have to judge situations which mathematically can be described in terms of Bayes' formula, they change their view in the right direction.

Later David M. Eddy (1982) found that judgments of medical doctors do not follow Bayes' formula when solving the following task:

The probability that a woman at age 40 has breast cancer (B) is 1%.
(P(B) = prevalence = 1%)

According to the literature, the probability that the disease is detected by mammography (M) is 80%.

(P(M+ |B) = sensitivity = 80%)

The probability that the test misdetects the disease although the patient does not have it is 9.6%.

(P(M+ |¬B) = 1 - specificity = 9.6%)

If a woman at age 40 tests positive, what is the probability that she indeed has breast cancer (P(B|M+)?

Bayes' formula yields the following result:

$$P(B|M+)=\frac{P(M+|B)\cdot p(B)}{P(M+|B)\cdot P(B)+P(M+|\neg B)\cdot P(\neg B)}=$$

$$=\frac{80\%\cdot 1\%}{80\%\cdot 1\%+9.6\%\cdot 99\%}=0.078$$

Thus, the probability of breast cancer is only 7.8%, while Eddy reported that 95 out of 100 doctors estimated this probability to be between 70% and 80% (Kahneman, Slovic, and Tversky, 1982). And even earlier Kahneman and Tversky (the latter a student of Edwards) had replied to Edwards: "In his evaluation of evidence, man is not a conservative Bayesian. He is not Bayesian at all" (1972, p. 450).

Kahneman and Tversky's "Heuristics and Biases" program determined the course of decision making theories in the last decades. The Italian researcher Massimo Piattelli-Palmarini, for instance indicated in 1991 that we are a species that is uniformly probability blind, from the humble janitor to the Surgeon General.

Gigerenzer and Hoffrage (1995) looked at the problem from another perspective: they focused on the aspect of representation of uncertainty. For instance, in Eddy's problem, quantitative information was given in probabilities. They argue that probabilities are not a natural format of information for the human mind. Observing the world we see samples and their subsets and keep records of the relations between their sizes. Only when we switch to quotients we are headed toward probabilities. The discussion goes even one step further: it is questionable, whether the single-event probability P (B | M+) exists at all. In fact, no one has ever seen a woman with a probability of .078 of breast cancer. There are only women *with* or women *without* breast cancer.

One can be a Bayesian without ever having heard of probabilities. Gigerenzer and Hoffrage presented Eddy's problem to medical doctors replacing probabilities with *frequencies*. In their formulation the task was:

> *100 out of 10000 women at age 40 who participate in routine screening have breast cancer.*
>
> *80 out of 100 women with breast cancer will get a positive mammography.*
>
> *950 out of 9900 women without breast cancer will also get a positive mammography.*
>
> *Here is a new representative sample of women at age forty who get a positive mammography in routine screening. How many of these women do you expect to actually have breast cancer?*

Although the task is – except for the information format – equivalent to Eddy's task, now nearly half (46%) of all doctors gave the Bayesian answer: 80 out of 1030 (7.8%).

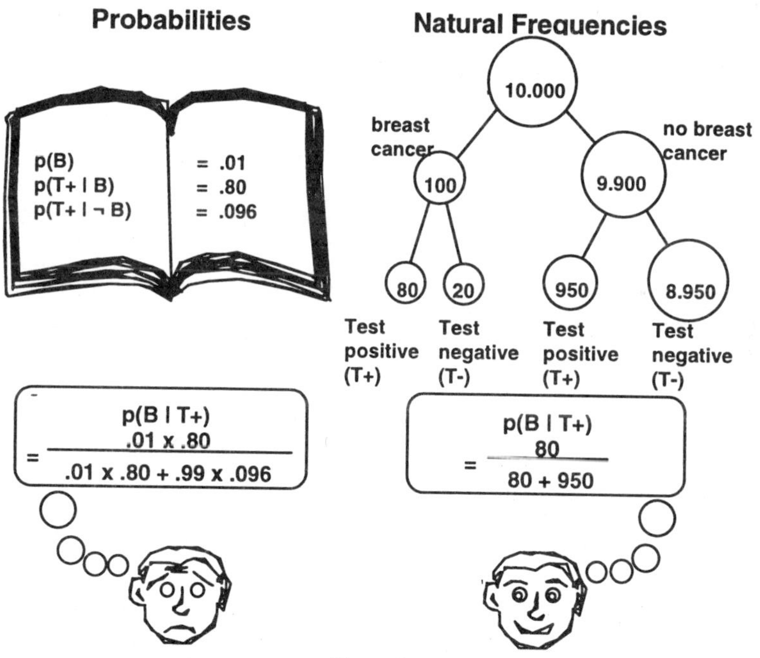

Figure 1.

Gigerenzer and Hoffrage call their frequencies "natural": imagine a doctor who does not know the sensitivity and the specificity of the test. After having applied mammographies to a large number of patients, he would "naturally" obtain numbers that he could place in the tree of Figure 1.

There is a lively discussion on these results. Recent articles confront, replicate or simply comment Gigerenzer and Hoffrage's findings, among others

Mellers and McGraw (1999), Lewis and Keren (1999) and Macchi and Mosconi (1998). What is the crucial property of natural frequencies, which helps one to find the Bayesian solution? To answer this question, it is helpful to consider a more general case. In real-life situations, decisions are usually based on several cues. A medical doctor, for instance, seldom diagnoses a disease based on a single test. The usual procedure after a mammography (M) is to perform an ultrasound test (U).

We tested empirically three versions of a Bayesian task with two cues (a medical diagnostic task based on a mammography and an ultrasound test), to find out how people deal with different information formats in this type of settings:

Empirical study

Version 1: "probability format", 41 subjects

The probability that a woman at age 40 has breast cancer (B) is 1%.

According to the literature, the probability that the disease is detected by mammography (M) is 80%.

The probability that the mammography (M) misdetects the disease although the patient does not have it is 9.6%.

The probability that the disease is detected by an ultrasoundtest (U) is 95%.

The probability that the ultrasound test (U) misdetects the disease although the patient does not have it is 4%.

What is the probablity that a woman at age 40 has breast cancer, given that she has a positive mammography and a positive ultrasound test?

Provided with this probability format, only 12.2% of our participants (5 out of 41) reached the correct solution (P(B | M+ & U+) = (2/3)).

1.2 Is there a "natural" frequency format for the two cue situation?

Dominic Massaro, who discussed Gigerenzer and Hoffrage's, findings in his book *Perceiving Talking Faces* (1998, pp. 174-179), gave an example describing the same situation with frequencies:[2]

[2] To integrate the research on this topic, we borrowed concepts from various sources and explored them in the breast cancer example. In fact, Gigerenzer and Hoffrage used a sample of 1.000 (not 10.000) women, Massaro speaks of symptoms instead of tests and we tested our subjects with "tuberculosis tasks" instead of "breast cancer tasks".

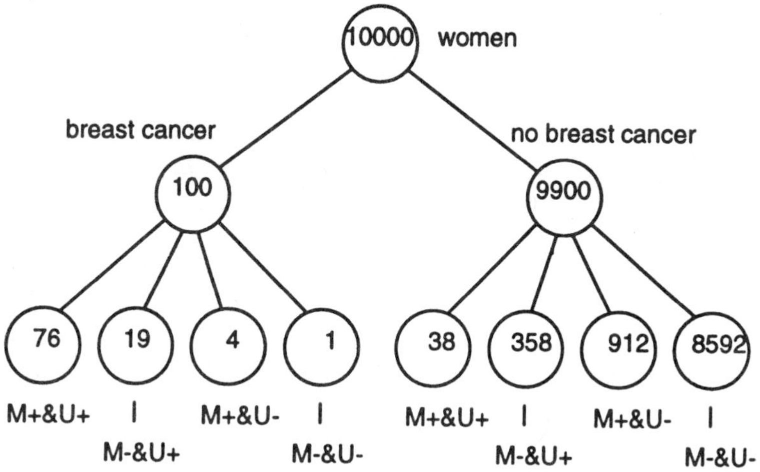

Figure 2.

Massaro states that in the case of two cues "a frequency algorithm will not work" and "it might not be reasonable to assume that people can maintain exemplars of all possible symptom configurations." However, his statements are not based on experimental evidence and his frequencies express the *combined* sensitivity P(M+ & U+ | B) and the *combined* specificity 1-P(M- & U- | ¬B). Because we only want to test frequency versions equivalent to our probability version we must find frequencies corresponding to the *single* sensitivities and specificities from our probabilistic format. One possibility is to formulate the situation with *two* trees in parallel for the *two* medical tests as one possible generalization of Gigerenzer and Hoffrage's *one* tree for *one* medical test.

From these trees we obtained the following frequency format, which we called "parallel frequency format". We used it as the second version in our empirical study (only the cover story was presented, no trees were drawn).

Empirical study

Version 2: "parallel frequency format", 41 subjects

100 out of 10000 woman at age 40 who participate in routine screening have breast cancer.

80 out of 100 women with breast cancer will get a positive mammography.

950 out of 9900 woman without breast cancer will also get a positive mammography.

95 out of 100 women with cancer will get a positive ultrasound test.

396 out of 9900 women, although they do not have cancer, nevertheless obtain a positive ultrasound test.

How many of the women who get a positive mammography and a positive ultrasound test do you expect to actually have breast cancer?

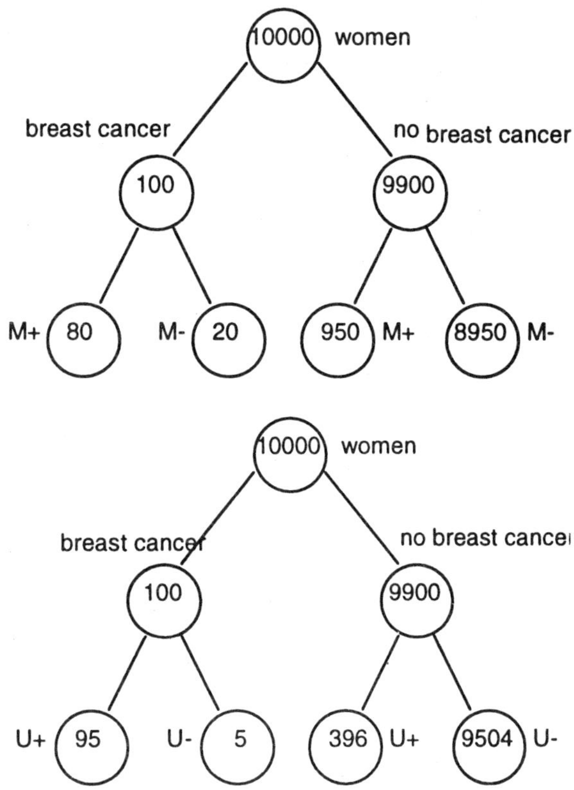

Figure 3.

Provided with the parallel frequency version only 14.6% of our participants (6 out of 41) solved the task correctly. This is not a frequency format that fosters insight into Bayesian inference. Note, that numbers like those of Figure 3 could only be obtained by testing disjoint groups of patients: one group with mammography test only (the tree on the left hand side), the other group with ultrasound test only (the tree on the right hand side). This situation is not realistic. The reality of medical testing is better represented by tests taken sequentially, one after the other. For a diagnosis of breast cancer a doctor usually first applies a mammography and then an ultrasound test. A possible numerical description of this type of repeated routine would be:

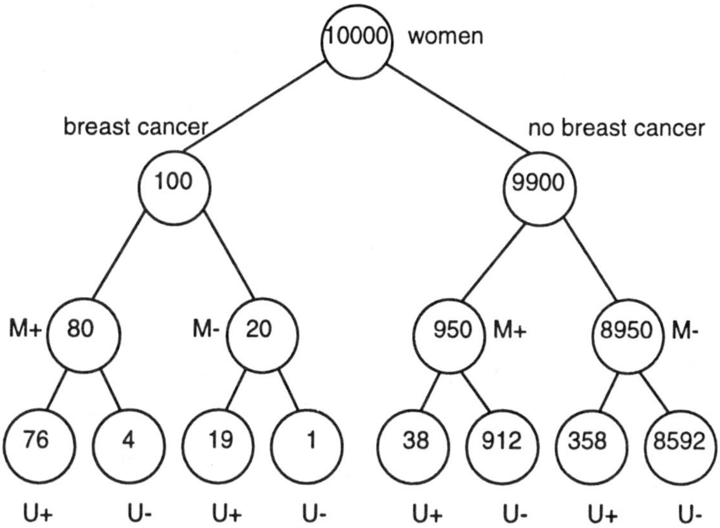

Figure 4.

We used a diagnostic task referring to the information in this tree as version 3 in our empirical study:

> Empirical study
>
> Version 3: "sequential frequency format", 41 subjects
>
> *100 out of every 10000 women at age 40 who participate in routine screening have breast cancer.*
>
> *80 of every 100 women with breast cancer will get a positive mammography.*
>
> *950 out of every 9900 women without breast cancer will also get a positive mammography.*
>
> *76 out of 80 women who had a positive mammography and have cancer also have a positive ultrasound test.*
>
> *38 out of 950 women who had a positive mammography, although they do not have cancer, also have a positive ultrasound test.*
>
> *How many of the women who get a positive mammography and a positive ultrasound test do you expect to actually have breast cancer?*

53.7% of our participants (22 out of 41) solved this task correctly, in spite of the fact that this version is equivalent to our probability version 1 (as it holds for version 2). The chart of Figure 5 summarizes our results.

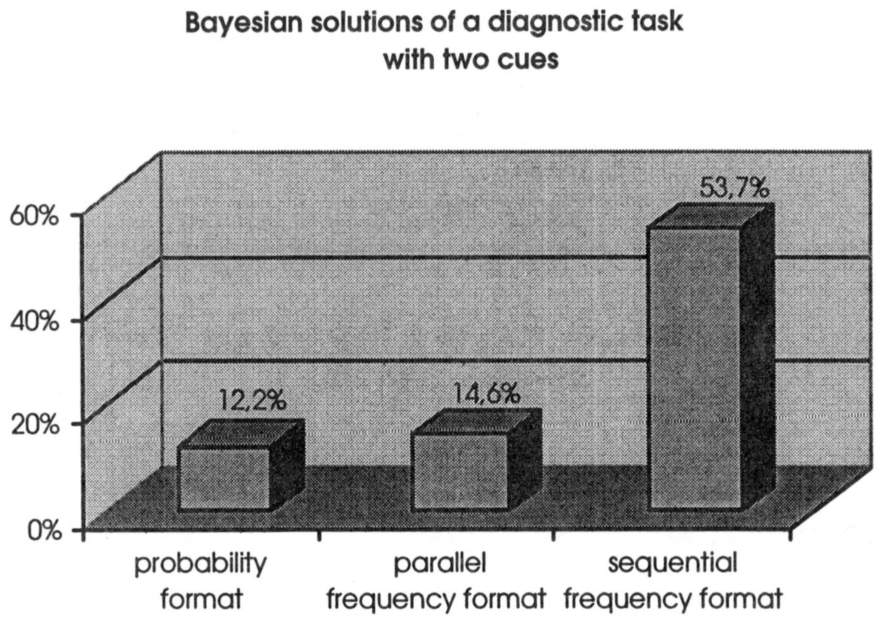

Figure 5.

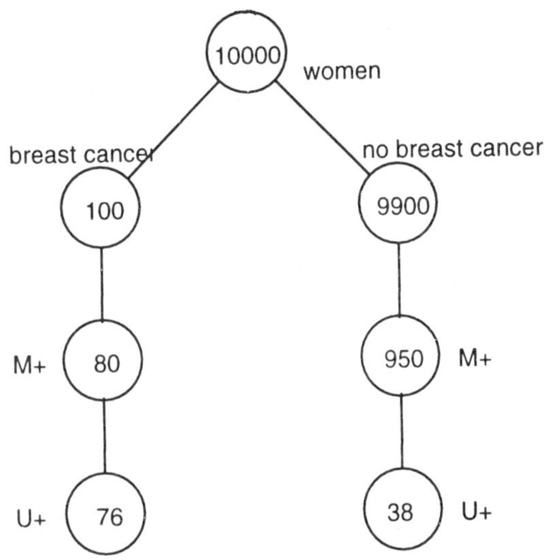

Figure 6.

Note, that not all frequencies in Figure 4 were actually used to formulate version 3. In order to visualize the breast cancer risk for a woman with two positive tests, it is possible to eliminate all frequencies irrelevant to this case (see Figure 6).

These frequencies provide the adequate visualization of general Bayesian situations. We call them *Markov frequencies* because of the analogy with Markov chains. In fact:

1. Our tree consists of two chains.
2. Each node corresponds to the reference class from which the label for the next node is extracted. As in a Markov chain, the frequency in each node depends only upon its predecessor, not upon other previous nodes.

2. REMARKS AND DISCUSSION

Through natural sampling we obtain frequencies all related to one reference class, namely the observed one. Markov frequencies correspond to this natural process and are therefore generalizations of Gigerenzer and Hoffrage's natural frequencies to the case of more than one cue. In order to grasp the crucial mathematical feature of Markov frequencies, let us briefly reconsider the one cue situation. Lewis and Keren (1999) stress that even in this simpler case not just *any* frequencies foster insight. They tested the following frequency version of Eddy's task (with slightly different numbers):

100 out of 10000 women at age 40 who participate in routine screening have breast cancer.

8000 out of 10000 women with breast cancer will get a positive mammography.

960 out of 10000 women without breast cancer will also get a positive mammography.

Here is a new representative sample of women at age forty who get a positive mammography in routine screening. How many of these women do you expect to actually have breast cancer?

Participants did not perform well when tested on this task. The frequencies of their task are equivalent to Eddy's probabilities, yet these frequencies do not correspond to natural observations. A doctor who had sampled Lewis and Keren's frequencies would have seen exactly as many women with cancer as without cancer. This relationship is very unlikely with the given prevalence (100 out of 10000 have breast cancer). There is a problem shared by Lewis and Keren's task and our version 2 (parallel frequency): the reported numbers correspond to disjoint reference classes. In natural observations the sampled numbers are connected to each other, because they all refer to subsets of the same observed reference class. Lewis and Keren's numbers do not relate to *one* reference class. Observe that the first information (about the prevalence) mentions 100 women with breast cancer, while the second (about the sensitivity) mentions 10000 women with breast cancer.

In 1995 Macchi had already recognized that even in a probabilistic format stressing the relationship between the provided informations can make a difference. She reached an improvement of the performance only by changing slightly the wording in the cover story. She introduced the term "partitive" in order to describe wordings that clarify the relationship between the percentages in the cover story. The following example is taken from Macchi (1995):

- (non partitive) "In a population of adolescents, 80% of *suicide attempts* are made by girls and 20% by boys. The percentage of *death by suicide* is three times higher among boys than among girls. What is the probability that an adolescent, selected at random from those who had died by suicide, was a boy?"
- (partitive) "In a population of adolescents, 80% of *suicide attempts* are made by girls and 20% by boys. The percentage of *suicide attempts* that result in death is three times higher among boys than among girls. What is the probability that an adolescent, selected at random from those who had died by suicide, was a boy?"

In Macchi's "partitive" version the percentage of suicide attempts is clearly a part of the 80%. Gigerenzer and Hoffrage's natural frequencies make this relationship even clearer, since the first number in the prevalence (*100* out of 10.000) is exactly the second number in the sensitivity (80 out of *100*).

In the case of two cues the *nested subsetting* must be maintained, because this is the only format of frequency data that corresponds to the outcome of real-life observations. Let us summarize the main features of Markov frequencies:

> *Markov Frequencies arise from successive partitioning from a unique reference class. They express more than the mere proportions for prevalence, sensitivity and specificity, because they reflect relationships between numbers which is a characteristic of the corresponding natural observations.*

One can easily imagine how Markov frequencies look like when more cues are added. For *n* cues one needs only check 2^n pieces of information, not 2^{n+1} as in Massaro's proposed frequency format. This is because Markov frequencies are *task oriented*: only branches and leaves relevant to the question "P (B | M+ & U+) = ?" have to be depicted in the tree (see Figure 5).

Clearly, Markov frequencies can describe the answers to all other possible questions. The Markov frequencies corresponding to the question "What is the probability that a woman with a positive mammography and a *negative* ultrasound test has breast cancer (P (B | M+ & U-) = ?)", for instance, is:

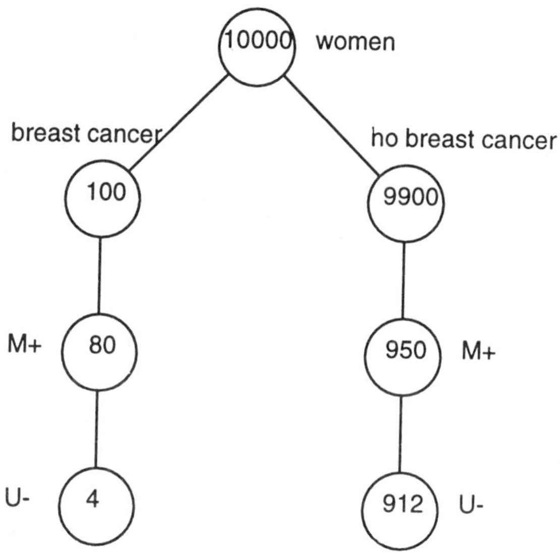

Figure 7.

Figure 4 collects the answers to all possible test configurations (M+ & U+, M+ & U-, M- & U+, M- & U-). In each case, the odds of the problem can be read directly from the last two leaves. These odds can also be computed by the well-known likelihood-combination rule (see, for instance, Spies, 1993, pp. 51-54):

prior odds • product of the likelihood ratios = posterior odds

For the (M+ & U+) configuration (used in our empirical study) we have:

the prior odds for breast cancer are $\frac{100}{9900}$.

Multiplying this with the likelihood ratio for the mammography,[3] we obtain $\frac{80}{950}$.

Again multiplying this with the likelihood ratio of the ultrasound test, we finally get $\frac{76}{38}$.

Macchi und Mosconi (1998) criticize that Gigerenzer and Hoffrage's natural frequencies eliminate all need for computation because their frequency formats contain the numbers that compose the answer. This is clearly also true of our Markov frequencies. Yet our claim is that the story has to be told in the opposite direction: probabilities introduced the need for computations. By observing samples in nature and assessing subsets' sizes we are naturally *performing and understanding* Bayesian inferences. Things first become cumbersome when the inference is expressed in terms of probabilities: from that moment on Bayesian inferences consist of distressing inversions. Side effects of probability formats are the famous base rate neglect and the confusion of different conditional probabilities. Performing Bayesian inference by means of Markov frequencies, instead, requires no inversions and the base rates cannot be neglected because the last two leaves of the tree carry the information about the base rates automatically.

Markov frequencies have another advantage: from a mathematical point of view one could criticize that our versions 1 and 2 contain no information on the conditional dependence of tests 1 and 2, given the disease. One could imagine that P(U+ | B & M+) ≠ P(U+ | B) = 95%. For instance a positive mammography test could reduce the probability that the ultrasound test is also positive (mammography and ultrasound tests actually do not influence each other, but the tests used for diagnosing prostate cancer, for instance, show interdependencies). Yet, even if P(U+ | B & M+) were equal to 70% rather than 95% a doctor would not have to worry about the dependency between tests in the presence of cancer. He would automatically sample 56 women (where in Figure 5 there used to be a 76). With the numbers 56 and 38 he can again compose the answer "56 out of 94 (= 59%)" to the question "P(B | M+ & U+) = ?", without ever having heard of the concept of stochas-

[3] The likelihood ratio *L(B, M+)* is defined by $\frac{P(M+/B)}{P(M+/\neg B)}$, which is $\frac{80\%}{9{,}6\%} \approx 8.3$

The likelihood ratio *L(B, U+)* therefore is $\frac{95\%}{4\%} = 23.75$

"P(B | M+ & U+) = ?", without ever having heard of the concept of stochastic dependency. Thus Figure 4 represents not just the risk of having breast cancer for all possible test configurations but also reflects all possible dependencies between cues.

Markov frequencies and their visualization can help subjects solve diagnostic tasks and doctors communicate risks. Furthermore, they reflect the outcomes of real observations. Saying that they "eliminate difficulties" which humans have dealing with uncertainty is only half of the story. These difficulties (base rate neglect, confusing conditional probabilities, struggling with the problem of stochastic dependency) were first created by the introduction of probabilities.

REFERENCES

Eddy, D.M., 1982, Probabilistic reasoning in clinical medicine: problems and opportunities, in: *Judgment under Uncertainty: Heuristics and Biases*, D. Kahneman, P. Slovic, and A. Tversky, eds., Cambridge University Press, Cambridge, pp. 249-267).

Gigerenzer, G., and Hoffrage, U., 1995, How to improve Bayesian reasoning without instruction: frequency formats, *Psychological Review* 102:684-704.

Hell, W., Fiedler, K., and Gigerenzer, G., eds., 1993, *Kognitive Täuschungen*, Spektrum Akademischer Verlag, Berlin-Oxford.

Kahneman, D, Slovic, P., and Tversky, A., eds., 1982, *Judgment under uncertainty: Heuristics and biases*, Cambridge University Press, Cambridge.

Kahneman, D., and Tversky, A., 1972, Subjective probability: a judgement of representativeness, *Cognitive Psychology* 3:430-454.

Kahneman, D., and Tversky, A., 1973, On the psychology of prediction, *Psychological review* 80:237-251.

Lauritzen, S.L., and Spiegelhalter, D.J., 1988, Local computations with probabilities on graphical structures and their application to expert systems (with discussion), *J. Roy. Statist. Soc. Ser. B* 50:157–224.

Lewis, C., and Keren, G., 1999, On the difficulties underlying Bayesian reasoning: comment on Gigerenzer and Hoffrage, *Psychological Review*, in press.

Macchi, L., 1995, Pragmatic aspects of the base-rate fallacy, *The Quarterly Journal of Experimental Psychology* 48A:188-207.

Macchi, L., and Mosconi, G., 1998, Computational features vs frequentist phrasing in the base-rate fallacy, *Swiss Journal of Psychology* 57:79-85.

Massaro, D., 1998, *Perceiving Talking Faces*, MIT Press, Boston.

Mellers, B.A., and McGraw, A.P., 1999, How to improve Bayesian reasoning without instruction: comment on Gigerenzer and Hoffrage, *Psychological Review*, in press.

Spies, M., 1993, *Unsicheres Wissen: Wahrscheinlichkeit, Fuzzy-Logik, neuronale Netze und menschliches Denken*, Heidelberg, Spektrum Akademischer Verlag, Berlin-Oxford.

Complexity versus Complex Systems: A New Approach to Scientific Discovery

F. Tito Arecchi
Department of Physics, University of Firenze and Istituto Nazionale di Ottica, Firenze, arecchi@firefox.ino.it

Abstract: Extraction of quantitative features from observations via measuring devices M means that the words of science are coded as numbers, and the syntaxis is a set of mathematical rules, thus all consequences should be worked in a purely deductive way. This characteristic of science displays two orders of drawbacks, namely, undecidability of deductive procedures, and intractability of complex situations. The way out of such a crisis consists in a frequent readjustment of M suggested by the observed events. This adaptive strategy differs from the adaptivity of a learning machine, which – inputted by a data stream – readjusts itself over a class of theoretical explanations in order to select the optimal choice. On the contrary, the scientist not only modifies the explanations for a fixed data set, but also explores different data sets by modifying M, that is, by selecting a different point of view. This M-adjustment is a pre-linguistic operation, not expressible by a formal language. Hence, the scientific endeavor can not be reduced to a machine task.

1. INTRODUCTION

It is important to distinguish between A) "Complexity" and B) "Complex Systems" (Arecchi and Farini, 1996). The first term has been differently defined in formal languages (Hopcroft and Ullman, 1979), computer science (Kolmogorov, 1965; Chaitin, 1966) and nonlinear signal analysis (Grassberger, et al., 1991; Abarbanel, et al., 1993), starting in the early 80's with the intrinsic non predictability associated with chaotic time series (Shaw, 1981). A) is associated with epistemic processes. Once a time series of data, coded in a given alphabet, has been assigned, can one retrieve the meaning of the message just by perusal of that sequence? This contextual analysis uncovers the grammatical rules which allow some symbol sequences (words) and forbid some others; it also attributes a likelihood to each word and hence it provides predictions about the future of the time series.

The A) approach has received different formulations, with different solutions leading to automatic procedures on complexity assignment (Grassber-

Model-Based Reasoning in Scientific Discovery, edited by L. Magnani, N.J. Nersessian, and P. Thagard, Kluwer Academic/Plenum Publishers, New York, 1999

ger, 1986; Crutchfield and Young, 1989; D'Alessandro and Politi, 1990; Badii and Politi, 1997).

On the other hand, natural scientists have focused rather on the ontological fact that reality uncovers lots of complex structures.

Two different tasks are thus to be considered, namely,

a) given an input, what is the optimal use we can make of it? We call "certitude" the subjective confidence that we have done the best in grasping the inner rules of the input;
b) facing a piece of world, can we express our knowledge of it in a suitable language, i.e. encode phenomena into symbol sequences from some alphabet (which later will be analyzed as in a))?

b) is "prior" to a), and it is dealt with by adaptive strategies, which we can later formalize as linguistic procedures, but which arise at a prelinguistic level and even determine the same choice of the most appropriate language (Simon, 1982). This more fundamental problem is that of "Truth", defined as *"adaequatio intellectus et rei"* that is "adjustment of our expectations to the changing world".

The two approaches are altogether different. In case a) a learning machine can be foreseen which automatizes the quest for complexity (Crutchfield and Young, 1989). On the other hand, case b) hints at the crucial role of a pre-linguistic stage where we still have to decide how to encode the stream of perceived phenomena into a linguistic sequence, which is then exposed to the inquiry of the complexity machine b).

Our main conclusion is that while there may be a complexity machine for a), it is in principle impossible to introduce a science machine for b). Hence b) remains a human endeavor not reducible to automatic procedures.

The paper is organized as follows.

Section 2 reviews current definitions of complexity showing the virtues and intrinsic limitations of a contextual analysis of a data stream. Section 3 is a dynamic approach to a complex situation. While in section 2 no apriori rules are requested but a data stream is preliminary, in section 3 we take for granted the general rules of dynamics and prospect a variety of possible data streams, that is, a variety of different physical situations. This variety appears as a natural implementation of the intuitive concept of complexity, as it is representative of what we call complex systems, from economy and sociology to biology and physics.

In section 4 we present an adaptive strategy recently introduced to recognize (Arecchi, et al., 1994) and control (Boccaletti and Arecchi, 1995) chaotic dynamics (Arecchi and Boccaletti, 1997). While an adaptive procedure was already incorporated in the complexity machine of section 2, in order to fit the theory to the input sequence, here, in a more radical way, we change the same structure of M in order to provide different sets of measurement sequences to be later analyzed.

We conclude in section 5 with some epistemological insight, recalling that a knowledge program based on assigned input data is how to make the best use of our mental representations according to a subjectivistic gnoseology started by Descartes and continued by Hume and Kant. On the contrary, knowledge based on readjustment of our measuring procedures appears as the most natural attitude of living beings.

2. COMPLEXITY OF SYMBOLIC SEQUENCES

In computer science, we define the complexity of a word (symbol sequence) as some indicator of the cost implied in generating that sequence. There is a "space" cost (length of the instruction stored in the computer memory) and a "time" cost (the CPU time for generating the final result out of some initial instruction).

A space complexity called AIC (Algorithmic Information Complexity) (Kolmogorov, 1965; Chaitin, 1966) is defined as the length in bit of the minimal instruction which generates the wanted sequence. This indicator is maximum for a random number, since there is no compressed algorithm (that is, shorter than the number itself) to construct a random number.

A time complexity called "logical depth" (Bennett, 1987) is defined as the CPU time required to generate the sequence out of the minimal instruction. It is minimal for a random number, indeed, once the instruction has stored all the digits, just command: PRINT IT.

Of course, for simple dynamical systems as a pendulum or the Newtonian two-body problem, both complexities are minimal.

While AIC refers to the process of building a single item, logical depth corresponds to finding the properties of all possible outputs from a known source.

In fact, the exact specification of the final outcome is too much for the ambition of the natural scientist, whose goal is more modest. It may be condensed in the two following items:

i) to transmit some information, coded in a symbol sequence, to a receiver in a compact way, possibly economizing with respect to the actual string length, that is, making good use of the redundancies (this requires a preliminary study of the language style);

ii) to predict a given span of future, that is, to assign with some likelihood a group of forthcoming symbols;

For this second goal, introduction of a probability measure is crucial (Grassberger, 1986; Gell-Mann, 1994) in order to design a complexity-machine, able to make the best informational use of a given data set.

In view of the difference between a) and b) introduced in section 1 let us sketch the essential elements of knowledge building in natural sciences. First of all, we realize a device, or measuring apparatus M, whose output is informationally equivalent to what is going on in the observed piece of world.

We thus attribute to knowledge two different meanings:

i) as we face a phenomenon, M captures (presumably) the relevant aspects of it, so that we can transfer sufficient information, either to another partner or to ourselves if we have to reflect in order to build a possible theory. Knowledge improvement implies trying with different M apparatuses by a suitable program that we will explore in section 4.
ii) As observer O_1 is exposed to a symbol stream, it has to transmit a compact explanation to O_2, so that O_2 is able to retrieve the same input sequence. The explanation consists of a tentative theory that we call model *m*.

The modeler O_1 is inputted by some sequence *s*, and it sends an explanation which should enable O_2 to reconstruct an output $s' = s$.

Notice that M is a whole class of possible instruments, and different M's will give rise to different data sequences (different words).

The explanation (Crutchfield, 1992) consists of a theoretical guess (model *m*) whose validity is tested by simulating an output and comparing it with the actual input data *s*. The difference yields an error signal *e*. Observer O_2 is provided with both *m* and *e* and it can reconstruct $s' = s$ upon this information.

The virtue of the explanation X is to have a bit length $\|x\| = \|m\| + \|e\|$ shorter than the sequence length $\|s\|$.

The explanatory machine is complex in so far as it spans over a whole class of models *m*. If one had access to a complete probabilistic description of the modeling universe, then the goal would be to maximize the probability of *m* conditional to the input *s*

$$\Pr(m \mid s) \qquad (1)$$

This ideal complete description is not available, but an approximation can be obtained by the Bayes' rule

$$\Pr(m \mid s) = \frac{\Pr(s \mid m) P(m)}{\sum_m P(s \mid m)} \qquad (2)$$

All these probabilities are conditioned on the choice of the model class. Furthermore, all terms on the right hand side refer to a *single* data stream *s*. Here $\Pr(s|m)$ is the probability that a model *m* produces the given data. With sufficient effort, $\Pr(s|m)$ can be estimated. Finally the normalization in the denominator depends only on the given data and so can be dropped as a constant.

The most likely explanation corresponds to the shortest code of length $\|x\| = -\log_2 \Pr(m|s)$.

There are two criteria for a good explanation:

i) *x* must explain s, that is, O_2 should resynthesize the original data $s' = s$
ii) the bit length

$$\|x\| = \|m\| + \|e\|$$

must be minimized.

The efficiency of an explanation is given by the compression ratio

$$C(m, s) = \frac{\|x\|}{\|s\|} = \frac{\|m\| + \|e\|}{\|s\|} \qquad (3)$$

C is a cost function of the "complexity" of the explanation. The optimal model minimizes this cost.

There are two limit cases. When the model is trivial ($\|m\| = 0$), the entire data are on the error channel: $\|e\| = \|s\|$. On the contrary a tautological model $m=s$ has no error: $\|e\| = 0$.

In this section we have explored the complexity of a given symbolic sequence either from the computational point of view, aimed at reconstructing the single item, or from the probabilistic point of view, aimed at selecting a model within a class.

However preliminary to that, the problem arises of how we have obtained a given sequence, and this implies a critical analysis of the measurement apparatus M.

3. A DYNAMICAL APPROACH TO COMPLEXITY

Modern science started from Galileo's rule: "do not try to grasp the 'essence' (i.e., the nature) of a thing, but stick only to its quantitative affections" (Galilei, 1932). Quantities are extracted by a measuring procedure that we denoted by M. We call the sequence of M outputs as the "words" of science.

The most crucial aspect of this procedure is the apparent arbitrariness of M, that is, the large number of different scientific words to be associated with an event usually denoted by a single word in any ordinary language.

This proliferation of M's has avoided for centuries the question of complexity. Suppose for instance that we look at the cell dynamics by M_1. Thereby, we build the specific language "cytology", with its own words. We may now realize that many cells form an organ, but the organ is observed by a different M_2 which provides different words and hence a different science, namely, "physiology". Similarly if we look at the biomolecules we build a "biochemistry".

We change language depending on the M apparatus.

A more economic criterion would consist in a reductionist approach. We can build a hierarchy from large to small and say that the behavior of smaller objects should determine that of larger ones. But here a perverse thing, already hinted by Anderson (1972), occurs. If our words were a global description of the object in *any* situation (as the philosophical "essences" in Galileo) then, of course, knowledge of elementary particles would be sufficient to make predictions on animals and society. In fact, Galileo's self-limitation to some "affections" is sufficient for a limited description of the event, but only from a narrow point of view. Even though we believe that humans are made of atoms, the affections that we measure in atomic physics are insufficient to make predictions on human behavior.

The fact that higher levels of organization display features not predictable from the lower ones, is what colloquially we call *complexity*.

This way, complexity is not a property of things (like being red or hot) but it is a relation with our status of knowledge, and for modern science it emerges from Galileo's self-limitation.

Reductionism from large to small as sketched above was accompanied by a logical reduction of the scientific explanation to a deductive task out of a set of axioms.

In this spirit, a scientific theory is considered as a set of primitive concepts (defined by suitable measuring apparatuses M) plus the mutual relations. Concepts and relations are the axioms of the theory. The deduction of all possible consequences (theorems) provides predictions which have to be compared with the observations. If the observations falsify the expectations, then one tries with different axioms.

The deductive process is affected by a Gödel undecidability like any formal theory, in the sense that it is possible to build a well formed statement, but the rules of deduction are unable to decide whether that statement is true or false.

Besides that, a second drawback is represented by *intractability*, that is, by the exponential increase of possible outcomes among which we have to select the final state of a dynamical evolution. Figure 1 sketches the bifurcation tree of a complex nonlinear dynamics, as one changes a suitable control parameter α. During the dynamical evolution of an open system, the control parameters $\{\alpha\}$ may assume different values, hence the cascade of bifurcations provides a large number of final states starting from a unique initial condition.

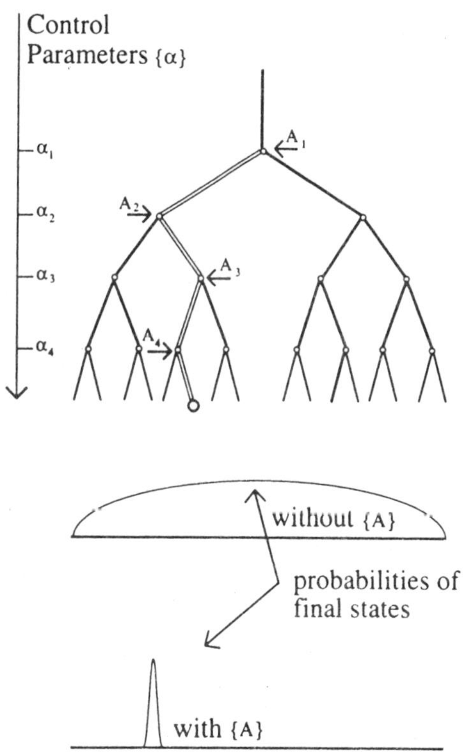

Figure 1. Bifurcation tree in nonlinear dynamics. As a control parameter α is tuned through different values, novel steady states appear. By tuning α from α_0 to α_N, the system goes from 1 to 2^N different states. We call dynamical complexity such an ambiguity. It is responsible for the failure of reductionism. In the absence of external gradients, all final outcomes are equally probable. We call "organization" the occurrence of just one event out of 2^N. This implies that at each α_i an external agent A_i has broken the bifurcation symmetry (see next figure).

Thus the reductionistic tentative of explaining reality out of its constituents yields an exponentially high number of possible outcomes, when only one is in fact that observed. This means that, while the theory, that is the syntaxis, would give equal probability to all branches of the tree, in reality we observe an *organization process*, whereby only one final state has a high probability of occurrence.

It is here necessary to recall some descriptive elements on the bifurcation of the stable branches of a nonlinear dynamics for different settings of a control parameter. Notice that dynamical bifurcations in a system of interactive identical particles display specific symmetries. Only external gradients break this symmetry (Figure 2).

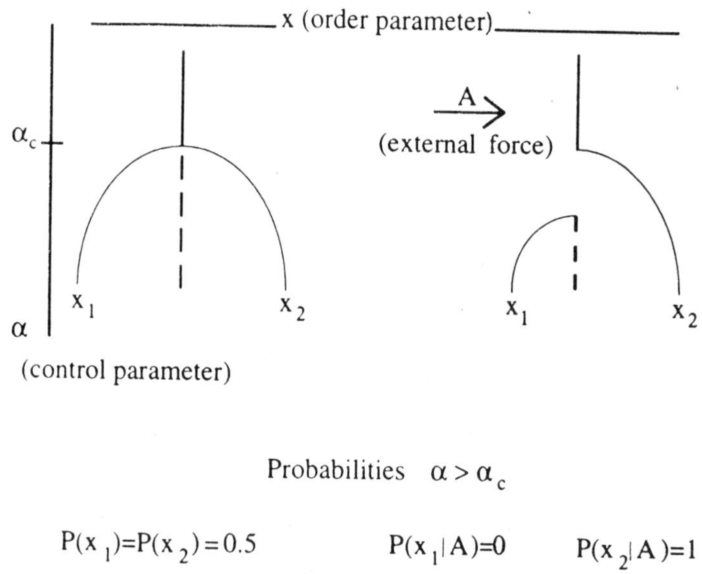

Figure 2. Examples of bifurcation diagrams. The dynamical variable x (order parameter) varies horizontally, the control parameter α varies vertically. Solid (dashed) lines represent stable (unstable) steady states as the control parameter is changed. Left: symmetric bifurcation with equal probabilities for the two stable branches. Right: asymmetric bifurcation in presence of an external field A. If the gap introduced by A between right and left branch is wider than the range of thermal fluctuations at the transition point α_c, then the right (left) branch has probability 1 (0).

Hence, there has been organization, that is a unique final state, this means that at each bifurcation vertex of Figure 1 the symmetry was broken by an agent external to the system under investigation.

We can thus stipulate the following things:

i) A set of control parameters

$$\alpha_1,\alpha_2,...,\alpha_N = \{\alpha\}$$

is responsible for successive bifurcations leading to an exponentially high number (in the example, of the order of 2^N) of final outcomes. If the system has no boundary effects (considered of infinite size) then all outcomes have comparable probabilities and we call complexity the impossibility of predicting which one is the state we will observe at the end of the chain of bifurcations.

ii) For a system of finite size embedded in an environment, there is a set of external forces

$$A_1, A_2,..., A_N = \{A\}$$

applied at each bifurcation point, which break the symmetries, biasing toward a specific choice and eventually leading to a unique final state.

We are in presence of a conflict between (i) syntaxis represented by the set of rules (axioms) $\{\alpha\}$ and (ii) semantics represented by the intervening external agents $\{A\}$. The syntaxis provides 2^N legal outcomes. But if the system is open to the external world, the presence of which is expressed by $\{A\}$, then it organizes to a unique final outcome. Once the syntaxis $\{\alpha\}$ is known, the final result is therefore an ascertainment that the set of external events $\{A\}$ must have occurred. Therefore we can take $\{A\}$ as the element of reality in which our system is embedded.

We define "certitude" the correct application of the rules $\{\alpha\}$, and "truth" the adaptation to the reality which is expressed by $\{A\}$.

However the same final outcome would be reached by a different set of rules $\{\beta\}$. In such a case, retracing back to the new tree of bifurcations, we would reconstruct a set $\{B\}$ of external agents. Thus, it seems that truth, $\{A\}$ or $\{B\}$, is language dependent!

Furthermore, the "emergence" of organization means that we can even build a set of axioms $\{\varepsilon\}$ which succeeds in predicting the correct final state without external perturbations, that is, $\{E\}=\varnothing$ (Figure 3). This is indeed the pretension of the so called "autopoiesis", or "self-organization" (Krohn, et al., 1990).

From a cognitive point of view, the theory $\{\varepsilon\}$ can be reputed to be a "petitio principi", a tricky formulation tailored for a specific purpose and not applicable to slightly different situations. Rather than explicitly detailing the elements of environment which break the symmetry, as e.g. $\{A\}$ for $\{\alpha\}$, the user of language $\{\varepsilon\}$ has already exploited at a pre-formalized level a detailed knowledge of the process in planning the axioms $\{\varepsilon\}$.

An "ad-hoc" model may be appropriate for a specific situation, but in general it lacks sufficient breadth to be considered as a general theory (think e.g. of the Ptolemaic model of the solar system). However, in describing the adaptive strategy of a living species, or a community etc., "self-organization" may be the most successful description. In other words, once the environmental influences have been known, it is better to incorporate that knowledge in the model, thus assuring the fast convergence to a given goal.

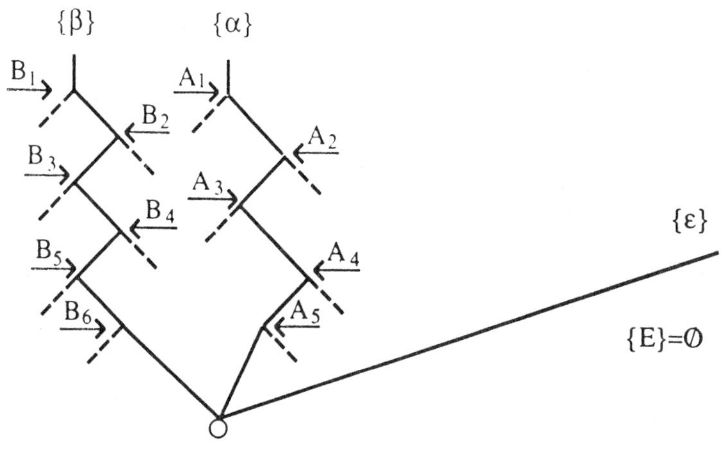

Figure 3. Different theoretical models may explain the same final state. The backward reconstruction of the dynamical history will then retrieve different classes of external agents.

These pieces of knowledge which preceed axiomatization have received different names, as "abduction" (Peirce, 1931) or "tacit dimension" (Polanyi, 1958). Some of them have been memorized as universal tools either in our genetic heritage, or during infancy in our learning age. This seems to be the cognitive valence of Jung's "archetypes" (Von Franz, 1988).

4. AN ADAPTIVE MEASUREMENT PROCEDURE

The measuring apparatus M can be modified by changing the number of probes, space or time resolution, or fuzziness (i.e., sharpness in localizing the data within a space-time box). Any change in M leads to a different set of output numbers, and hence to different sequences *s* (words). We can specifically tailor the M characteristics in order to emphasize a given set of dynamical properties and project out some other ones. Since M acts as a projector into the subspace of the variables measured by M itself, changing M means changing the "point of view" under which we observe the world, and hence making a different theoretical model.

In the previous section we show an indeterminacy in the reconstruction of the elements of reality $\{A, B, C, \ldots\}$ which modify the dynamical theories $\{\alpha, \beta, \gamma, \ldots\}$ of a mentally isolated system. As a result, the truth is represented by a combination of a model developed for the subsystem, plus the external gradients, or boundary conditions, once the subsystem is embedded in a suitable environment.

Which one is the most appropriate among the pairs

$$\{\alpha, A\}, \{\beta, B\}, \{\gamma, C\}, \text{etc.?}$$

The question is equivalent to asking: among all possible measuring apparatuses M applicable to an event, which one is the most appropriate?

If we prefer not to decide, we independently make use of different M's and correspondingly define different sciences.

In real life, we have to face problems overlapping different separate sciences. For instance, a cardiac disease may be due to a global offset of the pace-maker, or to some local cytopathology or even to a drug effect acting at a biomolecular level.

Figure 4 shows the difference between fixed and adaptive M. In the first case we sharply define three separate sciences. What can be exchanged among different specialists are not technical words, which are specific of each science, but just the residual metaphorical part, not filtered into the technical word. Should we say that two scientists of different areas always communicate by metaphors? A tentative way out (fuzzy logic, Zadeh, 1987) is to avoid sharp definitions, so that different disciplinary terms have regions of overlap.

However the most natural approach seems to start with a very low resolution, covering all the disciplinary areas, and then – depending upon a specific problem – to zoom toward one or the other narrow point of view.

A successful line of adaptive measurement has been worked out in my group (Arecchi and Boccaletti, 1997). It consists in developing a measuring procedure whereby we observe a system only at "almost equal" geometric separations in state space. As a consequence, observations are activated only for short time intervals at irregularly distributed times, separated from each other by unequal intervals. The stroboscopic sequence of these intervals has an information content which provides fast reliable answers to the following questions:

1. recognizing a chaotic dynamics (Lyapunov exponents, different unstable periodic orbits (UPO's);
2. discriminating determinism from stocastic noise;
3. controlling chaos, i.e. stabilizing one of the UPO's contained within a chaotic attractor.

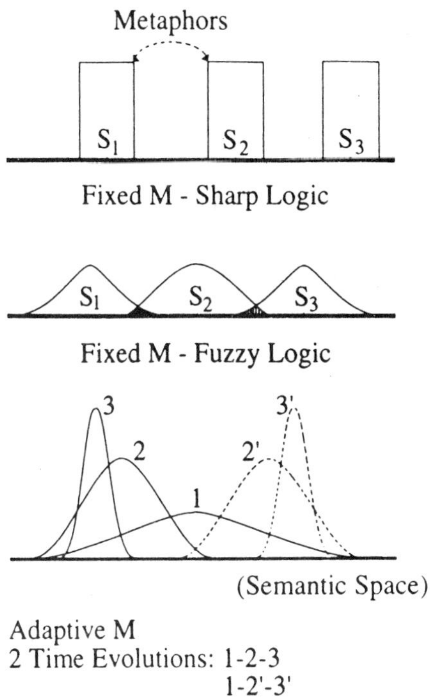

Figure 4. Fixed measuring apparatuses giving rise to different sciences. Inter communication is based on metaphors (upmost figure) unless by fuzzy logic one accepts ill defined terms with overlaps (center figure). When the measurement is no longer locked to a fixed position of semantic space, nor to a fixed resolution, then the scientific knowledge can cover, with different degrees of detail, different areas of the semantic space, allowing information exchange within a unique language.

When applied to the control of chaos (Boccaletti and Arecchi, 1995), this adaptive algorithm is effective for values of the stroboscopic times much larger than the elementary integration time step and smaller than the periods of all UPO's. In other words, the method introduces a natural adaptation time scale which is intermediate between the minimum resolution time of the dynamics and the time scale of the periodic orbits.

5. A NEW APPROACH TO THE SCIENTIFIC DISCOVERY: THE CARTESIAN CUT VS. THE REALISTIC APPROACH

In the scientific investigation, we select a quantitative feature by application of an apparatus M. The emerging description of reality represents an observation "from one point of view".

Due however to the variety of possible M's, we must justify at a metascientific level *why* we have selected that M rather than another one. This is a general question dealing with the role of those elements of reality which are preliminary to one particular program.

In section 4 we have seen that adaptation is a slalom among different sets of rules $\{\alpha\}$, $\{\beta\}$, ... under the guidance of a preferential set of external elements which has non zero intersections with $\{A\}$, $\{B\}$, ... but does not coincide with either of them.

When affronting of the truth problem we can assume three attitudes, namely,

i) assume the adaptive strategy and its associated reality set as a kind of privileged reference frame. Indeed, being the result of an optimization process, it appears more appropriate than any particular theory, $\{\alpha\}$, or $\{\beta\}$, etc.

ii) Consider the truth problem as a metatheoretical problem. At this metalevel, the set of all sets of truth values $\{A\}$, $\{B\}$, ... has to be considered as the truth, but with the stipulation that any individual set makes sense only if associated with the corresponding theory.

iii) A more fundamental approach recovers the polysemicity of the ordinary language as a virtue, not a drawback. More than questioning the power of any specific theory we question the same set-theoretical approach to physics. M provides a sharp connotation which allows to classify any observed entity within an appropriate set. Hence the set-theoretical character of all modern sciences, with the consequent antinomies of modern logic after Cantor, Russel, Gödel, etc., is transferred into the heart of the scientific language.

    An adaptive M means that the localization in semantic space is no longer as sharp as when it is defined by a precise stipulation as for the sets.

In the epistemology of scientific discovery, one usually relies on Bayes theorem to corroborate an hypothesis along the following procedure. We start hypothesizing a certain model x out of a possible set $\{x\}$ of models. Let us call $p(x)$ the apriori probability of x. As we collect some data d, the conditional probability $p(x|d)$ built upon a Bayes rule is a better approximation to $p(x)$. Thus we expect a convergent feedback loop as the data set $\{d\}$ increases.The Bayes rule was already given in (2); here we just replace m (models) by x (hypotheses) and the input sequence s with the data d. The rule presupposes the availability of the hypothesis set $\{x\}$ with a pre-assigned measure $p(x)$.

In fact, this approach overlooks the "theory ladenness" of the measuring procedure. Indeed, the apparatus M collecting the data $\{d\}$ is built upon the selected theoretical model. Thus, changing the probabilities $p(x|d)$ by conditioning them on the observations $\{d\}$ means also changing the structure of M, and hence perturbing the same set $\{d\}$. Hence, there are two feedback loops:

i) the standard one from data $\{d\}$ to models $\{x\}$; it assumes that the set $\{x\}$ is already given and one has just to choose within it;

ii) the new loop from $\{d\}$ to M, leading to $\{d'\} \neq \{d\}$.

The second loop cannot be in general formalized, since it works on different classes of M, and therefore it requires different operation concepts, or words. Changing point of view means changing language, and only a metalinguistic approach can cope with a variety of languages.

An intermediate compromise is what we called "adaptive procedure". Let us fix some finite bounds in the space of models $\{x\}$ as well in the space of apparatuses $\{M\}$. In such a bounded case, the procedure can be automatized as an algorithm. In the cases discussed by Arecchi and Boccaletti (1997), $\{M\}$ is the class of clocks set on different units of time, and we readjust the observation time upon the dynamical features.

Imposing apriori a bounded variation to M (i.e. limiting oneself to a finite number of "points of view") works fine only if you have tested its validity by a pre-formalized session. Indeed, before any scientific enterprise, we have already some non-scientific acquaintance with a problem, motivated by curiosity, practical interest, etc.

In a generic scientific endeavor, imposing a priori bounds, would be against wisdom, insofar as it cuts out ranges of phenomena which might be otherwise relevant.

Going back to the title, the truth versus certitude issue can be summarized by the following scheme (Figure 5). R stays for reality (whatever this means), S for a symbol interpreter (an intelligent being or even a Turing machine!), M is the measuring apparatus.

In modern science, M is usually not questioned, and the elaboration takes place on the output of M. This was called as the A) Complexity approach, leading eventually to certitude, when the S machine replicates exactly the M output sequence with the minimal amount of information. From a gnoseological point of view, if M are our senses, as suggested by Hume, then S (called by Descartes "res cogitans") has to face *not the world, but the representation* already coded by M. As shown in section 3, it has to face a grammatical problem inplementable by a machine. This means that Descartes' mind is equivalent to a Turing machine, as already suspected by many experts of artificial intelligence. This strong association of M on the side of R is equivalent to what Atmanspacher called "The Cartesian cut" (Atmanspacher, 1994).

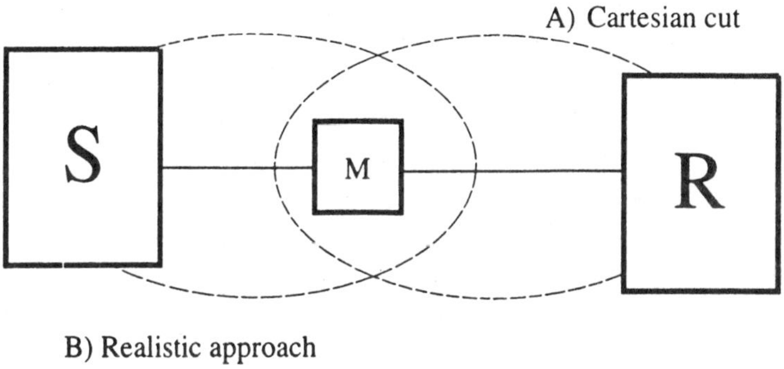

Figure 5. Knowledge interpretation: R=reality, M=symbol generator (measuring apparatus), S=symbol interpreter (model builder). Dashed line A (Cartesian cut): M+R provide representations as symbol sequences, which are interpreted by S. S can be replaced by Turing machine. Dashed line B (realistic approach): S+M globally face R. Before producing outputs, M is readjusted among a class of possible measuring apparatuses by a prelinguistic procedure not expressible within the formal language which later M provides to S.

On the contrary, the B) Complex system approach regards sciences as dealing with the world through an adaptive procedure M, for which however a linguistic foundation does not exist, because any linguistic formulation is subsequent to the operation of M. During the scientific operation, S+M act in an *entangled* way. A meta-level of investigation (psychology of cognition) is required to disentangle S from M.

In summary, there is a nonlinguistic residue in the scientific operation which then precludes a Turing machine from acting as a creative scientist.

REFERENCES

Abarbanel, H.D.J., Brown, R., Sidorowich, J.J., and Tsimring, L.S., 1993, The analysis of observed chaotic data in physical systems, *Rev. Mod. Phys* 65:1331-1365.

Anderson, P.W., 1972, More is different, *Science* 177:393-396.

Arecchi, F.T., Basti, G.F., Boccaletti, S., and Perrone, A., 1994, Adaptive recognition of a chaotic dynamics, *Europhys. Lett.* 26:327-331.

Arecchi, F.T., and Farini, A., 1996, *Lexicon of Complexity*, Studio Editoriale Fiorentino, Firenze.

Arecchi, F.T., and Boccaletti, S., 1997, Adaptive strategies for recognition, noise filtering, control, synchronization and targeting of chaos, *Chaos* 7:621-634.

Atmanspacher, H., 1994, Complexity and meaning as a bridge across the Cartesian cut, *J. of Consciousness Studies* 1:168-180.

Badii, R., and Politi, A., 1997, *Complexity*, Cambridge University Press, Cambridge.

Bennett, C.H., 1987, Dissipation, information, computational complexity and the definition of organization, in *Emerging Syntheses in Science*, D. Pines, ed., Addison Wesley, Redwood City, Ca.

Boccaletti, S., and Arecchi, F.T., 1995, Adaptive control of chaos, *Europhys. Lett.* 31:127-131.

Chaitin, G.J., 1966, On the length of programs for computing binary sequences, *J. Assoc. Comp. Math.* 13:547-560.

Crutchfield, J.P., 1992, Semantics and thermodynamics, in *Nonlinear Modeling and Forecasting*, M. Casdagli and S. Eubank, eds., Addison-Wesley, Redwood City, CA.

Crutchfield, J.P., and Young, K., 1989, Inferring statistical complexity, *Phys Rev. Lett.* 63:105-108.

D'Alessandro, G., and Politi, A., 1990, Hierarchical approach to complexity with applications to dynamic systems, *Phys Rev. Lett.* 64:1609-1612.

Galilei, G., 1932, Letter to M. Welser "On the sun spots", 1 December 1612, Italian original in: *Opere di G. Galilei*, Firenze, Vol. V, pp. 187-188.

Gell-Mann, M., 1994, *The Quark and the Jaguar*, Freeman, New York.

Grassberger, P., 1986, Toward a quantitative theory of self-generated complexity, *Int. J. Theor. Phys.* 25:907-919.

Grassberger, P., Schreiber, T., and Schaffrath, C., 1991, Nonlinear time sequences analysis, *Int. J. Bif. Chaos* 1:521-534.

Hopcroft, J.E., and Ullman, J.D., 1979, *Introduction to Automata Theory, Languages and Computation*, Addison-Wesley, Reading, MA.

Kolmogorov, A.N., 1965, Three approaches to the quantitative definition of information, *Problems of Information Trasmission* 1:4-20.

Krohn, W., Küppers, G., and Nowotny, H., 1990, *Selforganization, Portrait of a Scientific Revolution*, Kluwer, Dordrecht.

Peirce, C.S., 1931, *Collected Papers*, 1-6, C. Harshorne and P. Weiss, eds, Harvard University Press, Cambridge, MA; 7-8, A.W. Burks, ed., Harvard University Press, Cambridge, MA.

Polanyi, M., 1958, *Personal Knowledge: Towards a Post-critical Philosophy*, Routledge and Kegan Paul, London.

Shaw, R., 1981, Strange attractors, chaotic behavior, and information flow, *Z. Naturforsch,* 36a:80-97.

Simon, H., 1982, The architecture of complexity, *Proc. Amer. Philos. Soc.* 106:467-485.

Von Franz, M.L., 1988, *Psyche und Materie*, Einsiedeln, CH.

Zadeh, L.A., 1987, *Fuzzy Sets and Applications* (selected papers), J. Wiley, New York.

Creative Inferences and Abduction

Model-Based Reasoning in Creative Processes

Joke Meheus
Center for Logic and Philosophy of Science, Universiteit Gent, Gent, Belgium, Joke.Meheus@rug.ac.be

Abstract: Combining a contextual approach to problem solving with results on some recently developed (non-standard) logics, I present in this paper a general frame for the methodological study of model-based reasoning in creative processes. I argue that model-based reasoning does not require that we turn away from logic. I also argue, however, that in order to better understand and evaluate creative processes that involve model-based reasoning, and in order to formulate guidelines for them, we urgently need to extend the existing variety of logics.

1. AIM AND SURVEY

The aim of this paper is twofold. The first is to present a general framework for the methodological study of creativity and to show how model-based reasoning can be integrated in it. In line with Nersessian (1999a), I use the term "model-based reasoning" to refer to the construction and manipulation of analogue representations (representations that form structural analogues of real-world or imaginary situations). Typical forms of model-based reasoning are constructing and manipulating visual representations, thought experiments and analogical reasoning.

The second aim is to discuss some logical issues concerning model-based reasoning. Some readers may be surprised by the idea that logic has a role to play in the study of model-based reasoning. I shall argue, however, that logic and model-based reasoning seem to be incompatible *only* if one identifies the entire domain of logic with its existing systems. That the existing logical systems are inadequate to understand model-based reasoning does *not* imply that it is impossible to design such logics. Other readers may fear that I shall reduce model-based reasoning to a set of rules, and thus lose everything that is interesting about it. I can assure them that I have no intention to do so, and that I am even convinced that such a reduction is impossible. Contrary to what is sometimes believed, logic is not about formulating magic formulas that would turn entire domains into routine. It is about designing systems

Model-Based Reasoning in Scientific Discovery, edited by L. Magnani, N.J. Nersessian, and P. Thagard, Kluwer Academic/Plenum Publishers, New York, 1999

that allow one to distinguish between sound and unsound steps in specific types of reasoning processes. What I shall argue for is that we urgently need such systems for model-based reasoning.

I shall proceed as follows. First, I present the general framework (section 2). This frame is based on the contextual approach to problem solving that was developed by Batens (see his 1985, 1992a, and 1992b) and that was adapted to creative problem solving in Meheus (1997) and (1999a). Here, I shall show that it can account for model-based reasoning and that it moreover enables us to understand what is typical for model-based reasoning in creative processes, namely that they are inconsistent and/or that their manipulation involves ampliative steps. Next, I argue that, contrary to what is sometimes claimed, model-based reasoning does not require that we turn away from logic (section 3), and that it even has an important role to play in the normative study of model-based reasoning (section 4). I also argue, however, that in order to evaluate reasoning processes that involve analogue representations, and in order to design guidelines for them, the existing variety of logics has to be extended in three different ways: we need logics that work with non-verbal representations, that incorporate in a sensible way ampliative rules, and that can handle inconsistent representations (section 5). In section 6, I concentrate on reasoning processes that involve inconsistent representations, and I address the question which logic is appropriate for this type of reasoning. I end with some conclusions (section 7).

2. A CONTEXTUAL MODEL FOR CREATIVE PROCESSES

In the following paragraphs, I briefly outline a general model for the methodological study of creative processes. I shall only discuss some features of it that are specifically relevant to the present topic.

Creativity is commonly defined by referring to properties of products. In general, a person is considered as creative only if he or she produces something that is both novel and valuable. Similarly, a process is considered as creative only if it leads to novel and valuable products. In Meheus (1997) and Meheus and Batens (1999), however, it is argued that product-centered views not only lead to insurmountable difficulties, but moreover preclude the development of a methodological theory of creativity. It is typical for the model outlined here that it is thoroughly process-centered. Whether a person is considered as creative is dependent on the kind of actions this person has performed, *not* on the kind of products he or she produced. Whether a process is considered as creative is dependent on structural properties of that process, not on characteristics of the products obtained.

One of the chief tenets of the model is that creativity only occurs where ill-defined problems are attacked in a rational way. So, a person who merely solves well-defined problems is not considered as creative, even not if one of these problem solutions turns out to be highly innovative. Also, if a problem solver arrives at the solution of an ill-defined problem in a non-rational way (for instance, by modifying well-founded constraints in an unjustified way), he or she is not viewed as creative.

By an "ill-defined problem", I mean something that is rather different from what is usually understood by this in the literature. In order to understand this, I have to explain what is meant by a "context" (problem solving situation).

A context is composed of four elements: (i) a problem, (ii) a set of relevant items, (iii) a set of certainties, and (iv) a set of methodological instructions.[1] Elements (ii)-(iv) are considered as the constraints for (i). In line with Nickles (1980 and 1981), I use the term "constraint" in a broad sense to refer to any item of information that imposes a condition on the solution of the problem or that narrows down the solution path.[2]

The relevant items provide materials from which the solution can be obtained (if the context is adequate). They may be formulated as sentences, but also as diagrams or drawings, or as a combination of these. Methodological instructions specify the way in which the relevant items should or may be manipulated in order to reach the solution (or to come closer to it). Certainties determine the meaning of the entities that occur in the other elements of the context. In this sense, they determine which manipulations of the relevant items are justified. The certainties thus provide the "underlying logic" for a given context. More often than not, this underlying logic incorporates transformation rules for non – verbal elements. It may also contain ampliative rules (inductive rules, abductive rules, rules that govern analogical reasoning...). Note that I use "logic" here in a very broad sense (formal and informal systems, deductive as well as non-deductive systems). Note also that I do not require that the logic used in some context is explicit (usually it is not).[3]

It is worth pointing out that the context-elements vary from one context to the other. This means, first, that the constraints from one context may be

[1] The notion of a context was introduced in Batens (1985). There, the participants in the problem solving process constitute a fifth element of the context. In the present paper, I disregard complications involved in non-individual problem solving processes.

[2] The approach to problem solving presented here has several characteristics in common with Nickles' constraint-inclusion model. One of the differences is that several kinds of constraints are distinguished – see also Meheus and Batens (1998).

[3] Because of reasons of space, the discussion of the different context-elements is necessarily oversimplified. For a more detailed discussion of the different context-elements, see Batens (1985 and 1992a), and Meheus and Batens (1998).

absent in others. It also means that the same items of information may have a different function in different contexts. For instance, a certainty from one context may function as a relevant item or even as the problem in another.[4]

Let me give a simple example to make all this a bit more concrete. Suppose one is confronted with the following logical puzzle: [5]

> The country of Marr is inhabited by three types of people: liars, truth tellers and doubters. Liars always lie, truth tellers always tell the truth, doubters sometimes lie and sometimes tell the truth. At a local cocktail party, you are engaged in conversation with three citizens of Marr: Andrew, Bob and Charles. At some point, Andrew remarks that Charles is a truth teller. Bob replies that Andrew is a truth teller. Suppose you know that one of them is a liar, one a truth teller, and one a doubter. Can you determine who is who?

It is evident from the formulation of the puzzle what the problem is. What are the relevant items? Remember that the relevant items provide materials from which the solution to the problem should be derivable. Thus we have:

1. If Andrew is a truth teller, then Charles is a truth teller.
2. If Andrew is a liar, then Charles is not a truth teller.
3. If Bob is a truth teller, then Andrew is a truth teller.
4. If Bob is a liar, then Andrew is not a truth teller.

(1) follows from the remark Andrew made together with the definition of truth tellers. Analogously for (2)-(4).

We also know that exactly one of them is a liar, one a truth teller, and one a doubter. This piece of information is also needed to derive the solution to our problem. How can we represent it? One way would be to represent it in a purely verbal way. However, this will make our derivation long and complex. A more efficient way is to represent it implicitly by means of the following diagram

1.	Truth Teller	Liar	Doubter
2. Andrew			
3. Bob			
4. Charle Charles			

for which each of the following obtains:

[4] These characteristics of the model have various important implications which I cannot discuss here (but see Batens, 1984, and Batens and Meheus, 1996).

[5] The puzzle is inspired on one presented in Wickelgren (1974, pp. 118-119).

5. If there is a plus sign in a certain cell, it holds true that the corresponding object has the corresponding property; if there is a minus sign, it is false that the object has the property.
6. No cell can contain both a plus sign and a minus sign.
7. No column or row can contain more than one plus sign.
8. No column or row can contain only minus signs.

The certainties determine the meaning of the entities occurring in the other context elements. In this case, the certainties incorporate classical logic – determining the meaning of, for instance, the connective "if... then" in (1)-(4) – as well as interpretation and transformation rules for the diagram used. Examples of interpretation rules are (5)-(8). Typical examples of transformation rules are:

9. If a column (respectively a row) contains an empty cell, and all other cells in that column (respectively row) contain a minus sign, a plus sign may be put in the empty cell.
10. If a column (respectively a row) contains a plus sign, a minus sign may be put in all the other cells of that column (respectively row).

What are the methodological instructions? These may include heuristic rules such as

11. Look for a sentence of the form "If A then B" such that it is not possible to make both A and B true in the diagram; if such a sentence is found, A make false in the diagram.

Given all this, the problem can be easily solved. Consider (1) again. In view of (7), it is not possible to make both the antecedent and the consequent of (1) true in the diagram. Hence, in line with (11), we make the antecedent of (1) false in the diagram. Analogously for (3). In view of (9), a plus sign can now be put in the empty cell of the first column. But then, it is not possible to make both the antecedent and the consequent of (2) true in the diagram (because this would violate (6)). Hence, we put a minus sign in the top cell of the second column. Applications of (9) and (10) allow us to finish the diagram in the following way:

	Truth Teller	Liar	Doubter
Andrew	-	-	+
Bob	-	+	-
Charle Charles	+	-	-

In view of (5), we are now able to formulate the solution to our problem: Andrew is the doubter, Bob is the liar, and Charles is the truth teller.

Let us now come back to the general model. As I mentioned before, one of the central tenets of the model is that creativity only occurs where ill-defined problems are attacked in a rational way. The notion of a context allows us to classify problems into "well-defined" and "ill-defined". I say that a problem is well-defined, if and only if, given the set of certainties, the relevant statements and the methodological instructions determine a unique solution to the problem.

Two remarks should immediately be added to this. The first is that the model allows for two kinds of ill-defined problems. The constraints of a problem may fail to determine a unique solution because the relevant items are inconsistent (and the methodological instructions do not suggest a way to resolve the inconsistencies) or because the relevant items are incomplete (and the methodological instructions do not specify a procedure for completing them). Rudolf Clausius' problem of formulating a (consistent) theory for thermodynamic phenomena forms an example of the first type. As I have shown in Meheus (1996 and 1999b), this problem was defined by inconsistent constraints: the relevant items included both Carnot's theory (based on the conservation of heat) as well as experimental results (mainly from Joule) indicating that heat is not conserved. As an example of an ill-defined problem of the second type, one may think of the question of why the irregular openings in the foliage of trees cause circular light spots. When confronted for the first time with this question, most people have no idea what kind of information would allow them to answer this question.

The second remark is that both kinds of ill-defined problems require special logics and special mechanisms. When dealing with an ill-defined problem, classical logic does not allow one to determine (in a sensible way) which steps are valid and which are not. This is obvious in the case where the constraints are inconsistent. As is well known, classical logic allows one to derive *any* sentence from a contradiction. So, if you have reasons to believe that heat is conserved as well as reasons to believe that it is not conserved, then, according to classical logic, it follows that the moon is made of green cheese. It is clear that, under such circumstances, all sensible reasoning stops. But, also in the case where the relevant items are incomplete, classical logic is inadequate, or at least insufficient. The reason is that classical logic merely consists of deductive rules. It does not incorporate ampliative rules – rules that would allow one to *extend* the constraints. And precisely these are needed when the constraints of a problem are incomplete.

As I mentioned in the introduction, the model presented here can account for the practices that are generally considered as forms of model-based reasoning – such as constructing and manipulating visual representations, thought experiments, analogies, abduction, etc. The reason for this is that there is room for constructing and manipulating non-verbal representations as well as for non-deductive inferences (such as inferences based on analogies).

In line with the idea that there are only two kinds of ill-defined problems (and that creativity only occurs where ill-defined problems are attacked in a rational way), I assume that it is typical for the representations used in creative processes that they are "inconsistent" and/or that their manipulation involves ampliative steps. In section 5, I shall argue that, because of these characteristics, model-based reasoning in creative processes requires special kinds of logic. But, first I want to clear away some possible misunderstandings (section 3), and argue why logic should play an important role in the study of model-based reasoning (section 4).

3. SOME POSSIBLE MISUNDERSTANDINGS

It is sometimes claimed that the study of model-based reasoning requires that we turn away from logic. In order to justify this claim, one typically refers to the fact that model-based reasoning frequently involves the manipulation of non-verbal representations and that this cannot be understood from the point of view of logic. This argument is not tenable. It confuses the domain of logic with the very small subdomain for which logical systems are as yet available. It holds true that almost all logics available today are based on purely verbal representation systems. But, this does not mean that there would be something inherently verbal about logic.

Let me also approach this in a somewhat different way. As a discipline, logic is concerned with the study of correct reasoning. This means that logicians design systems that enable one to distinguish, in specific types of situations, between sound and unsound inference steps. Now, no one will deny that in model-based reasoning specific types of representations are manipulated in order to infer information from them. Hence, as soon as one accepts that it makes sense to distinguish between sound and unsound manipulations of these representations, model-based reasoning becomes a problem for logic.

I am evidently not claiming here that model-based reasoning would no longer be a problem for other disciplines, like cognitive psychology and philosophy of science. What I do claim is that logic, in its own way, can contribute to the study of model-based reasoning.

Does this imply that model-based reasoning is algorithmic, and hence, non-creative? Not at all. The fact that one is able to design a logical system that delineates the correct steps for a certain type of reasoning process, does not make the latter algorithmic. Actually, deductive and non-deductive reasoning processes are on a par in this respect. For a limited number of deductive reasoning processes, logical systems are available. And, for some of these systems, we have algorithmic proof search procedures. But, these procedures only allow one to check whether a *given* conclusion follows from a *given* set of premises. They neither form a method for retrieving interesting premises nor for deriving interesting conclusions from them. In the same way, we can design systems for specific types of model-based reasoning, for instance for analogical reasoning. And, for some of these systems, we shall be able to design algorithmic procedures. But, also these will only allow one to check whether some conclusion follows from a given representation. They will not enable one, for instance, to discover new and interesting analogies.

The upshot is that identifying "deductive" with "non-creative" and "non-deductive" with "creative" is erroneous. It holds true of course that many deductive reasoning processes are non-creative – for instance, those studied in Johnson-Laird and Byrne (1991). However, the same holds true for many non-deductive processes (think of simple analogy problems such as "planets/sun ≈ electrons/????"). In both cases, the reason why the process is non-creative is that the problem is well-defined (the premises are given beforehand, they are very simple, and the required inference takes very few steps).

There is a second possible misunderstanding, namely that the study of model-based reasoning requires that we turn away from proof-theoretic methods. The idea here seems to be that model-based reasoning involves "models", and hence, requires model-theoretic methods. Also this rests on a confusion. Models in the sense of analogue representations have little to do with models in the logical, model-theoretic sense. Moreover, they can be manipulated in a model-theoretic way (in which one tries to construct a representation that falsifies some conclusion) as well as in a proof-theoretic way (in which transformation rules are applied to them). An example of the latter is the puzzle presented above. In solving puzzles like these, we do not try to construct a representation in which all our premises are true and our conclusion is false. Instead, we simply apply transformation rules to a certain type of diagram. This process is entirely analogous to the construction of "ordinary" proofs in which transformation rules are applied to sentences.

So, the situation is this. Some representation systems are purely verbal, some are non-verbal, and some are mixed. For each of these types, it is possible to design logical systems with a proper syntax and a proper semantics. Once the systems are available, it is possible to design proof-theoretic methods for them as well as model-theoretic methods. I am personally convinced (but cannot argue for it here) that most of our reasoning proceeds in a proof-theoretic way. (In general, proof-theoretic methods are much more efficient and much more natural than model-theoretic methods). But, again, this does not imply that most of our reasoning would proceed in a verbal way.

4. WHY LOGIC IS IMPORTANT

If we take the study of model-based reasoning seriously, then we cannot restrict ourselves to purely descriptive accounts. No matter how important descriptive historical and psychological studies are to enhance our understanding of model-based reasoning, they do not offer an answer to the normative question of how one *should* proceed when confronted with a problem that asks for model-based reasoning. The importance of this question can hardly be overestimated. Indeed, one thing descriptive studies convincingly have shown is that, no matter how "natural" model-based reasoning is, laypersons do not seem to be very good at it.[6] As soon as the problems they are confronted with become a bit more complex, they tend to make a lot of mistakes.

There are two reasons why logic should play an important role in the normative study of model-based reasoning. The first is related to the fact that, as mentioned before, model-based reasoning involves the manipulation of representations in order to infer information from them. Needless to say, one's view on logic determines how one will evaluate such inference steps: they are considered as irrational (or non-rational) if and only if they cannot be accounted for by the logic *we* consider as appropriate. For instance, if a reasoning process that involves abduction is approached from the point of view of classical logic, the conclusion will necessarily be that the abductive steps are unsound. It follows that logic implicitly fulfils a central function in any normative account of model-based reasoning, and that this function be better made explicit. It also follows that in order to evaluate reasoning processes that involve analogue representations, we need insight in the kinds of logics that are appropriate for them. This will often mean that the design of new kinds of logic is required (see also the following section).

The second reason why logic is important for the normative study of model-based reasoning concerns the design of methods for the manipulation of various kinds of non-verbal representations. Model-based reasoning is often considered as "tricky". One of the reasons for this is that the representations used are often "overdetermined", and hence, the reasoning easily leads to erroneous conclusions. However, there is nothing inherently tricky about model-based reasoning. What we lack at the moment are methods for the sound ma-

[6] One may think here of the studies by Johnson-Laird and his collaborators.

nipulation of non-verbal representations. However, in order to design these methods, we first need insight in the appropriate logical systems.[7]

Someone may object here that the design of methods for model-based reasoning requires much more than insight in the appropriate logical systems. This holds true. In the case of analogical reasoning, for instance, we also need heuristics for the retrieval of analogies. And, as I explained before, logic has very little to say on this stage of the process. However, this does not change the fact that there are stages where logic can be of enormous help. In analogical reasoning, for instance, logic can help us to design methods for the evaluation of analogies as well as for the efficient transfer of information from the source domain to the target domain.

Someone else may object that logic cannot help us for the normative study of model-based reasoning in creative processes because the latter involve the invention of new kinds of logics. It indeed happens that problem solvers, especially when dealing with problems that ask for creativity, cannot rely on the existing variety of logics. To mention only one example: scientists had to find ways to reason sensibly from inconsistencies, long before the formal tools for this type of reasoning were designed. Let me add, however, two warnings to this.

The first is that, even if no logic exists for the specific type of problem someone is tackling, the greater the variety of logics available, the greater the chance he or she will find one that can be adapted for the problem at issue. Put differently, logicians are not able to offer a set of logics that is adequate for every problem that will be encountered in the future. However, the more the existing variety of logics is extended, the better problem solvers will be equipped to deal with novel problems.

The second remark is that, even in cases where problem solvers have to make inferences for which they cannot rely on the existing logics, designing these logics *post hoc* may help us to evaluate the inference steps involved. A logical analysis may reveal, for instance, that some of the steps (contrary to what the problem solvers themselves believed), necessarily lead to triviality, and hence, have to be considered as unsound.

[7] For more arguments and for examples of how insight in the logical systems can help us to overcome the fallacy of drawing unsound conclusions on the basis of overdetermined representations, I refer the reader to the different contributions in Allwein and Barwise (1996).

5. MODEL-BASED REASONING REQUIRES SPECIAL KINDS OF LOGIC

Thus far I argued that logic should play an important role in the normative study of model-based reasoning. However, in order to fulfil this role, the existing variety of logics needs to be extended along three dimensions.

The first dimension concerns the representational systems. At this moment, we have adequate logics for (a small number of) verbal representations. We also have adequate logical systems for some (elementary) diagrammatic systems (such as Venn diagrams[8]). For many other representational systems (for instance, the analogue systems for two-dimensional spatial reasoning as described by Johnson-Laird) the logics need urgently to be designed.

Someone may object here that in order to evaluate the steps in a concrete reasoning process, the representation system used is irrelevant. After all, the information contained in a non-verbal representation can always be expressed in a purely verbal way. Hence, we can simply use first-order predicate logic to evaluate whether the conclusion arrived at is valid. Two warnings are in order here.

The first is that non-verbal representations typically contain several implicit assumptions, and that the validity of the conclusion also depends on these. (For example, when spatial information is represented in a non-verbal way, the transitivity of the relations involved is not explicitly expressed). Hence, even if we want to "translate" arguments based on non-verbal representations, we have to be able to recognise the implicit assumptions involved. But this requires insight in the representation system at issue.

The second warning is that in order to design guidelines for the manipulation of specific representations, we need insight in the interpretation and manipulation rules appropriate for them. Hence, also here first-order predicate logic is of limited use.

The second dimension according to which the existing variety of logics has to be extended concerns the kind of steps that can be accounted for (deductive and ampliative). Especially in creative processes, the manipulation of analogue representations frequently involves a combination of deductive and ampliative steps (abductive steps, steps based on analogies, etc.). In section 2, I have related this to the fact that creative processes often involve problems defined by an incomplete set of constraints. In processes like these, one needs not only deductive rules that enable one to analyze the available constraints, but also ampliative rules that allow one to extend them. At this moment, several computational systems are available

[8] For examples, see the contributions in Allwein and Barwise (1996).

for specific types of ampliative steps.[9] What we urgently need, however, are logical systems that sensibly combine deductive with ampliative steps (see also my 1999a).

The third dimension concerns the kind of "abnormalities" a logic can handle. As I explained before, creative processes often involve inconsistent representations. To the best of my knowledge, the occurrence of inconsistent representations has largely been ignored in the literature on model-based reasoning.[10] I shall therefore consider this aspect in some more detail in the next section.

6. REASONING FROM INCONSISTENT REPRESENTATIONS

For some readers, it will not be evident that analogue representations can be inconsistent. It is important to realize, however, that analogue representations, just like verbal representations, are constructed in view of certain interpretation rules, and that in view of these, they can be inconsistent. Suppose, for instance, that we are given the following set of sentences

(11) There are three persons: Peter, An, and Tom
(12) Peter and An are in Room A
(13) Tom is not in the same room as An
(14) Tom is in Room A

and that we are asked to determine who is where. A natural way to represent the relevant items is by drawing two boxes (corresponding to the two rooms). That a certain person is present in a certain room can then simply be represented by writing this person's name in the corresponding box. An obvious interpretation rule for this representation is that no name can appear in more than one box. Hence, if a person's name appears in one of the boxes, we may conclude from this that the person is present in the corresponding room, but not in the other room. In view of these interpretation rules, the drawing that represents our relevant items is inconsistent

[9] For examples of computational systems that can handle abduction, see, for instance, Josephson and Josephson (1994). Interesting examples of computational systems for analogical reasoning can be found in Falkenhainer, Forbus, and Gentner (1989), Holyoak and Thagard (1995), and Long and Garigliano (1994).

[10] An important exception is Nersessian (1999b) in which a very interesting case study is presented on the role of inconsistencies in model-based reasoning.

(R1)

room A	room B
Peter An Tom	Tom

We can infer from it that Tom is present in room A and that he is not present in room A (but in room B). Similarly, we can infer that he is and is not present in room B.

Someone may object here that this cannot be an adequate representation of the situation at issue, because it is obvious that Tom cannot be in two places at the same time. This definitely holds true. However, the drawing represents the best possible knowledge we have about that situation. And, as long as we are uncertain about what is wrong with our information, removing Tom's name from either of the two boxes would be an arbitrary move.

An important observation should immediately be added to this. If the information we have about some situation or domain is inconsistent, it may nevertheless be important that we are able to work with that information. Case studies from the history of the sciences reveal, for instance, that scientists do not reject a theory simply because it turns out to be inconsistent. This seems reasonable. If the theory is the best possible (or the only available) alternative we have, rejecting it may cause a considerable loss of information. The same holds true in everyday life. We are not willing to give up all the information we have about some situation simply because it contains a mistake somewhere. Rather, we try to find out where the mistake is located. At the same time, we continue to make inferences from those parts of the information that we consider as reliable. However, all this requires that we are able to reason sensibly in the presence of inconsistencies. Precisely this is not possible on the basis of classical logic.

Some readers will object here that working with inconsistent representations can easily be avoided by considering different consistent alternatives. In the above example, for instance, one may consider two different alternatives: one in which Tom is in room A and one in which he is in room B. However, three things should be noted here. First, if we are dealing with inconsistent information, it is crucial that we remember that something is *wrong* with that information. By "splitting up" our information into consistent chunks we conceal this fact. Next, the example above only serves the purpose of providing the simplest possible example of an inconsistent representation. For this extremely simple case, "splitting up" the information in consistent alternatives seems indeed straightforward. However, in most (if not all) realistic examples, defining a set of consistent alternatives is far from evident, and requires that inferences are made from the inconsistent representations (see Meheus, 1996, for examples). Finally, in all interesting cases, resolving an inconsistency cannot be understood as making a selection among a set of previously defined consistent alternatives. It usually involves various kinds of conceptual change. However, finding such conceptual reinterpretations requires that one does not prematurely split up the theory (see Meheus, 1996).

Given that non-verbal representations can be inconsistent, and that it is often necessary to work with them, the question is now what kind of logic one should use for this. I have argued elsewhere that, of the logics available today, so-called inconsistency-adaptive logics[11] seem the most appropriate tools for this kind of purpose (see Meheus, 1996 and 1999b). At this moment, all inconsistency-adaptive logics work with purely verbal representation systems. However, the idea of an inconsistency-adaptive logic can also be applied to logics that work with non-verbal or mixed representations. Discussing the technical details of such a logic is not possible within the scope of this paper. Instead, I shall illustrate the idea of an inconsistency-adaptive logic by means of a simple example that involves a non-verbal representation.[12] But in order to do so, I first have to say something on the relation between non-verbal representations and models (in the logical, Tarskian sense).[13]

Consider (R1) again. I shall say that a representation like this is true in a model if and only if every item of information that is represented by it is true in that model.[14] So, (R1) is true in all models (and only in those models) in which each of the following is true: "Peter is in room A", "An is in room A", "Tom is in room A", "Tom is not in room A", "Tom is in room B", and

[11] The first inconsistency-adaptive logic was designed by Diderik Batens about 15 years ago (see his 1986 and 1989). Meanwhile, a whole variety of inconsistency-adaptive logics is available and the notion of an adaptive logic has been generalized to include other types of "logical abnormalities" – for an overview, see Batens (1999). In Meheus (1999c), an inconsistency-adaptive logic is presented that is extremely rich; in Meheus (1999d), it is argued that precisely because of this feature the logic is very close to natural thinking.

[12] My informal presentation may give the impression that inconsistency-adaptive logics are not logics in the proper sense. I can assure the reader, however, that they are well-defined: they have a proper syntax and a proper semantics, and for all of them soundness and completeness proofs are available or under construction (see Batens, 1998, for an introduction in the technical details).

[13] In the rest of this section, I use the term model to refer to a couple <D, *v*> in which D is a (non-empty) set and *v* is an assignment function.

[14] Evidently, what is represented by a diagram or drawing in a certain context is dependent on the interpretation rules in that context. Here and below, I assume that the interpretation rules for drawings like (R1) are such that, if a person's name occurs in a certain box, this not only represents that this person is present in the corresponding room, but also that he is not present in the other room.

"Tom is not in room B". It is important to note that in some cases, the information we have about a certain situation can only be represented by a *series* of drawings or diagrams. Consider, for instance, the following items

(15) Mary is in room A
(16) John is in room A or in room B

Using the same representation system as before, (15)-(16) cannot be represented in a single drawing. One way to solve this is as follows

	room A	**room B**		**room A**	**Room B**
(R2)	Mary John			Mary	John

in which we follow the convention that, if different drawings are put next to each other, they represent different alternatives. Again, I say that (compound) representations such as (R2) are true in a model if and only if all the items represented by it are true in that model. It is important to note that we have two different types of models for (R2): in some of them the drawing on the left is true, in the others the drawing on the right is true.

Let us now return to the question of how one should proceed when some representation turns out to be inconsistent. Consider, again (R1). As is well-known, classical logic only allows for consistent models. Hence, according to classical logic there is no model in which (R1) is true. Given the standard notion of semantic consequence (an item of information follows from a set of premises[15] if and only if it is true in all the models of those premises), this immediately entails that every possible item follows from (R1). (As there are no models of the premises, the criterion that an item should be true in "all" of them is trivially satisfied.) This strategy is clearly inadequate. Even if we have reasons to believe (11)-(14) (by lack of a better alternative), we surely do not want to conclude that two plus two is five.

An obvious way to resolve this difficulty is by allowing for inconsistent models. This is typically the strategy that is followed by paraconsistent logics. However, if we simply allow inconsistent models (in order to avoid triviality), we face another difficulty. In order to see this, consider the following variant of the problem presented above:

(17) There are five persons: Peter, An, Tom, John, and Mary
(18) John is in room A or in room B

[15] As should be clear from what is explained above, I allow that a set of premises is represented by a set of sentences, as well as by a series of drawings (diagrams, ¼).

(19) Mary is in the same room as John
(20) Mary is in room A
(21) Peter and An are in room A
(22) Tom is not in the same room as An
(23) Tom is in room A

Representing these items in the same system as before, we obtain

(R3)

room A	**room B**
John Mary Peter An Tom	Tom

room A	**Room B**
Mary Peter An Tom	John Mary Tom

If we simply allow for inconsistent models, there are two types of models for (R3): those in which the drawing on the left is true, and those in which the drawing on the right is true. Only in models of the first kind "John is in room A" is true. Hence, given the standard notion of semantic consequence, this item does not follow from (R3). From an intuitive point of view, this outcome seems unacceptable. We indeed have a problem with (21)-(23), but (18)-(20) are totally unrelated to this problem. Moreover, taken separately, (18)-(20) entail "John is in room A".

For this extremely simple example, one will probably not care about the the loss in inferential power. However, for more realistic examples, it is important that our logic is as strong as possible. For instance, if a scientific theory is confronted with an anomaly in a specific domain of application, we do not want it to lose its explanatory and predictive power in all other domains as well.

It is here that inconsistency-adaptive logics come in. Inconsistency-adaptive logics, like standard paraconsistent logics, allow for inconsistent models. In this way, they avoid that inconsistencies lead to triviality. However, they also have the property that they interpret a set of premises "as consistently as possible". What this comes to is that a selection is made from the set of models obtained by some (standard) paraconsistent logic. Intuitively, only those models are retained that are not more inconsistent than is necessary for the truth of the premises. In order to make this selection, one can follow different strategies (see Batens, 1999). However, one of the most natural ones is to eliminate a model *M* if it validates all the inconsistencies another model *M′* validates, but in addition validates some inconsistencies that are not validated by *M′*.

To make this all a bit more concrete, consider again (R3). All models in which the drawing on the left is true validate "Tom is in room A", "Tom is not in room A" as well as "Tom is in room B", "Tom is in not room B". The models in which the drawing on the right is true validate the same inconsistencies, but in addition to this also validate "Mary is in room A", "Mary is not in room A" and "Mary is in room B", "Mary is not in room B". The strategy mentioned in the previous paragraph allows us to eliminate all these models. The rationale for this is that the "extra" inconsistencies are not necessary for the truth of the premises (remember that they are not true in the models in which the drawing on the left is true). An important consequence of this is that, in all the models that are allowed for by an inconsistency-adaptive logic, the sentence "John is in room A" is true. Hence, according to an inconsistency-adaptive logic, this sentence *is* derivable from (R3).[16]

This result can be generalized. For "parts" of our information that behave consistently, adaptive logics allow one to derive everything that would have been derivable if the information would have been consistent. This is precisely what we want. If we have to work with inconsistent representations, we want to avoid triviality, but at the same time, we want that unproblematic "parts" of our information retain their full strength.

An important characteristic of inconsistency-adaptive logics is that some of their inference rules are only conditionally valid (they can only be applied provided specific items behave consistently). An implication of this is that arguments which are structurally the same, are in some cases considered as valid and in other cases as invalid. In Meheus (1999b), I discuss an example of a historical reasoning process that fits this pattern, namely Clausius' derivation of Carnot's theorem. This derivation was made from two incompatible approaches to thermodynamic phenomena (namely the theory of Sadi Carnot which was based on the heat is conserved, and the view that heat and work can be converted into each other). Against the background of this inconsistent set of beliefs, Clausius designed two different proofs for Carnot's theorem which are structurally similar (both are based on *reductio ad absurdum*). Nevertheless, Clausius considered only one of them as a valid derivation of Carnot's theorem. As I show in Meheus (1999b), this particular stage in Clausius's reasoning process appears to be irrational if one identifies rationality with classical logic, but comes out perfectly normal and rational from the point of view of inconsistency-adaptive logics.

[16] Remember that the purpose here was merely to illustrate the idea behind inconsistency-adaptive logics, not to present an inconsistency-adaptive logic that works (directly) with a specific non-verbal representation system. Such logics have to be designed yet. However, they are certainly possible, and can be formulated both at the semantic level as well as at the syntactic level.

7. CONCLUSIONS

I argued in this paper that model-based reasoning does not require that we turn away from logic, but that, on the contrary, the latter has an important role to play in the normative study of the former.

That logic has largely been ignored in the domain of model-based reasoning is understandable. Until very recently, logicians had very little to offer in this respect. At this moment, however, the situation is different. It still holds true that most of the existing logical systems are inadequate for model-based reasoning. But, the tools for designing adequate systems are now available. We have some logical systems for elementary non-verbal representation systems. These can form the basis for designing logical systems that work with more complex non-verbal representations. We also have a variety of (mainly computational) systems for different types of ampliative reasoning. These can function as an important source of inspiration for formulating logical systems that sensibly combine deductive and ampliative steps. Finally, various logics are available that can handle inconsistencies. And, as I have shown in this paper, some of these can easily be adapted to handle inconsistent non-verbal representations.

Evidently, designing these systems is a problem for the logicians. However, the impetus to address this problem will have to come from outside the domain of logic. Claiming that model-based reasoning goes beyond logic will not motivate the logicians. One better confronts them with the fact that model-based reasoning goes beyond the existing logics.

REFERENCES

Allwein, G., and Barwise J., eds., 1996, *Logical Reasoning with Diagrams*, Oxford University Press, Oxford.

Batens, D., 1985, Incommensurability is not a threat to the rationality of science or to the anti-dogmatic tradition, *Philosophica* 32:117–132.

Batens, D., 1986, Dialectical dynamics within formal logics, *Logique et Analyse* 114:161-173.

Batens, D., 1989, Dynamic dialectical logics, in: *Paraconsistent Logic. Essays on the Inconsistent*, G. Priest, R. Routley, and J. Norman, eds., Philosophica Verlag, München, pp. 187-217.

Batens, D., 1985, Meaning, acceptance and dialectics, in: *Change and Progress in Modern Science*, J.C. Pitt, ed., Reidel, Dordrecht, pp. 333–360.

Batens, D., 1992a, Do we need a hierarchical model of science?, in: *Inference, Explanation, and Other Frustrations. Essays in the Philosophy of Science*, J. Earman, ed., University of California Press, Berkeley, pp. 199–215.

Batens, D., 1992b, *Menselijke Kennis. Pleidooi voor een Bruikbare Rationaliteit*, Garant, Leuven.

Batens, D., 1998, Inconsistency-adaptive logics, in: *Logic at Work. Essays Dedicated to the Memory of Helena Rasiowa*, E. Orlowska, ed., Springer, Heidelberg, pp. 445-472.

Batens, D., 1999, A survey of adaptive logics, Kluwer, Dordrecht, to appear.

Batens, D., and Meheus, J., 1996, In-world realism *vs.* reflective realism, in: *Realism in the Sciences*, I. Douven and L. Horsten, eds., Universitaire Pers, Leuven, pp. 35–53.

Falkenhainer, B., Forbus, K., and Gentner, D., 1989, The Structure-mapping engine: algorithm and examples, *Artificial Intelligence* 41:1-63.

Holyoak, K.J., and Thagard, P.R., 1995, *Mental Leaps.* The MIT Press, Cambridge, MA.

Johnson-Laird, P.N., and Byrne, R.M., 1991, *Deduction*, Erlbaum, Hillsdale, NJ.

Josephson, J.R., and Josephson, S.G., eds., 1994, *Abductive Inference. Computation, Philosophy, Technology*, Cambridge University Press, Cambridge.

Long, D., and Garigliano, R., 1994, *Reasoning by Analogy and Causality*, Princeton University Press, Princeton.

Meheus, J., 1996, Adaptive logic in scientific discovery: the case of Clausius, *Logique et Analyse* 143-144:359-391.

Meheus, J., 1997, *Wetenschappelijke Ontdekking en Creativiteit*, Unpublished doctoral thesis.

Meheus, J., 1999a, Analogical reasoning in creative problem solving processes: logico-philosopical perspectives, Kluwer, Dordrecht, in press.

Meheus, J., 1999b, Inconsistencies in scientific discovery. Clausius's remarkable derivation of Carnot's theorem, Brepols, in press.

Meheus, J., 1999c, An extremely rich paraconsistent logic and the adaptive logic based on it, Kluwer, Dordrecht, in press.

Meheus, J., 1999d, How to reason from inconsistencies, to appear.

Meheus, J., and Batens, D., 1998, Steering problem solving between cliff incoherence and cliff solitude, *Philosophica* 58:153-187.

Meheus, J., and Batens, D., 1999, In defense of the rationality of creativity, to appear.

Nersessian, N., 1999a, Conceptual change: creativity, cognition, and culture, to appear.

Nersessian, N., 1999b, Inconsistency, generic modeling, and conceptual change in science, to appear.

Nickles, T., 1980, Can scientific constraints be violated rationally?, in: *Scientific Discovery, Logic, and Rationality*, T. Nickles, ed., Reidel, Dordrecht, pp. 285–315.

Nickles, T., 1981, What is a problem that we may solve it?, *Synthese* 47:85–118.

Wickelgren, W.A., 1974, *How to Solve Mathematical Problems*, Dover Publications, New York.

Model-Based Creative Abduction

Lorenzo Magnani
Computational Philosophy Laboratory, Department of Philosophy, University of Pavia, Pavia, Italy; also School of Public Policy, Georgia Institute of Technology, Atlanta, GA, USA, lorenzo@philos.unipv.it

Abstract: My contribution aims to introduce the distinction, not previously analyzed, between two kinds of *abduction*, *theoretical* and *manipulative*, in order to provide an integrated framework to explain some of the main aspects of both creative and *model-based reasoning* effects engendered by the practice of science. The distinction appears to be extremely convenient: creativity will be viewed as the result of the highest cases of theoretical abduction showing the role of the so-called *model-based abduction*. Moreover, I will delineate the first features of what I call *manipulative abduction* showing how we can find methods of constructivity at the experimental stage, where the recent epistemological tradition has settled the most negative effects of theory-ladenness.

1. THEORETICAL ABDUCTION: SENTENTIAL AND MODEL-BASED

In this section I will introduce the concept of *theoretical abduction*, in order to provide an integrated framework to explain some of the main aspects of both creative and *model-based reasoning* effects engendered by the practice of science. After having illustrated the *sentential models* together with their limitations (section 2), creativity will be viewed as the result of the highest cases of theoretical abduction, showing the role of the so-called *model-based abduction* (section 3) (cf. Figure 1). Finally, in section 4, I will delineate the first features of what I call *manipulative abduction* showing how we can find methods of constructivity at the experimental stage, where the recent epistemological tradition has settled the most negative effects of theory-ladenness.

Many reasoning conclusions that do not proceed in a deductive manner are *abductive*. For instance, if we see a broken horizontal glass on the floor[1] we might explain this fact by postulating the effect of wind shortly before: this is not certainly a deductive consequence of the glass being broken (a cat may well have been responsible for it). Hence, theoretical abduction is the process

[1] This event constitutes in its turn an *anomaly* that needs to be solved/explained.

Model-Based Reasoning in Scientific Discovery, edited by L. Magnani, N.J. Nersessian, and P. Thagard, Kluwer Academic/Plenum Publishers, New York, 1999

of *inferring* certain facts and/or laws and hypotheses that render some sentences plausible, that *explain* or *discover* some (eventually new) phenomenon or observation; it is the process of reasoning in which explanatory hypotheses are formed and evaluated.

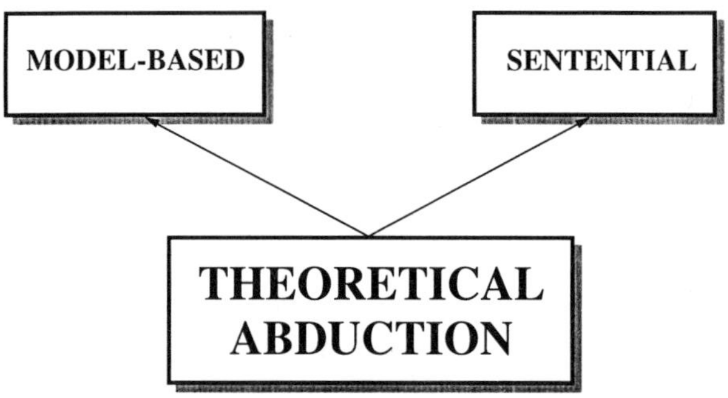

Figure 1. Theoretical abduction.

First, it is necessary to show the connections between abduction, induction, and deduction and to stress the significance of abduction to illustrate the problem solving process. I have developed with others (Magnani, 1992; Ramoni, et al., 1992) an epistemological model of medical reasoning, called the *Select and Test Model* (ST-MODEL) which can be described in terms of the classical notions of abduction, deduction and induction; it describes the different roles played by such basic inference types in developing various kinds of medical reasoning (diagnosis, therapy planning, monitoring) but can be extended and regarded also as an illustration of scientific theory change.

A hundred years ago, Peirce interpreted abduction essentially as an *inferential creative process* of generating a new hypothesis. There are two main epistemological meanings of the word abduction (Magnani, 1988): 1) abduction that only generates plausible hypotheses (*selective* or *creative*)[2] and 2) abduction considered as *inference to the best explanation*, which also evaluates hypotheses (see Figure 2). All we can expect of our "selective" abduction, is that it tends to produce hypotheses for further examination that have some chance of turning out to be the best explanation. Selective abduction will always produce hypotheses that give at least a partial explanation and

[2] To illustrate from the field of medical knowledge, the discovery of a new disease and the manifestations it causes can be considered as the result of a creative abductive inference. Therefore, *creative* abduction deals with the whole field of the growth of scientific knowledge. This is irrelevant in medical diagnosis where instead the task is to *select* from an encyclopedia of pre-stored diagnostic entities (Ramoni, et al., 1992).

therefore have a small amount of initial plausibility. In the syllogistic view (see below) concerning abduction as inference to the best explanation advocated by Peirce one might require that the final chosen explanation be the *most* plausible.

Induction in its widest sense is an ampliative process of the generalization of knowledge. Peirce distinguished three types of induction and the first was further divided into three sub-types. A common feature of all kinds of induction is the ability to compare individual statements: using induction it is possible to synthesize individual statements into general laws (types I and II), but it is also possible to confirm or discount hypotheses (type III). Clearly I am referring here to the latter type of induction, that in my model is used as the process of reducing the uncertainty of established hypotheses by comparing their consequences with observed facts.

Deduction is an inference that refers to a logical implication. Deduction may be distinguished from abduction and induction on the grounds that only in deduction is the truth of inference guaranteed by the truth of the premises on which it is based.

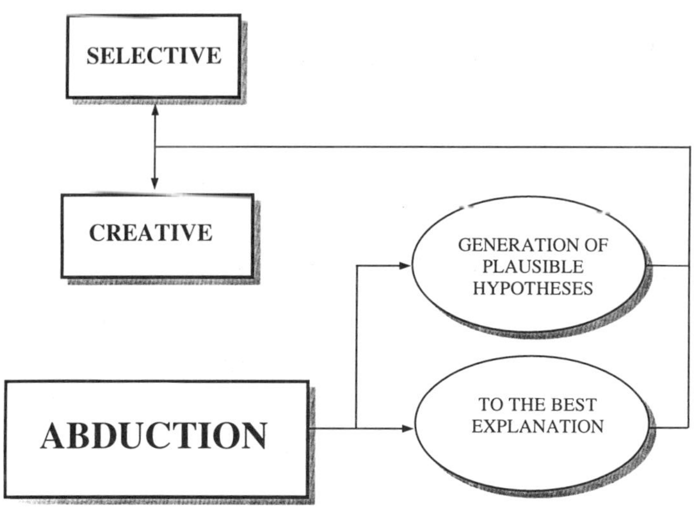

Figure 2. Creative and selective abduction.

All these distinctions need to be exemplified. To describe how the three inferences operate, it is useful to start with a very simple example dealing with diagnostic reasoning and illustrated (as Peirce initially did), in syllogistic terms:

1. If a patient is affected by a beta-thalassemia, his/her level of hemoglobin A2 is increased.
2. John is affected by a beta-thalassemia.
3. John's level of hemoglobin A2 is increased.

By deduction we can infer (3) from (1) and (2); by induction we can go from a finite set of facts, like (2) and (3), to a universally quantified generalization, like the piece of hematologic knowledge represented by (1). Starting from knowing – selecting – (1) and observing (3) we can infer (2) by performing a selective abduction. Such an inference is not affected by uncertainty, since the manifestation (3) is pathognomonic for beta-thalassemia. This is a special case, where there is not abduction because there is not "selection", in general clinicians very often have to deal with manifestations which can be explained by different diagnostic hypotheses: the abductive inference rule corresponds to the well-known fallacy called affirming the consequent

$$\frac{\begin{array}{c}\varphi \rightarrow \psi \\ \psi\end{array}}{\varphi}$$

Thus, *selective abduction* is the making of a preliminary guess that introduces a set of plausible diagnostic hypotheses, followed by deduction to explore their consequences, and by induction to test them with available patient data, (1) to increase the likelihood of a hypothesis by noting evidence explained by that one, rather than by competing hypotheses, or (2) to refute all but one.

If during this first cycle new information emerges, hypotheses not previously considered can be suggested and a new cycle takes place: in this case the *nonmonotonic* character of abductive reasoning is clear. As stated above, there are two main epistemological meanings of the word abduction: (1) abduction that only generates plausible hypotheses (*selective* or *creative*) – and this is the meaning of abduction accepted in my epistemological model – and (2) abduction considered as *inference to the best explanation*, that also evaluates hypotheses. In the latter sense the classical meaning of selective abduction as inference to the best explanation (for instance in medicine, to the best diagnosis) is described in my epistemological model by the complete abduction – deduction – induction cycle.

2. THE SENTENTIAL FRAMEWORK

Many attempts have been made to model abduction by developing some formal tools in order to illustrate its computational properties and the relationships with the different forms of deductive reasoning (see, for example, Bylander, et al., 1991). Some of the more recent formal models of abductive reasoning are based on the theory of the *epistemic state* of an agent (Boutilier and Becher, 1995), where the epistemic state of an individual is modeled as a consistent set of beliefs that can change by expansion and contraction (*belief revision framework*). Inconsistent beliefs do not describe any world, and so are unproductive; moreover, the changes of state have to be *epistemologically conservative*: the agent maintains as many of its beliefs as possible when it adjusts its beliefs to the new information, thus following Quine's idea of "minimum mutilation" (Quine, 1979).

Deductive models of abduction may be characterized as follows. An explanation for β relative to background theory T will be any α that, together with T, entails β (normally with the additional condition that $\alpha \cup T$ be consistent). Such theories are usually generalized in many directions: first of all by showing that explanations entail their conclusions only in a *defeasible* way (there are many potential explanations), so joining the whole area of the so-called nonmonotonic logic or of the probabilistic treatments; second, trying to show how some of the explanations are relatively implausible, elaborating suitable technical tools (for example in terms of modal logic) able to capture the notion of preference among explanations.

The idea of consistency that underlies some of the more recent deductive consistency-based models of selective abduction (diagnostic reasoning) is the following: any inconsistency (anomalous observation) refers to an aberrant behavior that can usually be accounted for by finding some set of components of a system that, if behaving abnormally, will entail or justify the actual observation. The observation is anomalous because it contradicts the expectation that the system involved is working according to specification. This types of deductive models go beyond the mere treatment of selective abduction in terms of preferred explanations and include the role of those components whose abnormality makes the observation (no longer anomalous) consistent with the description of the system (Boutilier and Becher, 1995; Magnani, 1999a).

This kind of *sentential frameworks* exclusively deals with selective abduction (diagnostic reasoning) and relates to the idea of preserving *consistency*. Exclusively considering the sentential view of abduction does not enable us to say much about creative processes in science. It mainly refers to the *selective* (diagnostic) and merely *explanatory* aspects of reasoning and to the idea that abduction is mainly an inference to the best explanation: when used to express the creativity events it is either empty or replicates the well-known *Gestalt* model of radical innovation. It is empty because the sentential view stops any attempt to analyze the creative processes. Already in the Peircian syllogistic and sentential initial conception of abduction – as the fallacy of affirming the consequent, described in the previous section, we immediately see it is perfectly compatible with the *Gestalt* model of discovery. In both cases the event of creating something new (for example a new concept) is considered so radical and instantaneous that its irrationality is immediately involved.

It is well known that for Peirce abduction is an *inferential process* (*CP* 5.188-189, 7.202) that includes all the operations whereby hypotheses and theories are constructed (*CP* 5.590) (see also Hintikka, 1998). Hence abduction has to be considered as a kind of *ampliative* inference that is not logical and truth preserving (in the sense of deductive): indeed valid deduction docs not yield any new information, for example new hypotheses previously unknown (on the possible abductive sides of some forms of deduction see also section 3.4 below).

If we want to provide a suitable framework for analyzing the most interesting cases of conceptual changes in science we do not have to limit ourselves to the *sentential* view of theoretical abduction but we have to consider a broader *inferential* one which encompasses both sentential and what I call *model-based* sides of creative abduction (see, for details, the following section).

3. MODEL-BASED ABDUCTION

3.1 Conceptual change and creative reasoning in science

I have analyzed elsewhere (Magnani, 1999a) some limitations of the sentential models of theoretical abduction in accounting for other reasoning tasks; for example they do not capture 1. the role of statistical explanations, where what is explained follows only probabilistically and not deductively from the laws and other tools that do the explaining; 2. the sufficient conditions for explanation; 3. the fact that sometimes the explanations consist of the application of schemas that fit a phenomenon into a pattern without realizing a deductive inference; 4. the idea of the existence of high-level kinds of *creative* abductions I cited above; 5. the existence of model-based abductions (for instance visual and diagrammatic, see section 3.2 below); 6. the fact that explanations usually are not complete but only furnish *partial* accounts of the pertinent evidence (Thagard and Shelley, 1997); 7. the fact that one of the most important virtues of a new scientific hypothesis (or of a scientific theory) is its power of explaining *new*, previously *unknown* facts: "[...] these facts will be [...] unknown at the time of the abduction, and even more so must the auxiliary data which help to explain them be unknown. Hence these future, so far unknown explananda, cannot be among the premises of an abductive inference" (Hintikka, 1998, p. 507), observations become real and explainable only by means of new hypotheses and theories, once discovered by abduction.

We will see in the following section that it is in terms of *model-based abductions* (and not in terms of sentential abductions) that we have to think for example of the case of a successful synthesis of two earlier theoretical frameworks which might even have seemed incompatible. The old epistemological view sees Einstein's theory as an attempt to "explain" certain anomalies and facts such as the Michelson-Morley experiment: "The most instructive way of looking at Einstein's discovery is to see it as a way of reconciling Maxwell's electromagnetic theory with Newtonian mechanics [...] it would be ridiculous to say that Einstein's theory 'explains' Maxwell's theory any more than it 'explains' Newton's laws of motion" (Hintikka, 1998, p. 510). This kind of abductive movement does not have that immediate explanatory effect illustrated by the sentential models of abduction: the new framework usually does not "explain" the previous ones but provides a very radical new perspective (just as illustrated in the limitation 7 above).

If we want to deal with the nomological and most interesting creative aspects of abduction we are first of all compelled to consider the whole field of the growth of scientific knowledge cited above. We may also see belief change from the point of view of *conceptual change*, considering concepts either cognitively, like mental structures analogous to data structures in computers, or, epistemologically, like abstractions or representations that presuppose questions of justification. Belief revision is able to represent cases of conceptual change such as adding a new instance, adding a new weak rule, adding a new strong rule (see Thagard, 1992), that is, cases of addition and deletion of beliefs, but fails to take into account cases such as adding a new part-relation, adding a new kind-relation, adding a new concept, collapsing part of a kind-hierarchy, reorganizing hierarchies by branch jumping and tree switching, in which there are reorganizations of concepts or redefinitions of the nature of a hierarchy.

Let us consider concepts as composite structures akin to frames of the following sort:

CONCEPT:
A kind of:
Subkinds:

A part of:
Parts:
Synonyms:
Antonyms:
Rules:
Instances:

It is important to emphasize (1) kind and part-whole relations that institute hierarchies, and (2) rules that express factual information more complex than simple slots. To understand the cases of conceptual revolutions we need to illustrate how concepts can fit together into conceptual systems and what is involved in the replacement of such systems. Conceptual systems can be viewed as ordered into kind-hierarchies and linked to each other by rules. Belief revision is able to represent cases of conceptual change such as adding a new instance, adding a new weak rule, adding a new strong rule (Thagard, 1992), that is, cases of addition and deletion of beliefs, but fails to take into account cases such as adding a new part-relation, adding a new kind-relation, adding a new concept, collapsing part of a kind-hierarchy, reorganizing hierarchies by branch jumping and tree switching, in which there are reorganizations of concepts or redefinitions of the nature of a hierarchy.

The last three types of conceptual change can be illustrated by the following examples. The Newtonian abandon of the Aristotelian distinction between natural and unnatural motion exemplifies the collapse of part of the kind-hierarchy. Branch jumping occurred when the Copernican revolution involved the recategorization of the earth as a kind of planet, when previously it had been considered special, but also when Darwin reclassified humans as a kind of animal. Finally, we have to say that Darwin not only reclassified humans as animals, he modified the meaning of the classification itself. This is a case of hierarchical tree redefinition: "Whereas before Darwin kind was a notion primarily of similarity, his theory made it a historical notion: being of common descent becomes at least as important to being in the same kind as surface similarity. Einstein's theory of relativity changed the nature of part-relations, by substituting ideas of space-time for everyday notions of space and time" (Thagard, 1992, p. 36).

These last cases are the most evident changes occurring in many kinds of creative reasoning in science, when adopting a new conceptual system is more complex than mere belief revision. Related to some of these types of scientific conceptual change are different varieties of *model-based abductions*.

3.2 Model-based abduction

The last cases of creative reasoning in science we have just illustrated demonstrate the radical *conjectural* character of the new concepts and the incommensurability as regarding previous ones, that is the cases in which "revolutionary" changes happen and the most "counterinductive" acts can become visible. The analysis of model-based conceptual change helps us to study the revolutionary changes of science: different varieties of what I call *model-based* abduction are related to some of these types of conceptual change.

Peirce stated that all thinking is in signs, and signs can be icons, indices, or symbols. Moreover, all *inference* is a form of sign activity, where the word sign includes "feeling, image, conception, and other representation" (*CP* 5.283), and, in Kantian words, all synthetic forms of cognition. That is, a great part of the thinking activity is *model-based.* Of course model-based reasoning acquires its peculiar creative relevance when embedded in abductive processes.

For Peirce a Kantian keyword is synthesis, where the intellect constitutes in its forms and in a harmonic way all the material delivered by the senses. Surely Kant did not consider synthesis as a form of *inference* but, notwithstanding the obvious differences,[3] I think synthesis can be related to the Peircian concept of inference, and, consequently, of abduction. After all, when describing the ways the intellect follows to unify and constitute phenomena through imagination Kant itself makes use of the term *rule* "Thus we think a triangle as an object, in that we are conscious of the combination of the straight lines according to a rule by which such an intuition can always be represented" (Kant, 1929, A140, B179-180, p. 182), and also of the term *procedure* "This representation of a universal procedure of imagination in providing an image for a concept, I entitle the schema of this concept" (Kant, 1929, A140-B179-180, p. 182). We know that rules and procedures represent the central features of the modern concept of inference (see also section 3.4 below). Moreover, according to Peirce, the central question of philosophy is "how synthetical reasoning is possible [...]. This is the lock upon the door of philosophy" (*CP* 5.348), and the mind presents a tendency to unify the aspects which are exhibited by phenomena: "the function of conception is to reduce the manifold of sensuous impressions to unity" (*CP* 1.545).

Most of these forms of constitution of phenomena are creative and, moreover, characterized in a model-based way. Let me show some examples of model-based inferences. It is well known the importance Peirce ascribed to

[3] For example Peirce considers space and time themselves as products of synthesis and not as forms of intuition, and consequently synthesis can not be seen as a process in time even if Kant would admit that "psychologically" mental phenomena are movements of the mind from one thing to another (Davis, 1972).

diagrammatic thinking, as shown by his discovery of the powerful system of predicate logic based on diagrams or "existential graphs". As we have already stressed, Peirce considers inferential any cognitive activity whatever, not only conscious abstract thought; he also includes perceptual knowledge and subconscious cognitive activity (Davis, 1972). For instance in subconscious mental activities visual representations play an immediate role.

We should remember, as Peirce noted, that abduction plays a role even in relatively simple *visual phenomena*. *Visual abduction*, a special form of non verbal abduction, occurs when hypotheses are instantly derived from a stored series of previous similar experiences. It covers a mental procedure that tapers into a non-inferential one, and falls into the category called "perception". Philosophically, *perception* is viewed by Peirce as a fast and uncontrolled knowledge-production procedure. Perception, in fact, is a vehicle for the instantaneous retrieval of knowledge that was previously structured in our mind through inferential processes. By perception, knowledge constructions are so instantly reorganized that they become habitual and diffuse and do not need any further testing: "[...] a fully accepted, simple, and interesting inference tends to obliterate all recognition of the uninteresting and complex premises from which it was derived" (*CP* 7.37). Many visual stimuli are ambiguous, yet people are adept at imposing order on them: "We readily form such hypotheses as that an obscurely seen face belongs to a friend of ours, because we can thereby explain what has been observed" (Thagard, 1988, p. 53). This kind of image-based hypothesis formation can be considered as a form of *visual* (or *iconic*) *abduction*. Of course such subconscious visual abductions of everyday cognitive behavior are not of particular importance but we know that in science they may be very significant and lead to interesting new discoveries (Magnani, et al., 1994; Shelley, 1996).

Peirce gives another interesting example of model-based abduction related to sense activity: "A man can distinguish different textures of cloth by feeling: but not immediately, for he requires to move fingers over the cloth, which shows that he is obliged to compare sensations of one instant with those of another" (*CP* 5.221). One more example is given by the fact that the perception of tone arises from the activity of the mind only after having noted the rapidity of the vibrations of the sound waves, but the possibility of individuating a tone happens only after having heard several of the sound impulses and after having judged their frequency. Consequently the sensation of pitch is made possible by previous experiences and cognitions stored in memory, so that one oscillation of the air would not produce a tone.

To conclude, all knowing is *inferring* and inferring is not instantaneous, it happens in a process that needs an activity of comparisons involving many kinds of models in a more or less considerable lapse of time. This is not in contradiction with the fact that for Peirce the inferential and abductive character of creativity is based on the instinct (the mind is "in tune with nature") but does not have anything to do with irrationality and blind guessing. All sensations or perceptions participate in the nature of a unifying hypothesis, that is, in abduction, in the case of emotions too: "Thus the various sounds made by the instruments of the orchestra strike upon the ear, and the result is a peculiar musical emotion, quite distinct from the sounds themselves. This emotion is essentially the same thing as a hypothetic inference, and every hypothetic inference involved the formation of such an emotion" (*CP* 2.643); this surely suggests that abductive movements have also interesting extratheoretical effects (see section 4).

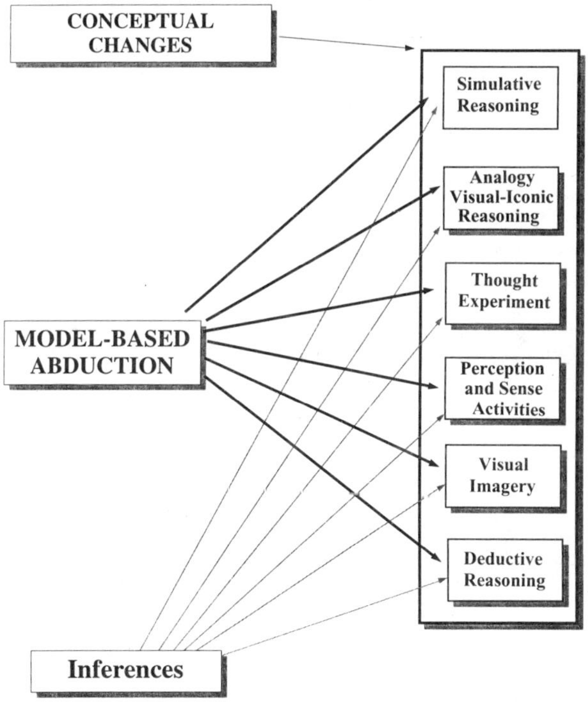

Figure 3. Model-based (theoretical) abduction.

In all these examples Peirce is referring to a kind of hypothetical activity that is inferential but not verbal, where "models" of feeling, seeing, hearing, etc. are very efficacious when used to build both habitual abductions of everyday reasoning and creative abductions of intellectual and scientific life (see Figure 3).

Following Nersessian (1995), I use the term "model-based reasoning" to indicate the construction and manipulation of various kinds of representations, not necessarily sentential and/or formal. Obvious examples of model-based reasoning are constructing and manipulating visual representations, thought experiment, analogical reasoning, but also the so-called "tunnel effect" (Cornuéjols, et al., 1999), occurring when models are built at the intersection of some operational interpretation domain – with its interpretation capabilities – and a new ill-known domain.

We have to remember that visual and analogical reasoning are productive in scientific concept formation too, where the role they play in model-based abductive reasoning is very evident; scientific concepts do not pop out of heads, but are elaborated in a problem solving process that involves the application of various procedures: this process is a *reasoned process*. Visual abduction, but also many kinds of abductions involving analogies, diagrams, thought experimenting, visual imagery, etc. in scientific discovery processes, can be just called *model-based*. We know that scientific concept formation has been ignored because of the accepted view that no "logic of discovery" – either deductive, inductive, or abductive algorithms for generating scientific knowledge – is possible. The methods of discovery involve use of *heuristic* procedures (Peirce was talking of creative abduction as the capacity and the "method" of making good conjectures); cognitive psychology, artificial intelligence, and computational philosophy have established that heuristic procedures are reasoned (see section 3.4 below). Analogical reasoning is one such problem solving procedure, and some reasoning from imagery is a form of analogical reasoning (Holyoak and Thagard, 1995).

How does this kind of analogical and/or imagery reasoning function in problem solving? Nersessian (1995, 1998, 1999b) has demonstrated that history of science abounds with instances of the use of imagery and of analogy to transform vague notions into scientifically viable conceptualizations of a domain. Her analysis deals with the important case of the use of imagery and analogy by Faraday and Maxwell in the construction of the concept of field. The concept of field had its origins in vague speculations about processes in the regions surrounding bodies and charges that might contribute to their action upon one another. In articulating a field representation for electric and magnetic actions, Faraday used primarily qualitative concepts and reasoned from imagery figures. He created a field representation for electric and magnetic actions by reasoning from an imagery representation of the "lines of force" that are formed when iron filings are sprinkled around a magnetic source. Many features of "lines" are incorporated into his field concept. He discussed many actions as "expanding", "bending", "being cut". All the forces of nature are unified and interconvertible through various motions of the lines of forces, and matter, itself, is nothing but point centers of converging lines of force. This representation enabled Faraday to express a quantitative relationship between the number of lines cut and the intensity of the induced force. At the end of this research Faraday introduced a pictorial representation that was to play an important role in Maxwell's construction of the quantitative field concept.

Use of analogy and imagery in ordinary and scientific problem solving is very complex. Nevertheless, we may observe in many cases all the features of a *productive*, *creative* mapping, where such "transfer of knowledge" is essential to the development of a new concept. Imagery representations appear to function analogically. The value of an imagery representation is that it makes some structural relations immediately evident.[4]

Hence the role of visual abduction and visual imagery in scientific discovery is very interesting. Let us consider the example above concerning the ambiguity of an obscurely seen face: more generally, we can face an *initial* (eventually) observed image in which we recognize a problem to solve. For example, given a visual or imagery datum, we may have to explain: 1) why an object is in a particular *position*; 2) the *presence/absence* of an object; 3) how an object can achieve a given *task* moving itself and/or interacting with the remaining objects in the scene/image; 4) how we can *recognize* an object as of *significance* (for instance the recognition of a stone as a tool, Shelley, 1996).

The generation of a "new" imagery hypothesis can be considered the result of the *creative abductive* inference previously described; in this respect we can consider how the imagery representations of new hypotheses lead to scientific discovery. The selection of an imagery hypothesis from a set of pre-enumerated imagery hypotheses, stored in long-term memory, also involves abductive steps, but its creativity is much weaker; of course we can call this type of visual abduction *selective* (Magnani, et al., 1994). Other interesting cases are given by cases that deal with more dynamic situations such as forming new images 1) changing in dimensions, 2) changing in shape; 3) changing in structure (discrete vs dense; bounded-unbounded; number of objects). We find these last kinds of image transformations in many cases related with *creative* visual abduction, as involved in scientific discovery, where we have also to establish relevant strategies for passing from the *propositional-verbal* level to the *visual* one and vice versa.

[4] The so-called *constructive* and *generic* modeling in scientific discovery are illustrated in Nersessian (1995) and Nersessian, et al. (1997).

3.3 Model-based abduction and inconsistencies

We can see abductive inferences "as answers to inquirer's explicit (or usually) tacit questions put to some definite source of answers (information)" (Hintikka, 1998, p. 519) stressing the interrogative features of this kind of reasoning. If abduction is the making of a set of possible answers, the choice of the possible questions is also decisive (and this choice of course is not indifferent as regards the further process of finding answers). At present, the cognitive sciences lack a good theory of how fruitful questions are produced, able to involve also emotional, ethical, and pragmatic aspects of cognition (Thagard and Croft, 1999).

I think that the role of *inconsistencies* in scientific reasoning could be very important to delineate the first features of such a theory. Surely surprise, curiosity, need, are related to the detection of inconsistencies. Model-based ways of generating a hypothesis that explains some phenomenon or conceptual problem that produced the question are heuristically linked to the activity itself of "finding" that certain puzzling phenomenon or that particular conceptual problem. In its turn, the deductive component and the inductive testing of hypothesis of the whole cycle of reasoning illustrated above – abduction as inference to the best explanation – clearly shows the presence of a further interrogative feature (further generation of questions).

We have now to notice some limitations of the formal models (section 2 above) in accounting for other kinds of inconsistencies embedded in many reasoning tasks and the contribution given by the analysis in terms of model-based abduction. The logical accounts of abduction certainly elucidate many kinds of inconsistency government, which nevertheless reduce to the act of finding contradictions able to generate the abduction of some hypotheses, beliefs, reasons, etc.; these contradictions always emerge at the level of data (observations), and consistency is restored at the theoretical level. This view may distract from important aspects of other kinds of reasoning that involve intelligent abductive performances. *Empirical anomalies* result from data that cannot currently be fully explained by a theory. They often derive from predictions that fail, which implies some element of incorrectness in the theory. The detection of these points involves defining which theoretical constituents are employed in the explanation of the anomaly. Thus, the problem is to investigate all the relationships in the explanatory area.

Empirical anomalies are not alone in generating impasses, there are also the so-called *conceptual anomalies*. To illustrate the first features of a theory of the role of inconsistencies in model-based abduction we can present the case of conceptual problems as triggers for hypotheses. The so-called conceptual problems represent a particular form of anomaly. In addition, resolving conceptual problems may involve satisfactorily answering questions about the nature of theoretical entities. Nevertheless such conceptual problems do not arise directly from data, but from the nature of the claims in the principles or in the hypotheses of the theory. It is far from simple to identify a conceptual problem that requires a resolution, since, for example, a conceptual problem concerns the adequacy or the ambiguity of a theory, and yet also its incompleteness or (lack of) evidence.

The formal sciences are especially concerned with conceptual problems. Let's consider an example deriving from the well-known case of the non-Euclidean revolution, which plays a remarkable role in illustrating some actual transformations in rational conceptual systems. The discovery of non-Euclidean geometries involves some interesting cases of model-based abductions. It demonstrates a kind of visual abduction, as a strategy for *anomaly resolution* related to a form of *explanatory* and *productive visual thinking*.

Since ancient times the fifth postulate has been held to be not evident. This "conceptual problem" (just an anomaly) has caused much suspicion about the reliability of the whole theory of parallels, consisting of the theorems that can be only derived with the help of the fifth postulate. The recognition of this anomaly was fundamental to the development of the great non-Euclidean revolution. Two thousand years of attempts to resolve the anomaly have generated many more-or-less fallacious demonstrations of the fifth postulate (for example, a typical attempt was that of trying to prove the fifth postulate from the others), until the discovery of non-Euclidean geometries (Greenberg, 1980).

I have presented elsewhere (Magnani, 1999b) some examples derived from the historical discovery of non-Euclidean geometries which illustrate the relationships between strategies for anomaly resolution and explanatory and productive visual thinking: I consider how Lobachevsky's strategy for resolving the anomaly of the fifth postulate was to manipulate the symbols, rebuild the principles, and then to derive new proofs and provide a new mathematical apparatus. The failure of the demonstrations of his predecessors induced Lobachevsky to believe that the difficulties that had to be overcome were due to causes other than those which had until then been focused on. I have shown how some of the hypotheses created by Lobachevsky were mostly *image-based* trying to demonstrate that *visual abduction* is relevant to hypothesis formation and scientific discovery.

Other epistemological settings show cases in which anomalies and inconsistencies are not triggers for new hypotheses and theories but have to be maintained for a certain period of time. When a scientist introduces a new hypothesis, especially in the field of natural sciences, he is interested in the potential rejection of an old theory or of an old knowledge domain. Consistency requirements in the framework of deductive models, governing hypothesis withdrawal in various ways, would arrest further developments of the new hypothesis reached by abduction. In the scientist's case there is not the deletion of the old concepts, but rather the coexistence of two rival and competing views. Newtonian mechanics is forced to cohabit with the anomaly of perihelion of Mercury until the development of the theory of relativity, but it also has to stay with its false prediction about the motion of Uranus. In diagnostic reasoning too, it is necessary to make a diagnosis even if many symptoms are not explained or remain mysterious. As noted above, when we create or produce a new concept or belief that competes with another one, we are compelled to maintain the derived inconsistency until the possibility of rejecting one of the two becomes feasible.

To delineate a constructive formal framework that can handle the coexistence of inconsistent theories (and unify many of the themes concerning the theory of abduction previously illustrated) we first of all have to be able to deal with the treatment of non verbal representations (that is model-based representations, see above, section 3.2).

I think that the problem of coexistence of inconsistent scientific theories and of reasoning from inconsistencies in scientific creative processes leads to analyze the characters of what I call the *best possible information* of a situation. It is also necessary to distinguish between the dynamic and the static sides of the exploitation of the best possible information. If we stress the sequential (dynamic) aspects we are more oriented towards analyzing anomalies as triggers for hypotheses as for example illustrated by the traditional deductive models of abduction (see section 2 above): the problem concerns the sentential abductive steps of the sequential comprehension and integration of data into a hypothetical structure that represents the best explanation for them. Analogously, as we have seen in the case of conceptual anomaly in geometry, the "impasse" can also be a trigger for a whole process of model-based abduction. On the contrary, if we consider the holistic (static) aspects we are more interested in the coexistence of inconsistencies as potential sources of different reasoned creative process. In this last case we have to deal with a particular behavior of model-based abduction that can be elucidated by its formal treatment; I plan to derive some suggestions from the area of paraconsistent and adaptive logic (Meheus, 1999), for instance handling hierarchies of inconsistent models of a given representation.[5]

Very interesting cases present a set of (scientific) hypotheses at a certain time, where the lack of inconsistencies and anomalies weakens the possibility of producing questions and also of withdrawing hypotheses by falsification. Elsewhere (Magnani, 1999a) I have considered a kind of "weak" hypotheses that are hard to negate and the ways for making it easy. In these cases the subject can rationally decide to withdraw his hypotheses even in contexts

[5] See also the analysis of the relationships between inconsistency, generic modeling, and conceptual change given in Nersessian, 1999a.

where it is impossible to find "explicit" contradictions; moreover, thanks to the new information reached simply by finding this kind of negation, the subject is free to find new questions and to abduce new hypotheses. I have explored whether *negation as failure* can be employed to model hypothesis withdrawal in Freudian analytic reasoning and in Poincaré's conventionalism of the principles of physics, showing how conventions can be rationally abandoned.

3.4 Model-based heuristic and deductive reasoning

As already said at the end of section 2 the kind of reasoned inference that is involved in creative abduction goes beyond the mere relationship that there is between premises and conclusions in valid deductions, where the truth of the premises guarantees the truth of the conclusions, but also beyond the relationship that there is in probabilistic reasoning, which renders the conclusion just more or less probable.

On the contrary, we have to see creative abduction as formed by the application of *heuristic procedures* that involve all kinds of good and bad inferential actions, and not only the mechanical application of rules. It is only by means of these heuristic procedures that the acquisition of *new* truths is guaranteed. Also Peirce's mature view on creative abduction as a kind of inference seems to stress the strategic component of reasoning.

Many researchers in the field of philosophy, logic, and cognitive science have sustained that deductive reasoning also consists in the employment of logical rules in a heuristic manner, even maintaining the truth preserving character: the application of the rules is organized in a way that is able to recommend a particular course of actions instead of another one. Moreover, very often the heuristic procedures of deductive reasoning are performed by means of a model-based abduction. A most common example of creative abduction is the usual experience people have of solving problems in geometry in a *model-based* way trying to devise proofs using diagrams and illustrations: of course the attribute of creativity we give the abduction in this case does not means that it has never been made before by anyone or that it is original in the history of some knowledge.

Hence we have to say that theoretical model-based abductions also operate in deductive reasoning (see Figure 3 above). Following Hintikka and Remes's analysis (1974) proofs of general implication in first order logic need the use of instantiation rules by which "new" individuals are introduced, so they are "ampliative". In ordinary geometrical proofs auxiliary constructions are present in term of "conveniently chosen" figures and diagrams. In Beth's method of semantic tableaux the strategic "ability" to construct impossible configurations is undeniable (Hintikka, 1998).

This means that also in many forms of deductive reasoning there are not trivial and mechanical methods of making inferences but we have to use *models* and *heuristic procedures* that refer to a whole set of strategic principles. All the more reason that Bringsjord (1999) stresses his attention on the role played by a kind of "model based deduction" that is "part and parcel" of our establishing Gödel's first incompleteness theorem, showing the model-based character of this great abductive achievement of formal thought.[6]

4. MANIPULATIVE ABDUCTION AND SCIENTIFIC CREATIVITY

The problem of the incommensurability of meaning has distracted the epistemologists from the procedural, extra-sentential and extra-theoretical aspects of scientific practice. Since Kuhn, the problem of translating between languages and of conceptual creativity has dominated the theory of meaning. *Manipulative* abduction (Figure 4) happens when we are thinking *through* doing and not only, in a pragmatic sense, about doing. So the idea of manipulative abduction goes beyond the well-known role of experiments as capable of forming new scientific laws by means of the results (the nature answer's to the investigator's question) they present. It refers to an extra-theoretical behavior that aims at creating communicable accounts of new experiences to integrate them into previously existing systems of experimental and linguistic (theoretical) practices. In manipulative abduction the suggested hypotheses are inherently ambiguous until articulated into configurations of real or imaginated entities (images, models or concrete apparatus and instruments). In these cases only by experimenting we can discriminate between possibilities: they are articulated behaviorally and concretely by *manipulations* and then, increasingly, by words and pictures.

Gooding (1990) refers to this kind of concrete manipulative reasoning when he illustrates the role in science of the so-called "construals" that embody tacit inferences in procedures that are often apparatus and machine based. This embodiment is of course an expert manipulation of objects in a highly constrained experimental environment, and is directed by abductive movements that imply the strategic application of old and new *templates* of behavior mainly connected with extra-theoretical components, for instance emotional, esthetical, ethical, and economic.

[6] Many interesting relationships between model-based reasoning in creative reasoning and its possible deductive models are analyzed in Meheus (1999), we have already said to be also related to the formal treatment of inconsistencies.

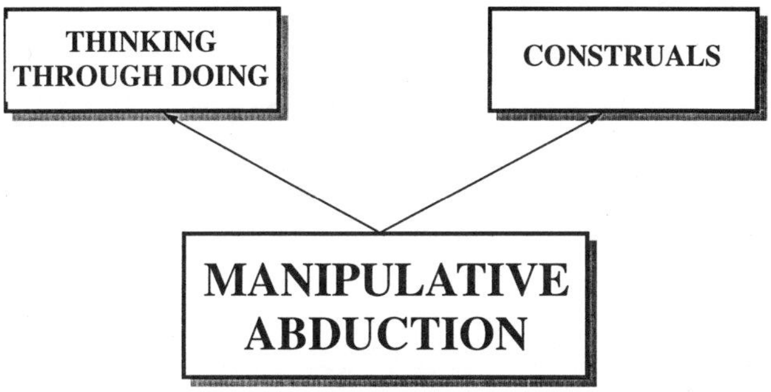

Figure 4. Manipulative abduction.

The hypothetical character of construals is clear: they are provisional creative organization of experience and some of them become in their turn hypothetical *interpretations* of experience, that is more theory-oriented, their reference is gradually stabilized in terms of established observational practices. Step by step the new interpretation relates to more "theoretical" modes of understanding (visual, diagrammatic, symbolic, conceptual, simulative), closer to the constructive effects of theoretical abduction. When the reference is stabilized the effects of incommensurability with other stabilized observations can become evident. But it is just the construal of certain phenomena that can be shared by the sustainers of rival theories. Gooding (1990) shows how Davy and Faraday could see the same attractive and repulsive actions at work in the phenomena they respectively produced; their discourse and practice as to the role of their construals of phenomena clearly demonstrate they did not inhabit different, incommensurable worlds in some cases.

REFERENCES

Boutilier, C. and Becher, V, 1995, Abduction as belief revision, *Artificial intelligence* 77:43-94.

Bylander, T., Allemang, D., Tanner, M.C., and Josephson, J.R., 1991, The computational complexity of abduction, *Artificial Intelligence* 49:25-60.

Bringsjord, S., 1999, Is (Gödelian) model-based deductive reasoning computational?, Special Issue "Abduction and scientific discovery", L. Magnani, N.J. Nersessian, and P. Thagard, eds, *Philosophica*, forthcoming.

Cornuéjols, A., Tiberghien, A., and Collet, G., 1999, A new mechanism for transfer between conceptual domains in scientific discovery and education, Special Issue "Model-Based Reasoning in Scientific Discovery: Learning and Discovery", L. Magnani, N.J. Nersessian, and P. Thagard, eds., *Foundations of Science*, forthcoming.

Davis, W.H., 1972, *Peirce's Epistemology*, Nijhoff, The Hague.

Gooding, D., 1990, *Experiment and the Making of Meaning*, Kluwer, Dordrecht.

Greenberg, M.J, 1980, *Euclidean and Non-Euclidean Geometries*, Freeman and Company, New York.

Hintikka, J., and Remes, U., 1974, *The Method of Analysis. Its Geometrical Origin and its General Significance*, Reidel, Dordrecht.

Hintikka, J. 1998, What is abduction? The fundamental problem of contemporary epistemology, *Transactions of the Charles S. Peirce Society* 34:503-533.

Kant, I., 1929, *Critique of Pure Reason*, translated by N. Kemp Smith, MacMillan, London; originally published 1787.

Holyoak, H.J., and Thagard, P., 1995, *Mental Leaps. Analogy in Creative Thought*, The MIT Press, Cambridge, MA.

Magnani, L., 1988, Epistémologie de l'invention scientifique, *Communication & Cognition* 21:273-291.

Magnani, L., 1992, Abductive reasoning: philosophical and educational perspectives in medicine, in: *Advanced Models of Cognition for Medical Training and Practice*, D.A. Evans and V.L. Patel, eds., Springer, Berlin, pp. 21-41.

Magnani, L., 1999a, Creative abduction and hypothesis withdrawal in science, in: *Proceedings of International Congress on Discovery and Creativity (ICDC)*, J. Meheus, ed., Kluwer, Dordrecht, in press.

Magnani, L., 1999b, Visual abduction in mathematical discovery, Technical Report, Department of Philosophy, University of Pavia.

Magnani, L., Civita, S., and Previde Massara, G., 1994, Visual cognition and cognitive modeling, in: *Human and Machine Vision: Analogies and Divergences*, V. Cantoni, ed., Plenum Press, New York, pp. 229-243.

Meheus, J., 1999, Model-based reasoning in creative processes, this book.

Nersessian, N.J., 1995, Opening the black box: cognitive science and history of science, Technical Report, Cognitive Science Report Series, Georgia Institute of Technology, Atlanta, GA. Partially published in *Osiris* 10:194-211.

Nersessian, N.J., 1998, Kuhn and the cognitive revolution, *Configurations* 6:87-120.

Nersessian, N.J., 1999a, Inconsistency, generic modeling, and conceptual change in science, in: *Inconsistency in Science*, J. Meheus, ed., Kluwer, Dordrecht, in press.

Nersessian, N.J., 1999b, Model-based reasoning in conceptual change, this book.

Nersessian, N.J., Griffith, T.W., and Goel, A., 1997, Constructive modeling in scientific discovery, Technical Report, Georgia Institute of Technology, Atlanta, GA.

Peirce, C.S, 1931-35, 1958, *Collected Papers* 1-6, C. Hartshorne and P. Weiss, eds.; 7-8, A. Burks, ed., Harvard University Press, Cambridge, MA.

Quine, W.V.O., 1979, *Philosophy of Logic*, Prentice-Hall, Englewood Cliffs, NJ.

Ramoni, M., Stefanelli, M., Magnani, L., and Barosi, G., 1992, An epistemological framework for medical knowledge-based systems, *IEEE Transactions on Systems, Man, and Cybernetics* 22(6):1361-1375.

Shelley, C., 1996, Visual abductive reasoning in archaeology, *Philosophy of Science* 63(2):278-301.

Thagard, P., 1988, *Computational Philosophy of Science*, The MIT Press, Cambridge, MA, 1988.

Thagard, P., 1992, *Conceptual Revolutions*, Princeton University Press, Princeton.

Thagard, P. and Shelley, C.P., 1997, Abductive reasoning: logic, visual thinking, and coherence, in: *Logic and Scientific Methods*, M.L. Dalla Chiara, K. Doets, D. Mundici, and J. van Benthem, eds., Kluwer, Dordrecht, pp. 413-427.

Thagard, P. and Croft, D., 1999, Scientific discovery and technological innovation: ulcers, dinosaur extinction, and the programming language Java, this book.

Abduction and Geometrical Analysis. Notes on Charles S. Peirce and Edgar Allan Poe

Ilkka Niiniluoto
Department of Philosophy, University of Helsinki, Helsinki,Finland, ilkka.niiniluoto@helsinki.fi

Abstract: The method of analysis and synthesis in Greek geometry is perhaps the most significant key idea in the history of heuristic reasoning. In theoretical analysis, reasoning goes backward from a theorem to the axioms from which it deductively follows. In problematical analysis, the desired thing is supposed to be given (this defines the so-called "model figure"), and then the reasoning again goes backward looking for possible constructions from which the sought thing results. This paper states that Charles S. Peirce's description of abduction, as a retroductive inference of a cause from its effects, is an instance of the propositional interpretation of analysis. What Jaakko Hintikka calls the "analysis of figures" view of problematical analysis is illustrated by detective stories of Edgar Allan Poe. The same idea is repeated in Poe's essay "Philosophy of Composition" where he tells that he wrote his poem *The Raven* (1845) analytically by starting at the end.

1. PEIRCE ON ABDUCTION: A BRIEF HISTORY

Throughout his career Charles S. Peirce (1839-1914) defended the thesis that, besides deduction and induction, there is a third mode of inference which he called "hypothesis", "retroduction", or "abduction".

In 1865, influenced by Aristotle and Kant, the young Peirce characterized the third mode of inference as "reasoning *a posteriori* to a hypothesis", or "inference of a cause from its effect" (*W* 1:180). Starting with an Aristotelian Barbara syllogism as the paradigm of deduction, Peirce noted that it can be inverted in two ways: induction is the inference of the major premise, and hypothesis the inference of the minor premise. More precisely, a typical deduction (syllogism of the first figure) infers a result (conclusion) from a general rule (major premise) and a case (minor premise), induction infers the rule from the case and the result, while hypothesis infers the case from the rule and the result:

Model-Based Reasoning in Scientific Discovery, edited by L. Magnani,
N.J. Nersessian, and P. Thagard, Kluwer Academic/Plenum Publishers, New York, 1999

All *Fs* are *G*
These are *Gs* (1)
∴ These are *Fs*

In 1867, inspired by John Venn, Peirce suggested the idea of probable arguments with a truth-frequency, i.e., probability expresses the proportion of situations in which the argument carries truth from the premises to its conclusion. In his 1878 paper, Peirce gave probabilistic formulations of the deductive syllogism and its inversions by replacing the major premise with a statistical generalization "Most Fs are G". An argument of this form, which of course is not, in general, logically valid, may nevertheless have a high truth-frequency.

The deductive syllogism was called "explanation" by Peirce (*W* 1:267, 428, 440). Anticipating C.G. Hempel's 1962 model of inductive-probabilistic explanation (see Niiniluoto 1993), Peirce used this terminology consistently also in the case of "statistical syllogisms" (*CP* 2.716).

Peirce's own pioneering work in logic led him to think that he had been "too much taken in considering syllogistic forms" (*CP* 2.102). In papers from 1901 to 1910, Peirce then redefined induction as "the operation of testing a hypothesis by experiment" (*CP* 6.526) and abduction as "the first starting of a hypothesis and the entertaining of it, whether as a simple interrogation or with any degree of confidence" (*CP* 6.525). Abduction starts from facts and seeks a theory to explain them (*CP* 7.218), but it infers only a "may-be" (*CP* 8.238). Its general form is as follows (*CP* 5.189):

The surprising fact *C* is observed;
But if *A* were true, *C* would be a matter of course. (2)
Hence, there is a reason to suspect that *A* is true.

A historical case illustrating this mode of reasoning is Kepler's investigation of the orbit of Mars (*CP* 1.71-74).

After N.R. Hanson's work on scientific discovery in 1958, and Gilbert Harman's essay on inference to the best explanation in 1965, the Peircean notion of abduction has become a major theme within the contemporary philosophy of science, both in the study of scientific creativity and patterns of non-deductive reasoning (for a survey and some references, see Niiniluoto, 1998). To put Peirce's views in a historical perspective, it is useful to relate them to another important idea of heuristic reasoning, viz. the method of analysis and synthesis in Greek geometry. It not only turns out that abduction can be viewed as a special case of the method of analysis, but we shall also see that

Peirce's ideas about heuristic reasoning were interestingly anticipated by the poet Edgar Allan Poe.

2. PEIRCE AND ARISTOTLE

Peirce's account of abduction as the inference of the minor premise of a syllogism, or the initial condition of an explanation, resembles Aristotle's examples of "quick wit", defined as the power of "hitting upon the middle term without a moment's hesitation" (*An. Post.* I, 34). For example, we see someone talking to a rich man, and immediately decide that it is because he is trying to borrow money.

Aristotle's treatment of *epagoge* in *Analytica Priora* (II, 23) involves the inversion of syllogisms, but for him this process in inseparable from concept formation. William Whewell, in *Philosophy of Discovery* (1860), followed Aristotle closely, as in his "inductions" there is always "a new conception", suggested by "the sagacity of discoverers" (see Niiniluoto, 1984, 1995). Peirce, instead, argued in 1883 that in a valid induction the attribute to be detected in the sample has to be "predesignated" or "agreed upon in advance", so that the discovery of laws in science is accomplished by induction, and the discovery of causes by hypothetic inference (*CP* 2.736).

In his 1898 Cambridge Lectures, Peirce suggested that induction corresponds to Aristotle's *epagoge* and hypothesis (abduction) to Aristotle's *apagoge*. The latter term is usually translated as "reduction". A historically important form of *apagoge* is the argument known as *reductio ad absurdum*: consequences are derived from a given statement, and if a false consequence is reached, the original statement must be false as well. But if the new facts are always found to match the derived consequences, the hypothesis has been tested. This kind of inference was called "induction" by Peirce after 1901.

Peirce's characterization of deductive explanation as proceeding from causes to effects has its roots in Aristotle's doctrine in *Analytica Posteriora* (see I, 13, 34) that a "scientific syllogism" proceeds from causes to their effects. The medieval followers of Aristotle, from Robert Grosseteste in the 12th Century to Jacopo Zabarella in the 16th Century, called this inference *demonstratio propter quid* or *compositio*. The inverse inference from effects to their causes was called *demonstratio quia* or *resolutio* (see Crombie, 1953). Taking the clue from Aristotle's distinction between "knowledge of a fact" and "knowledge of the reason of a fact" (*An. Post.* II, 13), the mode of presenting science was assumed to follow a "regressive" method: starting from something that is observable, better known to us, a singular and complex effect, we first proceed to something which is better known to nature, a simple, and general cause, and then back to the original phenomenon, but now with "knowledge-that" replaced by "knowledge-why". Some Aristotelians believed that resolution can proceed by deductively valid syllogisms while some identified this inference with induction.

The main idea of the regressive method, including the terminology with "resolutions" and "compositions", was followed by such great figures of modern science as Galileo Galilei, Isaac Newton, and Karl Marx. But in modern science, these procedures were regarded as modes of inquiry leading to new knowledge – rather than as modes of teaching and organizing old knowledge.

In this light, the novelty in Peirce's theory of abduction is not in the claim that there is a type of inference from effects to their causes. Peirce's originality can be seen in his attempt to generalize Aristotelian logic, including the statistical and probabilistic forms of inference.

Another interesting aspect of Peirce is related to his dynamic conception of laws. For Aristotle, the laws of nature expressed essential connections between qualitative properties, and this was true of his description of causes and effects as well. But the modern conception takes causal laws to be dynamic or relational invariances – "laws of succession" in J.S. Mill's sense. Thus, one way of understanding the general rule of a syllogism is that it expresses a causal law. As causes temporally precede their effects, the inverse inference (1) from effects to their causes involves reasoning backwards in time. This temporal interpretation of hypothetic inference motivates Peirce's term *retroduction* and some of his examples: perceptual judgments as "an extreme case of abductive inferences" (*CP* 5.181-185) and our knowledge of history (*CP* 2.714).

3. PAPPUS ON GEOMETRICAL ANALYSIS

The Latin terms *resolutio* and *compositio* are translations of the Greek terms *analysis* and *synthesis*. In opposition to J.H. Randall's (1961) thesis, that Galileo and Newton adopted the method of science from the regressive method of the Paduan Aristotelians, it has been argued that their main inspiration was from the method of analysis and synthesis of Greek geometry (see Hintikka and Remes, 1974). The most famous description of this "secret" of the Greek geometers, or the "Treasury of Analysis", was given by Pappus (ca. 300 AD), whose work was translated into Latin in 1566, and was known to the great champions of the new mathematical physics.

The method of analysis and synthesis was already known in Plato's school. In *Nicomachean Ethics* (1112b15-29), Aristotle compares it to the structure of deliberation: in the process of planning, a decision-maker searches for means to obtain given ends, and further means to obtain intermediate means, etc., until this reasoning "backward" comes to something that can be done or is impossible. In the former case, the synthesis reverses the steps in order to reach the desired goal. Hence, analysis goes here from ends to means, and synthesis conversely from means to ends.

The famous description by Pappus (in A. Jones' translation, see Pappus, 1986, p. 82) is as follows:

> Now analysis is the path from what one is seeking, as if it were established, by way of its consequences, to something that is established by synthesis. That is to say, in analysis we assume what is sought as if it had been achieved, and look for the thing from which it follows, and again what comes before that, until by regressing in this way we come upon some one of the things that are already known, or that occupy the rank of a first principle. We call this method "analysis", as if to say "reduction backward". In synthesis, by reversal, we assume what was obtained last in analysis to have been achieved already, and setting now in natural order, as precedents, what before were following, and fitting to each other, we attain the end of the construction of what was sought. This is what we call "synthesis".

Pappus goes on to explain that there are two kinds of analysis: "theoretical" which seeks truth, and "problematic" which tries to find or construct what was requested.

> In thc case of the theorematic kind, we assume what is sought as a fact and true, then, advancing through its consequences, as if they are true facts according to the hypothesis, to something established; if this thing that has been established is a truth, then that which was sought will also be true, and its proof the reverse of the analysis; but if we should meet something established to be false, then the thing that was sought too will be false. In the case of the problematic kind, we assume the proposition as something we know, then, proceeding through its consequences, as if true, to something established; if the established thing is possible and obtainable, which is what the mathematicians call "given", the required thing will also be possible, and again the proof will be reverse of analysis; but should we meet with something established to be impossible, then the problem too will be impossible. Diorism is the preliminary distinction of when, how, and in how many ways the problem will be possible.

There is a long-standing and lively debate on the meaning of this passage and its interpretation (see, e.g., Heath, 1956, vol. 1, pp. 137-142; Hintikka and Remes, 1975; Mäenpää, 1997). The "standard" interpretation of theoretical analysis is as follows: from a proposed theorem T_0 a logical consequence T_1 is deduced, from that T_2, etc. until a statement T_n known to be false or true is reached. In the former case, T_0 is false as well, in the latter case, T_0 can be proved in synthesis from T_n, provided all the previous deductive steps from T_i to T_{i+1} are reversible, i.e., T_i and T_{i+1} are in fact logically equivalent (see, e.g., A.K. Szabó's essay in Hintikka and Remes, 1974). This is called "downward propositional interpretation" of analysis by Hintikka and Remes (1974). The "upward propositional interpretation", instead, suggests that analysis seeks premises from which T_0 can be deduced, so that it is always required that T_{i+1} entails T_i for all $i = 0, 1,..., n\text{-}1$.

We can now see that with these propositional interpretations, the method of analysis and synthesis closely resemble the regressive method: analysis reasons backwards from a hypothetical theorem (T_0) to some axioms of geometry (T_n), and synthesis gives the proof of T_0 as a theorem on the basis of T_n.

We also see that Peirce's description (2) of abduction corresponds to the upward propositional interpretation of analysis: a surprising fact is given, and a theory from which it follows (by deduction or probability) is sought. However, in the case of natural science, we do not have an independent or direct way of verifying the theory, and there may be several rival theories which could all explain the given fact. Therefore, abduction only leads to potential theoretical explanations which are testworthy, and a correct selection requires the performance of further "inductive" tests by their other consequences.

Hintikka and Remes (1974) make important objections to the propositional interpretations of analysis. One of their observations is that theorems in geometry are typically general statements (e.g., "for all triangles, the sum of their angles equal 180 degrees") or universal-existential statements (e.g., "for all geometrical figures *x*, if *x* is a square, then *x* will have two diagonals and these diagonals bisect each other"). Proof of such general implications proceeds through their instantiations and by attempting to derive the consequent from the antecedent by suitable axioms or rules of inference. As worked out to some extent by Hintikka and Remes, one particularly effective way of attempting to construct such proofs is Beth's method of semantic tableaux.

E.W. Beth himself suggested that his semantic tables might be understood as an explanation of the method of analysis. It is interesting that these same methods have recently been used to give a model of abduction by Atocha Aliseda-Llera (1997). This in a sense reconfirms the intimate connection between analysis and Peircean abduction.

Another observation by Hintikka and Remes is that proof of general implications in ordinary predicate logic may need the use of instantiation rules by which new individuals are introduced into the inference. Hintikka has used the amount of such needed additional individuals as a measure of "non-triviality" of deductive inference. Again, Peirce anticipated this idea in his discussion of trivial or "corollarial" deductive reasoning (*CP* 7., p. 124; see Hintikka, 1998). In geometrical proofs, such new individuals correspond to auxiliary constructions that are drawn in figures and illustrations. In this view, the heuristic power of analysis lies precisely in these auxiliary constructions: as first-order logic is not effectively decidable, there are no mechanical methods of making such constructions or even knowing their number in a particular case. Analysis is not a "foolproof" algorithm for discovery, but as soon as the necessary constructions have been found by the wit or imagination of the researcher, it specifies steps for making genuine discoveries.

These considerations lead Hintikka and Remes to replace the propositional interpretation of analysis by what they call the "analysis of figures" or "configurational analysis". Instead of being a method which seeks deductive connections between propositions, analysis and synthesis consider geometrical figures and the functional interconnections between their elements. Following Kant, Peirce also argued that such "observation and experiment" of "a diagram of our own creation" is an essential feature of all "necessary reasoning" in geometry (*CP* 3.554-560).

The configurational idea is especially clear in the case of problematic analysis which attempts to solve geometrical problems. An elegant formalization of this method is given in Petri Mäenpää's 1993 doctoral dissertation (see Mäenpää, 1997): instead of using predicate logic, Mäenpää employs Per Martin-Löf's type theory to account for constructions and their results.

In 20th Century textbooks of school mathematics, the method of solving geometrical problems by analysis and synthesis was sometimes called "the method of model figures". The solution of a problem starts by assuming that the desired construction has already been accomplished. Then some auxiliary constructions are added to this model picture and the relations of its parts are studied. Eventually a condition for the thing sought is obtained in this "backward solution", and the number of ways of realizing this condition is counted as the number of possible solutions.

The configurational interpretation of analysis was also the basis of the analytic geometry of Descartes (see Mäenpää, 1997):

> If, then, we wish to solve any problem, we first suppose the solution already effected, and give names to all the lines that seem needful for its construction, - to those that are unknown as well as to those that are known. Then, making no distinction between known and unknown lines, we must unravel the difficulty in any way that shows most naturally the relations between these lines, until we find it possible to express a single quantity in two ways. This will constitute an equation [...] (Descartes, 1954, p. 6.)

So the trick is again to draw a model figure, maybe with additional constructions, identify the known elements, and mark the unknown as "x". Then, treating x as if it were known, an equation is established, and finally the equation is solved with respect to x.

4. POE'S STORIES OF RATIOCINATION

Thomas A. Sebeok and Jean Umiker-Sebeok (1980) tell a true story about an incident in 1879, when Charles Peirce was able to guess the person who stole his watch and overcoat on a boat trip. They argue that this inference, which was later verified by a house search, was an example of Peircean abduction.

Sir Arthur Conan Doyle (1859-1930) published his first Sherlock Holmes story, *A Study in Scarlet* in 1888. There is no direct evidence that Peirce had read Holmes stories, but Edgar Allan Poe (1809-49) seems to be one of his favorite writers. Poe published his first detective story *The Murders in the Rue Morgue* in 1841, followed by *The Mystery of Marie Rogêt* in 1843 and *The Purloined Letter* in 1845 (see Poe, 1977). These "stories of ratiocination", with their hero C. Auguste Dupin, mark the birth of the new detective genre in the world literature.

Dupin and Holmes are master detectives who solve crime cases with their powerful methods of reasoning. Carefully observed facts about the case serve as clues for further investigations and questions, but when the necessary information has been collected, the solution is obtained by an intellectual operation. Agatha Christie's Hercule Poirot follows this genre, while Raymond Chandler's Philip Marlowe operates with straightforward action and intervention. Indeed, Chandler's "The Simple Art of Murder" (1950) questions the whole tradition of "logic-and-deduction-novels" with its artificial treatment of "formal and exact problems".

Reasoning in classical detective stories is usually called "deduction". There is no doubt that deduction in the strict sense of logically valid inference has an important role. But it also easy to see that many of the inferential steps and "guesses" that Dupin and Holmes perform during their stage of investigation are abductive or retroductive in Peirce's sense, i.e., they infer a cause from an effect. (This is observed also by Sebeok and Umiker-Sebeok, 1980; Truzzi, 1983, p. 69; Harrowitz, 1983, pp. 185-194.) For example, Holmes notes that a person has heavier and bigger muscles in his right arm than in his left arm and infers that he is a manual worker, or Holmes observes the tanned skin of a gentleman and infers that he has served in the British army in India.

More generally, the solution of a murder is a paradigmatic example of retroduction. Given facts about the present that are traces of earlier events, the task of the detective is to give a historical reconstruction of the historical process that led to the death of the victim. It is no surprise that inferential steps leading to this reconstruction are abductive in the Peircean sense.

For the purposes of this paper, it is also important to observe that this kind of retroduction involves "solution backward" in the classical geometrical sense of analysis. This is not greatly emphasized by Doyle, who was educated in medicine and patterned the character and skill of Sherlock Holmes after his professor Joseph Bell of the Royal Infirmary of Edinburgh (see Sebeok and Umiker-Sebeok, 1980, p. 54). There are two well-known passages where Holmes describes his method in terms that are linked to the method of analysis (cf. *ibid.* p. 64, 66), but we shall see that they are directly influenced by Poe. In *A Study in Scarlet*, Holmes states:

> In solving a problem of this sort, the grand thing is to be able to reason backward. That is a very useful accomplishment, and a very easy one, but people do not practice it much. In the everyday affairs of life it is more useful to reason forward, and so the other comes to be neglected. There are fifty who can reason synthetically for one who can reason analytically [...]. There are few people, however, who, if you told them a result, would be able to evolve from their own inner consciousness what the steps were which led up to that result. This power is what I mean when I talk of reasoning backward, or analytically.

In *The Five Orange Pip*, Holmes refers to Cuvier's study of a whole animal "by the contemplation of a single bone", and concludes that

> the ideal reasoner [...] would, when he had once been shown a single fact in all its bearings, deduce from it not only the chain of events which led up to it but also the results which would follow from it.

A link between detective stories and abduction is provided by Jaakko Hintikka's suggestion that information-seeking through questioning, or the treatment of inferences as answers to tacit questions, serve as a formalization of Sherlock Holmes's logic (see Hintikka and Hintikka, 1983) and Peirce's abduction (see Hintikka, 1998). Moreover, a further indirect link is established by the fact that Hintikka's tool for studying such inferences both in the case of geometrical analysis (see Hintikka and Remes, 1974) and Sherlock Holmes (see Hintikka and Hintikka, 1983) has been Beth's semantic tableaux. As noted above, the same tool has been used to investigate the Peircean abduction structure in science by Aliseda-Llera (1997).

It is remarkable that Edgar Allan Poe was aware that "the analytic power" of Monsieur C. Auguste Dupin is historically associated with the mathematical tradition. Poe, who had studied at the University of Virginia, had a wide interest in the scientific and philosophical literature of his time, as vividly illustrated by his science fiction stories (see Poe, 1976). This interest covered also mathematics and its applications, such as cryptography and chess-players (see Poe, 1977).

In the opening paragraph of *The Murders in the Rue Morgue*, Poe discusses "the mental features discoursed of as the analytical" (Poe, 1977, pp. 332-333):

> The faculty of re-solution is possibly much invigorated by mathematical study, and especially by that highest branch of it which, unjustly, and merely on account of its retrograde operations, has been called, as if *par excellence*, analysis.

Poe is referring here to the basic area of modern mathematics, Differential Calculus, known as "the Analysis" and "the Calculus". He objects immediately that "to calculate is not in itself to analyze". Calculation, such as chess-playing, is not yet analysis, since the analytic ability requires imagination which goes beyond mere ingenuity (*ibid.*, p. 336). In *The Purloined Letter*, mathematical reasoning is characterized as "merely logic applied to observation upon form and quantity" (*ibid.*, p. 453). The criminal Minister in the story is described by Dupin as being both a poet and a mathematician, and therefore he could reason well (*ibid.*, p. 453). Similarly, Dupin himself is both creative and resolvent. Dupin's criticism of the French "deception" of speaking about "analysis" in connection with algebra suggests, however, that Poe was not acquainted with the historical connection between geometrical analysis and analytical geometry.

In the story about Rue Morgue, two women are found murdered in an unusually brutal way: one in a locked room, strangled and thrown in the chimney, the other in a yard in the rear of the building with her throat cut. The men making these horrible findings hurried to the place alarmed by terrible shrieks from the house. When they rushed to the stairs, they heard an angry contention by two voices, one by a Frenchman and the other by a foreigner with a shrill voice. The murders seemed to have no motive, and the police found no clues for solving them.

After investigating the scene of crime and the testimonies of the witnessess, Dupin solves the mystery by an astounding chain of inference. After excluding the possibility of suicide or supernatural explanations, he concludes that the murders are committed by a third person. But how could the assassin escape from the room? "Fortunately, there is but one mode of reasoning upon the point, and that mode *must* lead us to a define decision", Dupin explains. As other means of egress are excluded, the murderer must have passed through the windows of the back room. In showing that this "apparent impossibility" is a real possibility, Dupin concludes *a posteriori* that there must be something wrong in the nail which fastened the sashes of the window.

Dupin then puts together the "very unusual" strength and agility, brutal cruelty, startling absence of motive, peculiar shrill and unequal voice "about whose nationality no two persons could be found to agree and in whose utterance no syllabification could be detected". With the additional clues of a tuft of hair in the hand of one victim and the facsimile of the finger nails in the throat of the other victim, Dupin infers – by appealing to Cuvier's zoological study – that the murder has been accomplished by an Ourang-Outang of the East Indian Islands.

The structure of Dupin's inference resembles the backward solution of a mathematical problem by means of the method of analysis. His starting point is a situation with a murder (cf. the geometrical model figure), with the so far unidentified murderer (cf. the sought, unknown entity *x*). The question is how this situation has been generated (cf. the needed constructions for creating the unknown *x*). Effectively using the information about the situation, he develops the solution step by step backward (cf. the retroductive direction of analysis). Progress often requires additional questions and observations (cf. the auxiliary constructions in the model figure), which also serve to eliminate some of the alternative explanations. After the mystery has been solved, the succession of events is reconstructed in its temporal order (cf. the synthesis after an analysis).

The story about Rue Morgue has been designed so that there is only one reasonable, though imaginative solution which explains all the details about the case. It thus corresponds to a geometrical problem with a unique solution. In such a situation the backward analysis can proceed only in one way. As Dupin states about his "legitimate deductions": "the deductions are the *sole* proper ones, and [...] the suspicion arises *inevitably* from them as the single result". Still, this analysis requires creative imagination, as the analyst has to know what to ask, observe, and search for.

Poe also lets his hero verify the inferred conclusion by putting an advertisement in *Le Monde*, stating that "a very large Ourang-Outang of the Bornese species" has been caught in a park and its owner may have it again by paying a few charges. This kind of action might be compared to experiments or manipulations of nature that have an important role in Newton's application of the method of analysis in science.

Poe knew that all problem-solving cannot be so straightforward or compelling as in the story of Rue Morgue, since there may exist alternative backward paths and the task of the investigator is to find the most probable among them. This topic – which of course was crucial in Peirce's probabilistic formulations of abduction as well – is discussed in *The Mystery of Marie Rogêt*. Poe's understanding of the Calculus of Probabilities is not entirely adequate, since he seems to believe in the phenomenon of the "maturity of chances": the occurrence of two sixes in succession with a dice is for him a sufficient cause for betting that sixes will not be thrown in the third attempt (Poe, 1977, p. 438). More interestingly, Poe observes that the probability of a hypothesis is increased in the ratio of the "peculiarity" or "unusualness" of the facts it explains (cf. *CP* 7.182). Such corroborative evidence does not increase probability additively but geometrically: each new proof or piece of evidence multiplies the earlier probability (*ibid.*, pp. 405-406). These ideas could be easily formalized by Bayes' Theorem: if hypothesis H entails evidence E, then the posterior probability P(H/E) of H given E equals the prior probability P(H) of H multiplied by the factor 1/P(E).

Poe's story *The Gold-Bug* (1843) illustrates the principle that the most probable explanation deserves to be investigated in a heuristic search. The topic of the story, secret writing or cryptography, was discussed by Poe also in an essay two years earlier (see Poe, 1977, pp. 538-546). In the story, William Legrand is able to decode a letter of the pirate Captain Kidd by constructing the key on the basis of the statistical frequencies of letters in ordinary English texts. This method could be presented in any modern textbook of information theory (cf. Smith, 1955).

5. POE'S PHILOSOPHY OF COMPOSITION

Poe's fascination with mathematical analysis extends to his work as a poet. The construction of works of art requires, besides ingenuity, "the faculty of analysis" (Poe 1927, p. 326). As Poe remarks on Longfellow, someone may sometimes write brilliant poems "by accident" (*ibid.*, p. 269). But the proper way of composing poetry must be based upon knowledge of the aims of poetry - which is the creation of beauty rather than truth (*ibid.*, pp. 271, 279) - and of its general principles. The poetry of words is defined by Poe as "the rhythmical creation of beauty" (*ibid.*, p. 99), and two-thirds of the principles of versification belong to the mathematics (*ibid.*, p. 333).

Poe's essays "The Poetic Principle", "The Rationale of Verse", "Fancy and Imagination", and "Rhyme" contain many interesting observations. Anticipating modern information-theoretical accounts of aesthetics, he remarks that "perfection of rhyme is attainable only in the combination of the two elements, Equality and Unequality", and "unexpectedness must arise from expectedness" (*ibid.*, p. 311). Anticipating later doctrines of emergent qualities, Poe points out "the range of Imagination is unlimited", as the admixture of two elements may result in something that has no qualities of either (*ibid.*, p. 316). And anticipating modern semiotic studies, Poe points out that in a poetic text there may lie "beneath the transparent upper-current of meaning, an under or suggestive one" (*ibid.*, p. 285).

The essay "The Philosophy of Composition" gives a surprising description of the birth of Poe's perhaps best-known poem "The Raven" (1845):

> It is my design to render it manifest that no one point in its composition is referable either to accident or intuition - that the work proceeded, step by step, to its completion with the precision and rigid consequence of a mathematical problem (*ibid.*, pp. 165-166).

That Poe's account here corresponds to the problematic analysis in the geometrical sense becomes evident when he states that his poem has its beginning "at the end, where all works of art should begin" (*ibid.*, p. 171). The poet should start by selecting some desired effect, and then find aids in "the construction of the effect" (*ibid.*, p. 164). The artistic means are then chosen so that "in the whole composition there should be no word written, of which the tendency, direct or indirect, is not to the one pre-established design" (*ibid.*, p. 188). "It is only with the *dénouement* constantly in view that we can give the plot its indispensable air of consequence [...]" (*ibid.*, p. 164).

Poe starts from the intention of composing a poem that suits "at once the popular and the critical taste". From the principles that a poem should not be "too long to be read at one sitting", that beauty is "the sole legitimate province of the poem", and that melancholy is "the most legitimate of all the poetical tones", Poe concludes that the length of the intended poem should be about one hundred lines and it should deal with beauty with the tone of sadness. Using "ordinary induction" Poe then chooses refrain as his artistic effect. The best refrain, he decides, is a single word forming the close of each stanza. It should include the long *o* as the most sonorous vowel and *r* as the most producible consonant. The first word, expressing the melancholy tone, that occurred to Poe is "Nevermore". But what would the occasion for the repeated use of the word "nevermore"? As a rational human being would not fit the purpose, and a parrot would not be in keeping with the intended tone, a raven as "the bird of ill omen" was chosen.

According to Poe, death is the most melancholy of all melancholy topics, and the death of a beautiful woman is "unquestionably the most poetical topic in the world". The only intelligible way of combining these ideas is to imagine the raven employing the word 'nevermore' in answer to the queries of a lover. The last of these self-torturing questions should involve "the utmost conceivable amount of sorrow and despair". At this moment Poe puts his pen to paper and writes the climax of the poem:

> "Prophet", said I, "thing of evil! prophet still if bird or devil!
> By that heaven that bends above us – by that God we both adore,
> Tell this soul with sorrow laden, if within the distant Aidenn,
> It shall clasp a sainted maiden whom the angels name Lenore –

Clasp a rare and radiant maiden whom the angels name Lenore."
Quoth the raven "Nevermore."

The writing of the poem thus starts from the end, more precisely, from the sixteenth stanza. After writing the first fifteen stanzas gradually approaching the *dénouement*, where the raven destroys the lover's final demand to meet his mistress in another world, Poe still adds two stanzas at the end to create "suggestive meaning" by metaphorical expressions.

According to Poe's "philosophy of composition", the construction of a poem follows an equally definite pattern as Dupin's inferences in solving crime mysteries. Both are "analytic" in the traditional sense that they start from the end and seek the solution by backward steps. In this sense, Poe generalizes to the realm of poetry and artistic creation the principle that Isaac Newton expressed in his *Opticks* (1704) as follows: "As in Mathematics, so in Natural Philosophy, the Investigation of difficult Things by the method of Analysis, ought ever to precede the Method of Composition", where such Analysis proceeds "from Compounds to Ingredients, and from Motions to the Forces producing them; and in general, from Effects to their Causes". Peirce once remarked that physics owes all its triumphs to the "Analytic Method" (*CP* 1.64). Essentially the same method has been applied in engineering design (where construction problems are analyzed into subproblems until we reach something that can be done), and in computer science for the construction of programs (Mäenpää, 1997). Analysis was advertised as a general method of problem-solving already by George Polya in his *How to Solve It* (1945).

Not all commentators have taken seriously Poe's account of his poetic construction "with the precision and rigid consequence of a mathematical problem", since it seems to clash with established views about creativity in art. A hallmark of creation is its non-purposiveness, i.e., "its outcome can be a surprise even to its originator" (Hintikka, 1975, p. 196). But it seems to me that this aspect of creativity can find its place also in Poe's method of composition.

The first part of Poe's description consists in the discovery of the climax of the poem. This stage involves artistic inspiration and imagination, and – in spite of some inferential steps concerning the kind of effect that is desired – it is not yet present within the initial intention of writing a poem.

The second part finds a construction in order to produce this *dénouement*. Here this starting point may be compared to the model figure in geometrical analysis, observable phenomena in an Aristotelian resolution, a physical experimental set-up in the Newtonian analysis, or the scene of crime in a detective story. In all of these cases the desired effect has to be built up from its causes. But in the case of natural and historical phenomena, at least if we are ontological realists about nature and history, the causal ingredients to be found in abduction in a sense already exist in the given totality, and the problem is to find or reveal them, and then the synthesis puts this totality together in an organized way. In the case of poetic creation, the "totality" which serves as the starting point of analysis does not yet contain the ingredients from which it can be composed, but they have to be invented by the artist. Mathematics may be ontologically closer to art than to nature. But the possible differences in ontology do not seem to make a difference in the logical structure of analysis and synthesis within these domains.

NOTE

The prehistory of this paper, presented in the conference on "Model-Based Reasoning in Scientific Discovery", goes back to the year 1975 when I acted as one of the opponents of Unto Remes's (1942-75) doctoral dissertation on the geometrical method of analysis and synthesis (see Hintikka and Remes, 1974). During the disputation, I suggested that Edgar Allan Poe's detective stories follow the method of analysis which has also a close resemblance to Charles S. Peirce's account of abductive inference. Two years later by accident I found a short article, published by my father Yrjö Niiniluoto as a young man in 1925, which referred to Poe's fascinating essay on the "Philosophy of Composition". My paper was published in Finnish in 1978, and later with some revisions in 1990. In the meantime, Thomas A. Sebeok had published his article on Peirce and Sherlock Holmes in 1980, followed by a collection of essays on abduction and detective stories (see Eco and Sebeok, 1983). Therefore, I did not hurry to write an English version of my paper, but waited for a suitable occasion for its presentation.

REFERENCES

Aliseda-Llera, A., 1997, Seeking explanations: abduction in logic, philosophy of science and artificial intelligence, *ILLC Dissertation Series*, University of Amsterdam, Amsterdam.

Chandler, R., 1964, The simple art of murder, in: *Pearls are a Nuisance*, Penguin Books, Harmondsworth.

Crombie, A., 1953, *Robert Grosseteste and the Origins of Experimental Science (1100-1700)*, Oxford University Press, Oxford.

Descartes, R., 1954, *Geometry*, Dover, New York.

Doyle, A.C., 1952, *The Complete Sherlock Holmes*, Doubleday, Garden City, N.Y.

Eco, U., and Sebeok, T.A., eds., 1983, *The Sign of Three: Dupin, Holmes, Peirce*, Indiana University Press, Bloomington.

Harrowitz, N., 1983, The body of the detective model: Charles S. Peirce and Edgar Allan Poe, in: U. Eco and T.A. Sebeok, eds., pp. 179-197.

Heath, T.L., 1956, *The Thirteen Books of Euclid's Elements*, vols. 1-3, Dover, New York.

Hintikka, J., 1975, *The Intentions of Intentionality*, Reidel, Dordrecht.

Hintikka, J., 1998, What is abduction? The fundamental problem of contemporary epistemology, *Transactions of the Charles S. Peirce Society* 34:503-533.

Hintikka, J., and Hintikka, M.B., 1983. Sherlock Holmes confronts modern logic, in: U., Eco, and T.A., Sebeok, eds., pp. 154-169.

Hintikka, J., and Remes, U., 1974, *The Method of Analysis. Its Geometrical Origin and Its General Significance*, Reidel, Dordrecht.

Mäenpää, P., 1997, From backward reduction to configurational analysis, in: *Analysis and Synthesis in Mathematics*, M. Otte and M. Panza, eds., Kluwer, Dordrecht, pp. 201-226.

Niiniluoto, I., 1984, *Is Science Progressive?*, Reidel, Dordrecht.

Niiniluoto, I., 1993, Peirce's theory of statistical explanation, in: *Charles S. Peirce and the Philosophy of Science*, E.C. Moore, ed., The University of Alabama Press, Tuscaloosa, pp. 186-207.

Niiniluoto, I., 1995, Hintikka and Whewell on Aristotelian induction, *Grazer Philosophische Studien* 49:49-61.

Niiniluoto, I., 1998, Defending abduction, *PSA 1998*.

Pappus, 1986, *Synagoge VII*, English trasl. by A. Jones, Springer-Verlag, New York.

Peirce, C.S., 1931-35, 1958, *Collected Papers* 1-6, C. Hartshorne and P. Weiss, eds.; 7-8, A. Burks, ed., Harvard University Press, Cambridge, MA.

Peirce, C.S., 1982-, *Writings of Charles S. Peirce: A Chronological Edition* 1-, M. Fisch, et al., eds., Indiana University Press, Bloomington.

Poe, E.A., 1927, *Poems and Essays*, Everyman's Library, Dent, London. (Reprinted, 1977.)

Poe, E.A., 1976, *The Science Fiction of Edgar Allan Poe*, H. Beaver, ed., Penguin Books, Harmondsworth.

Poe, E.A., 1977, *The Portable Poe*, P. van Doren Stern, ed., Penguin Books, Harmondsworth.

Polya, G., 1945, *How to Solve It*, Princeton University Press, Princeton.

Randall, J.H., 1961, *The School of Padua and the Emergence of Modern Science*, Padova.

Sebeok, T.A., and Umiker-Sebeok, J., 1980, *"You Know My Method": A Juxtaposition of Charles S. Peirce and Sherlock Holmes*, Gaslight Publications, Bloomington, Indiana. (Reprinted in U. Eco and T.A. Sebeok, eds., pp. 11-54.)

Smith, L.D., 1955, *Cryptography: The Science of Secret Writing*, Dover, New York.

Truzzi, M., 1983, Sherlock Holmes: applied social psychologist, in U. Eco and T.A. Sebeok, eds., pp. 55-80.

Whewell, W., 1860, *The Philosophy of Discovery*, London. (New Edition: Lenox Hill, Burt Franklin, New York, 1971).

The Hierarchy of Models in Simulation

Eric Winsberg
Department of History and Philosophy of Science, Indiana University, Bloomington, Indiana, ewinsber@indiana.edu

Abstract: This paper focuses on the role of the computer in helping to manage mathematically unsolvable sets of equations that arise within models in the physical sciences. The study of these models often consists in developing representations of the underlying physics on a computer, and using the techniques of computer simulation in order to learn about the behavior of these systems. I argue that this process of transformation, which involves a hierarchy of models, is also a process of knowledge creation, and it has its own epistemology. Accordingly, I urge philosophers of science to examine more carefully the process of *theory articulation*, the process by which a general theory is made to conform to a particular application. It is a relatively neglected aspect of scientific practice, but it plays a role that is often as crucial, as complex, and as creative as theorizing and experimenting. Indeed, my conclusion will be that we now need a new philosophy of simulation to complement recent work on the philosophy of experiment.

1. INTRODUCTION

There has been a growing interest, in the philosophy of science, on the role of models and modeling techniques in scientific practice. This philosophical literature on modeling, moreover, is quite diverse; there are many different notions of model that are covered in the literature. In particular, there has been a great deal of recent interest in what has come to be called "mediating models" – models that serve to extend scientific theories into new domains and applications.

Many such applications, however, involve mediating models whose constitutive equations are analytically unsolvable. The study of these applications often consists in developing representations of the underlying physics on a computer, and using the techniques of computer simulation in order to learn about the behavior of these systems. In many instances, these computer simulations are not simple number-crunching techniques. They involve a complex chain of inferences, involving elaborate modeling practices at all levels, that serve to transform the initial theoretical model into a model designed to provide a representation of the actual behavior of a physical system.

Model-Based Reasoning in Scientific Discovery, edited by L. Magnani,
N.J. Nersessian, and P. Thagard, Kluwer Academic/Plenum Publishers, New York, 1999

In this paper I argue that this process of transformation is also a process of knowledge creation, and it has its own epistemology. It is an epistemology that is unfamiliar to most philosophy of science, which has traditionally concerned itself with the justification of theories, not in their application. I also argue that the complex and motley nature of this epistemology suggests that the end results of some simulations do not bear a simple, straightforward relation to the theories from which they stem. Accordingly, I urge philosophers of science to examine more carefully the process of "theory articulation," the process by which a general theory is made to conform to a particular application. It is a relatively neglected aspect of scientific practice, but it plays a role that is often as crucial, as complex, and as creative as theorizing and experimenting. Indeed, my conclusion will be that we now need a new philosophy of simulation to complement recent work on the philosophy of experiment.

The structure of this paper will be as follows. In section 2, I will attempt to characterize the kind of simulation research that I will be focussing on. This section will serve to limit the scope of my claims throughout the paper to a specific kind of scientific research using the computer. In section 3, I will outline the methodological structure of this kind of simulation research. I will also introduce, in this section, a "hierarchy of models", which presents the various kinds of models that play important roles as intermediate steps in the process of simulation. In section 4, I argue for the unique character of the epistemology of simulation, and I outline how this epistemology fits (or fails to fit) into the philosophical literature. In section 5, I explore the relationship that models of phenomena bear to theoretical structures, arguing that the relationship is more complex than one might suppose. In section 6, I compare the role played by the hierarchy of models in simulation to that of the hierarchy of theories in Patrick Suppes' "models of data" (1962). In the final section I offer some concluding remarks, and I urge philosophers of science to pay more attention to the role of "calculation" in science.

2. SIMULATION MODELING: AN OVERVIEW

The use of computational techniques in the sciences has become more and more widespread, and the range of techniques and applications is enormous. I will focus on one very particular type of application of these techniques – the practice of modeling very complex physical phenomena for which there already exist good, well-understood theories of the processes underlying the phenomena in question. Though the underlying theories are well understood, because of the complexity of interactions involved in generating the phenomena, the phenomena themselves are not well understood.

These computational techniques, involving the brute force solution of intractable differential equations, via simulation modeling, have as their aim an understanding of complex phenomena they model.

Let me begin by giving a rough sketch of the kinds of techniques. As an illustration, let us suppose that we are confronted with a physical system of which we would like to gain a better understanding: a severe storm, a gas jet, or the turbulent flow of water in a basin.[1] The system in question is made up of certain underlying components that behave according to a strict set of physical laws. We begin with the assumption that we know what these components are and that we know the laws that govern them.

The assumptions we have made so far about our system allow us write down a set of partial differential equations. These differential equations represent the exact determination of how the system will evolve through time that is given by the physical model. The problem is that when these underlying components of the system, whether they be solid particles, parcels of fluid, *etc.*, interact in the manner that we suppose that they do in our physical system, the mathematical equations of our model have an unfortunate feature. In the types of systems that the simulation modeler is concerned with, it is mathematically impossible to find an analytic solution to these equations – the model is said to be nonintegrable. That is, it is impossible to write down closed form equations, equations given in terms of known mathematical functions, which represent an exact solution to the set of differential equations and will thereby tell us what the system will do over time.

The approach that the simulationist takes to this problem is turn differential equations, which relate continuous rates of change over infinitesimal intervals, into difference equations, which relate rates of change over finite, or discrete, intervals. The values that these difference equations give can then be calculated by a digital computer, from one discrete moment of time to the next.

3. A HIERARCHY OF MODELS

Why, should simulation – a form of calculation – require an epistemology? We need an epistemology of simulation because simulation modeling is a set of scientific techniques that produces *results*. When science produces results, we often like to have standards for deciding whether or not these results have some degree of reliability. Even though simulation is fundamentally about replacing analytical solution with calculation, merely a mathe-

[1] Cf. Wilhelmson (1990) on the simulation of a severe storm, Smarr (1985), and Kaufmann and Smarr (1993) on the simulation of intergalactic gas jets, and Moin and Kim (1997) on computer simulation in the study of turbulence.

matical transformation, the question of the reliability of the results of simulation modeling goes beyond mere worries about the reliability of the calculation, and reaches out to the entire simulation process and the conclusions scientists reach using these techniques.

For an adequate discussion of this point, we need to understand more about the simulation process, how complex it is, and the number of diverse factors involved. The heart of the practice of simulation is the construction of a hierarchy of models. This hierarchy includes a mechanical model, a dynamical model, *ad hoc*, a computational model, and finally, a model of the phenomena.

The first step in any simulation study is to identify the theoretical structure under whose domain the phenomena of interest lie. This will form the basis for the simulation. For example, we might start with mechanical laws, force laws, or equations of state (Kaufmann and Smarr, 1993). By itself, the theoretical structure tells us very little about anything but the most idealized systems. To apply them to real world systems requires a *mechanical model.* A mechanical model is a bare bones characterization of a physical system that allows us to use the theoretical structure to assign a family of equations to the system. The *locus classicus* of philosophical discussion of these kinds of models is Nancy Cartwright's simulacrum account of models in *How the Laws of Physics Lie* (Cartwright, 1983).

Even though a mechanical model provides a foothold for the application of a theory to a set of real world problems, a mechanical model by itself is still a very general entity, it is not about any particular system. The next step for a simulationist is to specify a class of parameters, boundary values, and initial conditions that restricts the theoretical model to a specific class of phenomena. This conjunction of the theoretical model with parameters, boundary values, and initial data make a concrete *dynamical model* (or really, a family of models) for a highly specific class of phenomena. The specification of these values for a simulation is rarely a straightforward process, and is often a delicate balancing act between accuracy and tractability (Smarr, 1985).

Next comes the construction of a *computational model.* Some dynamical models are analytically tractable; their differential equations can be solved and a mathematical function can be given which provides a good representation of the dynamics of the system. But in systems of interest to simulationist, this approach is not possible. The dynamical model needs to be transformed into a computational model so that computational techniques can be used to overcome the problem of analytical intractability.

There are two steps to this process. First of all, the continuous differential equations of the dynamical model need to be converted into discrete algebraic equations that can be cranked out by the computer. Even though this solves the problem of analytical intractability, the new model must be *computationally* tractable. Simulationists use *ad hoc* modeling assumptions to help make their computational models more tractable and manageable.

Ad hoc modeling assumptions include such techniques as simplifying assumptions, removal of degrees of freedom, and even substitution of simpler empirical relationships for more complex, but also more theoretically founded laws. This model making can be eliminative or creative. The modeling can involve eliminating considerations from the dynamical model, or making up new ones. Sometimes the simulationist will ignore important factors or influences from their computational models because limitations of computational power make their inclusion impractical. This is what I refer to as eliminative *ad hoc* modeling. In this case, the simulationist has one of two options: either to assume that the effects of this neglected factor are negligible or to make use of some sort of empirical "fudge factor" – creative *ad hoc* modeling – to make up for the absence of the neglected factor.

Ad hoc models typically involve some kind of relatively simple mathematical relationship which is designed to approximately capture some physical effect in nature. When "coupled" to the more theoretical equations of a simulation, they allow the simulation to produce results that are more realistic than they could have been without *some* consideration of that physical effect.

Once the computational model is implemented on a computer in the form of a particular algorithm, the algorithm produces results in the form of a data set; often a very large one. This data sct requires interpretation. For this, the data can be visualized, subjected to mathematical analysis, and used in conjunction with other sources of knowledge, including observation, in order to arrive at the final goal of a simulation study; what I call a *model of the phenomena*.

A model of the phenomena is a manifold representation that embodies the relevant knowledge, gathered from all relevant sources, about the phenomena. It can consist of mathematical relations and laws, images, both moving and still, and textual descriptions. The construction of the model of the phenomena is an attempt to summarize the basic robust qualitative features of a whole class of structurally similar phenomena. It might include such features as:

- An emergent high-level mathematical relationship among certain aspects of the system, such as a scaling law.
- A transport mechanism.
- Threshold values of parameters.
- Characteristic coherent structures.
- Characteristic geometries of flow.
- Patterns of interaction and competition among coherent structures.

4. A UNIQUE EPISTEMOLOGY

Each of these steps in the simulation modeling process is a step on the way toward an inference. Figure 1 illustrates the different layers of models involved in simulation and the resources used in each inferential step. Via these steps, the simulationist hopes to infer, from *existing* theoretical knowledge, *new* knowledge about the system being simulated. While it is typical in philosophy of science to talk about deducing results from theoretical structures, the inferential moves described above are patently *not* deductive. They have neither the inevitability nor the epistemic certainty associated with deduction.

If one of the central tasks of a philosophy of science is applied epistemology, then any philosophical look at simulation modeling needs to ask when, and why, these nondeductive inferential steps are acceptable – that is, under what sorts of conditions do they produce reliable knowledge, and why. But another task of philosophy of science is interpretation. So we will also need to ask what the nature of the inferred knowledge is, and in what relation it stands to the theoretical structure from which it was inferred.

In trying to apply the resources of philosophy of science to this problem, we face a fundamental obstacle. Beyond the fact that there is little philosophy of science literature on simulation, there is moreover very little on applied science in general. The typical focus of philosophy of science is on the most theoretical aspects of science, and much less attention is paid to the subsequent application of this theoretical knowledge in the modeling of higher level, more complex phenomena. When attention is paid to the application of theories, it is almost always in the context of a discussion of the testing of the theory.

Probably the first philosopher of science to emphasize the importance of *theory articulation* as a creative process was Thomas Kuhn. In "The function of measurement in modern physical science", Kuhn illustrates the point that much of "normal science" consists of theory articulation. He notes that most novel theories, including Newton's theories of motion and gravitation, are capable, upon their first presentation, of few novel predictions.

> The new order provided by a revolutionary new theory in the natural sciences is always overwhelmingly a potential order. Much work and skill, together with occasional genius, are required to make it actual (Kuhn 1977, p. 184).

"Complex mathematical manipulation," together with "extensive approximations," is often necessary in order to get theories to yield experimental predictions (Kuhn, 1977, p. 185). Kuhn also argues that none of these ought to be construed as an attempt at confirmation, because as he points out, failure does not count as disconfirmation.

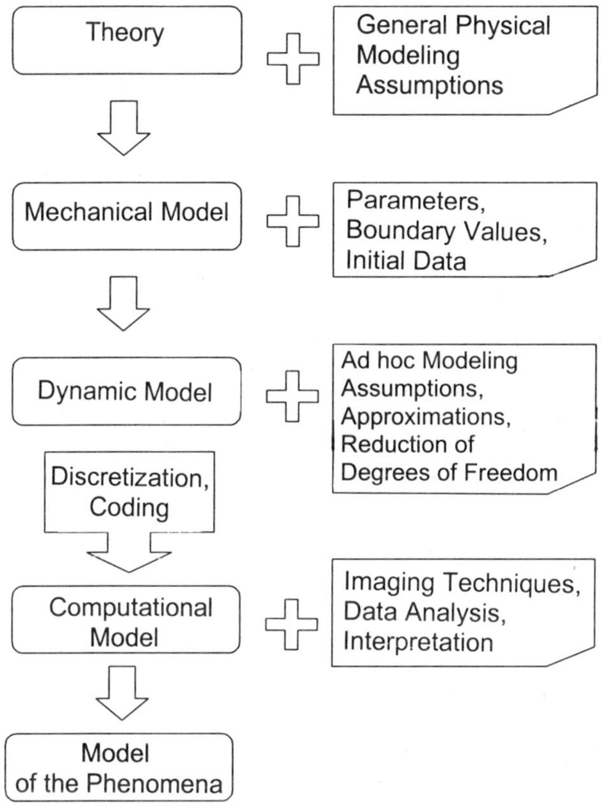

Figure 1. The hierarchy of models.

Nevertheless, Kuhn resists going so far as to say that theory articulation is truly creative in the sense that it produces new knowledge. "His [the scientist's] success [at opening up new areas of application for a theory] lies only in the explicit demonstration of a previously implicit agreement between theory and the world. No novelty in nature has been revealed" (Kuhn, 1977, p. 192).

So Kuhn's insights are helpful, yet somewhat insufficient. The concept of theory articulation, the idea that bringing theory into contact with the world is often a nontrivial and nondeductive process, is obviously very helpful in understanding what goes on in complex situations involving simulation. His assumption, however, that "no novelty in nature is revealed" is somewhat inhibiting. Simulation modeling is clearly a case of theory articulation in the spirit of Kuhn. It is a nontrivial process of bringing a theoretical structure into resonance with some phenomena that are "implicitly" in its domain. But simulation modeling, when successful, does reveal novel aspects of nature. Often simulation will enable us to produce a representation of an aspect of nature that is extremely difficult to observe. Even if the system in question can be observed in detail, often the simulation will bring a level of mathematical order where before there was only seemingly random detail. In Kuhn's own words, there is *actual order* where before there was only *potential order*.

This is a persistent problem in the philosophy of science literature on theory articulation. Most philosophers of science assume that approximations, idealizations, and other such transformations serve only to allow us to compare theoretical predications with observed results, usually in order to test the theory. As such, they have not been concerned with the autonomous assessment of the reliability of these inductive steps; autonomous in the sense of being independent of our satisfaction with the outcome.

Jeffrey Ramsey has made some telling criticisms of prevailing philosophical accounts of approximation. In "An expanded epistemology for approximations" (Ramsey, 1992), Ramsey argued that all prior accounts take as the sole criteria of epistemic adequacy the degree to which the *end result* of the approximation *resembles* some empirical structure. He prefers, instead, to look at the act of approximating, and argues that scientists often have to evaluate these moves independently of how well the outcome matches up with the world.

When we look specifically at the techniques employed in simulation, we find this point to be especially on the mark. It is precisely on this point that an epistemology of simulation modeling must focus: what *are* the factors that contribute to the notion that a computational model constructed on a set of approximations and idealizations are valid? Here, Ramsey's analysis of approximations is very important for us. The realization that techniques of approximation are in fact inferences that require an epistemology internal to their own structure is exactly the jumping off point that we need.

This is the project of an epistemology of simulation. It is the study of the means by which we sanction belief in the outcome of simulation studies despite their motley methodological structure. I will argue that in order to really understand the relationship between models of phenomena and scientific theories, we need to understand the process by which these results are get sanctioned. One of the benefits of gaining some insight into the epistemological foundations of the simulation practice is that it facilitates a better understanding of this relationship. Naturally, the complexity of this task is proportional to the complexity of the methodological structure of simulation. This structure includes, among its components:

- A calculational structure for the theory.
- Techniques of mathematical transformation.
- A choice of parameters, initial conditions, and boundary conditions.
- Reduction of degrees of freedom.
- *Ad hoc* models.
- A computer and a computer algorithm.
- A graphics system.
- An interpretation of numerical and graphical output coupled with an assessment of their reliability.

It is beyond the scope of this paper to provide a detailed analysis of the role of each of these components and to examine how a skilled simulationist can manage each of their potential contributions as sources of error. Nevertheless, this is what a thourough epistemology of simulation requires. Here, I will have to be satisfied to note a few crucial features of the process of sanctioning that such an examination reveals.

The first thing worth noting is the diverse range of elements that need to be subjected to epistemological scrutiny. Each of the elements listed above mediate between our theoretical models and our simulation results. If the influences and possible pitfalls of each element are not properly understood and managed by the simulationist, then they present a potential emergence of a threat to the credibility of simulation results

Understanding and managing these diverse influences is no simple matter. In particular, doing so requires reliance upon an equally diverse range of sources of knowledge and skills. A great deal of this knowledge is not contained in the theoretical knowledge that formed the original basis for the simulation.

Every step down the hierarchy of models represents a transformation of knowledge in one form into another form, to be used either as a tool, or as a representation. Each transformation introduces the possibility, or even the likelihood, or introducing artifacts into the results.

For example, in Wilhemlson's storm simulation, it was very important for the researcher to keep track of the movement of kinetic energy in the system. According to theory, of course, kinetic energy is determined entirely by the masses and motions of the particles, and parcels of fluid, of the system. But in any simulation that employs a discreet space-time grid, there is always the need to consider the impact of motions that occur "inside" the cells of the grid.

In order to avoid losing energy to these local motions, the storm simulation keeps track of a variable called "sub-grid-scale kinetic energy" that is calculated for each cell. But theory does not dictate what mathematical function should be used to "couple" this variable to the rest of the system. Therefore, the particular choice of how to calculate the interaction between this variable and the others of the simulation needs to be justified on some other basis. In general, these kinds of moves are evaluated on a variety of fronts, and they can be justified based on considerations coming from theory, from data, or from experience in modeling similar phenomena in other contexts.

It is also important to note the role of the "observer" in the sanctioning of simulation results. The observer plays a pivotal role. On the one hand, simulations are quite often performed to learn about systems for which data are sparse. As such, and as I have tried to make clear, comparison with real data can never be the autonomous criterion on which simulation results can be judged. Nevertheless, calibration does play a very important role. Simulation results can be calibrated in three different ways: by experiment, by analysis, and by other simulations. The first criterion that a simulation must meet is to be able to reproduce known analytical results. No matter how complex a system, there often exist, under highly constrained conditions, limited analytical results for the full equations of a theoretical model. Results can also be compared to the output of other simulations in virtue of the fact that the other simulation uses a different algorithm, or even better, if the other simulation is of a small local region within the broader system and makes use of a more complete dynamical model.

Simulation results are also calibrated against experimental results. Unfortunately, this kind of comparison is sometimes not as easy as it might seem, since data from these different sources may come in different forms. For example, simulation and experimental data are not always obtained at the same spatial position within a system; the grid points of the simulation do not often correspond to the location of probes within the experimental setup. Moreover, it is typically less interesting to compare detailed data, piece by piece, than it is to compare the characteristic features of the simulation results and empirical results, especially when the empirical results come in the form of flow visualization (Shirayama and Kuwahara, 1990, p. 67).

Most of all, absent an observer who can compare images against images, there is no metric of similarity between the different data sets that need to be compared. Visualization is by far the most effective means of identifying characteristic features out of complex dynamical data sets, and so it is the most, if not only, effective means of judging the degree of calibration a simulation enjoys with other data sets and with analytic results. Thus, visualization plays a crucial role in sanctioning as well as in analyzing simulation results. Not only does the epistemology of simulation call upon resources that are empirical, and come from outside of the theory, it also calls upon the faculties of the observer.

5. WHAT IS THE RELATION OF MODELS TO THEORIES?

Once we understand that simulation has its own epistemology, and that epistemology necessitates the scrutiny of all of the elements I have listed, we are forced to take a fresh look at what we take to be the nature of the relationship between scientific theories and their empirical consequences.

One thing that is clear is that our models of phenomena do not stand in an actual logical relation to our theories. Our models are approximate, and they are often rich with detail that we know is not robust. What we might propose though, is that if all has gone well, the results of a particular simulation closely resemble an ideal structure that stands in a purely logical relation to the theory. The structure itself and the exact nature of the relationship are unfortunately inaccessible to us, but we should be confidant that it is there and that we have successfully approximated it. The fact that we lack the certainty of deduction merely reflects the fact that there is some chance that we are wrong and that our results do not bear any relation to the ideal structure. In other words, there is some structure that is a logical consequent of our theories, and what we have is something which we strongly believe approximates it. Is this a view that we should adopt?

The relevant questions are the following: should we indeed think of the models of phenomena that simulationists provide to us as qualitative approximations of the actual (inaccessible) "solutions" of the theoretical equations we began with? Are these "models of phenomena" really (approximate) *models* of our theories in the strict formal parlance of the semanticists? Simply put, are these models direct consequences of our best theories?

I would argue that they are not. Given the rich complexity of the process of deriving warrant for simulation results, and the extent to which this process focuses on elements *external* to the theoretical structure itself, it would be unrealistic to interpret this warranting process as being about the results' relationship to some formal modal. That is, if we think carefully about the epistemological steps that go into warranting the reliability of simulation results, we find that they have everything to do with ensuring that the results match up well with the world, and little to do with ensuring that they resemble some ideal structure.

In other words, the confidence we have in our simulation results depends on several factors being in place, none of which is guaranteed by our theoretical knowledge. It depends facts we know about our computers, and about our graphical techniques. It depends on the confidence we have in the various *ad hoc* models we use – confidence we derive from laboratory and observational experience.

For example, consider Wilhelmson's own words describing his simulation:

> A very simple model is used to account for the development of rain. In many studies such simple models are sufficient for studying storm dynamics. Although simple, they provide the key storm-driving forces, namely, warming due to the release of latent heat as water vapor condenses and cooling due to evaporation of cloud and rain drops in unsaturated regions (Wilhelmson, 1990, p. 22).

It depends on our ability to calibrate models against empirical results. And finally, it depends on the confidence we have in our tacit abilities as observers to make judgements about the degree of resemblance between different classes of images; often abilities acquired in the role of skilled experimenters and observers, as well as in the role of skilled simulators. The epistemology of simulation is very much an empirical epistemology, and not merely a mathematico-logical one.

6. MODELS OF DATA – THE HIERARCHY OF THEORIES

Devotees of the semantic view of theories may wonder what all the hubbub is about here. After all, wasn't it Patrick Suppes who, more than thirty years ago, recognized the role of what he called the "hierarchy of theories?"

According to the views he advocated early on in the history of the semantic view, an isomorphism between model and data can only be established with the help of a wide range of other scientific theories and with the help of theories of methodology, experiment and inference. Why then, should adherents of the semantic view worry about what I call the motley character of simulation inferences, and their motley epistemology? Isn't this exactly what Suppes was talking about along time ago? And wasn't it exactly these characteristic that led people to prefer the semantic view to the received view in the first place?

In the semantic conception, the relation between model and phenomena, the isomorphism, is not described within the theory. In the other words, the determination of whether or not an empirical substructure is isomorphic to a set of appearances can not be carried out with the help of the theory alone. An isomorphism between model and data can only be established with the help of a wide range of other scientific theories and with the help of theories of methodology, experiment and inference.

The syntactic view of theories places the responsibility of determining the empirical application of the theory on a portion of the theory itself; the correspondence rules. But with the semantic view, we can draw on anything we need from outside the theory, because we are able to use whatever is available in order to establish an isomorphism between the substructure and the phenomena. The job of applying the theory to the world can be farmed out to whatever mathematical theories we have at our disposal, they need not be included in the theory in question in order to be available to us. And just as we have at our disposal whatever mathematical apparatus we need for establishing isomorphisms, we also have at our disposal any scientific theory that we take to be useful and accurate. This allows for what Patrick Suppes calls the "hierarchy of theories" which is often made use of in actual scientific practice.

For Suppes, the hierarchy of theories can be used for constructing a model of the data, and for establishing an isomorphism between the model of the data and a designated model of the theory. But no external resources, on Suppes' account, should be needed for determining what is to be suitably considered a legitimate model of the theory. For Suppes, what counts as a suitable model of the theory is a logical question, to be determined by the semantic structure of the theory itself.

Nevertheless, it is difficult to resist the intuition that Suppes, with his notions of models of data and hierarchy of theories, was onto something that could be quite useful for us. Even though it is clear that Suppes' characterization of models of data does not describe the *role* played by the models involved in simulations, it appears that Suppes' models of data better describe the *character* of what is going on in simulation.

As we have seen at length, simulation modeling is many layered, it draws on vast regions of extra-theoretic knowledge, and its epistemology is dynamic and autonomous. In these respects, simulation is a lot like Suppes' descriptions of modeling data. But of course, simulation has nothing to do with getting from raw data to models of data, or with establishing isomorphisms. In fact, it plays a role analogous to the task, in the semantic view of theories, of choosing an empirical substructure; a task that, according to the semantic view, is supposed to be remanded exclusively to the governing theory.

7. SPECULATION, CALCULATION, AND EXPERIMENTATION

Ultimately, even though all simulation modeling of the kind described in this paper fundamentally begins with a theoretical model, and even though we think of simulation as an attempt to "solve" the mathematical equations of this theoretical structure, our theoretical knowledge is just one of several ingredients that simulationists use to produce their end product – a model of the phenomena. So the relation that theory bears to these models is simply this: our confidence in our physical theories is one of the things that, in combination with other elements, *warrants* our rational belief that the models we construct are reliable and cognitively useful. This conclusion is simply the result of a fundamental limitation of our cognitive power. When it comes to complex systems, we simply cannot bend our theories to our cognitive will – they will not yield results with any mechanical turn of a crank. The models that we need to construct in order to do our science need to be constructed delicately, and from as many sources as are available. Consequently, these models are no mere "solutions" to our theoretical structures, though they are the results of a form of calculation, they are rich physical constructs that mediate between our theories and the world.

What is the moral of all of this? Here is one way to look at it. Ian Hacking's book *Representing and Intervening*, is often credited with having created a whole new kind of philosophy of science by writing a book on the philosophy of experiment. There is certainly a great deal of truth this. However, there are a couple of interesting anomalies relative to that claim.

The first one should be obvious from the discussion above. Patrick Suppes' discussion of models of data *is* philosophy of experiment and it goes back to 1960. But perhaps more salient is Hacking's own characterization of the situation.

In chapter 12 of his book, "Speculation, Calculation, Models, Approximations", Hacking argues against the traditional dichotomy of theory and experiment. He wants to replace it, he says, with a tripartite division; speculation, calculation, and experimentation.

> Kuhn remarked that normal science is a matter of what he calls articulation. We articulate theory to make it mesh better with the world, open to experimental verification. Most initial speculations hardly mesh with the world at all. This is for two reasons. One is that one can seldom directly deduce from a speculation consequences that are even in principle testable. The other is that even a proposition which is in principle testable is often not testable, simply because no one knows how to conduct the test.
>
> Thus Kuhn's articulation must denote two kinds of thing: the articulation of theory and the articulation of experiment. I shall arbitrarily call the more theoretical of the these two activities "calculation." I do not mean mere computation, but the mathematical alteration of a given speculation, so that it brings it into closer resonance with the world (Hacking, 1983, pp. 213-214).

The structural similarity between the hierarchy of models in simulation, and the hierarchy of theories in models of data would seem to suggest that, at times, the articulation of theories, is as rich, as layered, and as epistemologically complex as the articulation of experiment. *Representing and Interven-*

ing was widely read as a call for a philosophy of experiment. Perhaps it is now time to call for a philosophy of calculation.

REFERENCES

Cartwright, N., 1983, *How the Laws of Physics Lie*, Essays 7 and 8, Oxford University Press, Oxford.

Hacking, I., 1983, *Representing and Intervening*, Cambridge University Press, Cambridge.

Kaufmann, W.J., and Smarr, L.L., 1993, *Supercomputing and the Transformation of Science*, Scientific American Library, New York.

Kuhn, T., 1977, The function of measurement in modern physical science, in: *The Essential Tension: Selected Studies in Scientific Tradition and Change*, The University of Chicago Press, Chicago.

Moin, P., and Kim, J., 1997, Tackling turbulence with supercomputers, *Scientific American* 276, 1:62-68.

Ramsey, J.L., 1992, An expanded epistemology of approximations, in: *PSA 1992*, vol. 1, D. Hull, M. Forbes, and K. Okruhlik, eds., Philosophy of Science Association, East Lansing, pp. 154-66.

Shirayama, S., and Kuwahara, K., 1990, Flow visualization in computational fluid dynamics, *The International Journal of Supercomputer Applications* 4/2:66-80.

Smarr, L., 1985, An approach to complexity, *Science* 228:403-408.

Suppes, P., 1962, Models of data, in: *Logic, Methodology and Philosophy of Science*, E. Nagel, P. Suppes, and A. Tarski, Stanford University Press, Stanford.

Wilhelmson, R.B., et. al., 1990, A study of the evolution of a numerically modeled severe storm, *The International Journal of Supercomputer Applications* 4/2:20-36.

Abducting Explanation

Vincent F. Hendricks and Jan Faye
Department of Philosophy, University of Copenhagen, Denmark, vincent@coco.ihi.ku.dk, faye@coco.ihi.ku.dk

Abstract: This paper examines the nature of three well-known but distinct discovery engines: Inductive discovery (*ID*), Inference to the best explanation (*IBE*) and Abductive discovery (*AD*). Any discovery process requires modifications in the background knowledge: these belief revisions take place in four different dimensions. The three discovery engines will be compared with respect to these dimensions. Special attention will paid to (*AD*) since it constitutes the most troublesome of the three.

1. INTRODUCTION

This paper concerns the study of scientific of discovery as opposed to the study of scientific justification. Indeed, prominent epistemologists and philosophers of science have argued that justification methods are the real objects of research for methodology and the philosophy of science. This lead Reichenbach to formulate the now classical distinction between the *context of justification* and the *context of discovery*. Whether a conjectured hypothesis or theory is verified, refuted or decided by the evidence is strictly a logical matter which can be settled "out of court" in a reliable fashion. At least when the Quine-Duhem thesis of global underdetermination does not present impediments. Conversely, no guidelines nor methodological prescriptions can be given for the discovery of new hypotheses from finite sequences of evidence. The discovery of new hypotheses is like art, aided by intuition and perhaps divine insights – a thoroughly *unreliable* exercise. This line of thought is also emphasized by Hempel's later speak of the *logic* of justification while retaining only Reichenbach's context of discovery.

Since scientific discovery is an unreliable activity it should, if studied at all, be studied by empirical psychology, sociology and history, Popper concluded. This classical dichotomy is nowadays challenged by a number of different philosophers and methodologists. From the point of view of logical *reliabilism* (Kelly, 1996), formal (Osherson, 1986) and modal (Hendricks, 1999) learning theory the dictum is simply false. These methodological dis-

Model-Based Reasoning in Scientific Discovery, edited by L. Magnani,
N.J. Nersessian, and P. Thagard, Kluwer Academic/Plenum Publishers, New York, 1999

ciplines study the intrinsic solvability of inductive problems for both ideal and Turing computable agents in either a topological, recursive or modal logical setup. First of all, justification or assessment may be viewed as a simple form of discovery in which the scientist is asked to discover whether the hypothesis currently held is either true or false. Secondly, if the charge against engaging in issues of discovery is based on their supposedly immanently unreliable nature, then it can be shown that both discovery and justification present problems that may or may not have reliable solutions. The lack of reliable solutions may be due to either the intrinsic difficulty of the inductive problem presented or due to too demanding methodological prescriptions the method is asked to satisfy and still get to the truth. In short, to say that scientific discovery is unreliable *simpliciter* is incorrect, but to say that different discovery methods can fail on complex inductive problems or become unreliable due to the methodological prescriptions they are required to satisfy is definitely correct.

The above argument works neatly in a well-defined formal setup and with a particular understanding of what scientific discovery amounts to. We call the understanding of scientific discovery presented by logical reliabilism and formal/modal learning theory for *inductive* discovery (*ID*). The characteristic feature of (*ID*) is that the hypothesis conjectured always lies within the boundaries of the inductive vocabulary specified by the evidence language. One may also envision however a discovery engine which in some sense outputs hypotheses transcending the vocabulary of the evidence language. Call this for *abductive* discovery (*AD*). And finally, consider a discovery mechanism which enumerates competing hypotheses in accordance with some adequate yardstick and subsequently picks the one with the highest score. Call this *inference to the best explanation* (*IBE*). Especially between the latter two there seems to be some confusion which needs to be resolved.

This paper attempts to characterize (1) the defining features of (*ID*), (*IBE*) and (*AD*) with respect to a certain set of principles, goals, prescriptions and definitions inherent in any discovery problem; and (2) outlines the relations between these principles and the aim of getting to the correct answer. Since (*ID*) and (*IBE*) are the inference engines primarily discussed in the literature, special tribute will be paid to (*AD*).

2. PRINCIPLES, GOALS, PRESCRIPTIONS AND DEFINITIONS

Any discovery process requires modifications to the background knowledge. How radical these changes are essentially separate the three different discovery algorithms from each other. In general such a belief revision seems to take place within four different dimensions: 1) ontological principles (OP); 2) cognitive goals (CG); 3) methodological prescriptions (MP); and 4) semantical definitions (SD).

OP	CG	MP	SD
Space and Time	Truth	Reliability	Invariance
Causality	Empirical Adequacy	Simplicity	Variance
Determination	Simplicity	Consistency	Inclusion
Continuity	Coherence	Unification	Disjointedness
Conservation	Unity	Conservatism	Expansion
Indexicality			Correspondence

2.1 Ontological principles

Inference takes place on a background of ontological principles which make up the presuppositions of the general practice of individuation and identification buried in our pre-scientific and/or scientific language. Now, within the sciences of nature the principles of individuation are: (i) the principle of space and time, i. e., the assumption (a) that physical objects exist separately in space and time in a way so that they are localizable and countable because spatially separated objects do have real separated and definite properties, and (b) that causal processes connecting changes of these properties happen in space and time; (ii) the principle of causality, i. e., the assumption that every change of a property has a cause; (iii) the principle of determination, i. e., the assumption that the values of every later state of a system are uniquely determined by those of any earlier state; (iv) the principle of continuity, i. e., the assumption that all processes be distinguished by a difference in the initial and the final state have to go through every intermediary state; (v) the principle of the conservation, i. e., the assumption that certain basic properties of an isolated system can be transformed into various modes but are never gained, lost or destroyed; and finally (vi) the principle of indexicality, i. e., the assumption that the cause and the effect belong to the same realm of reality.

A physical system can be seen as a conceptual abstraction and idealization from a common-sense notion of a thing in which the thing, now considered as a physical object, is merely characterized by a state of properties as mass, position, velocity, energy, and electric charge. Moreover, a physical system may contain one physical object or a collection of physical objects. One presupposition for the individuation of a physical system is the assumption that two spatially separated systems do have real separated states with definite, well-defined properties, and that the joint state of the two systems is wholly determined by their separate states. This presupposition corresponds to the above principle (i). But in classical physics the presuppositions for being a physical system are also the principles (ii) to (v). A physical system continues in the same state, unless it is acted upon by an external (or internal) cause. It means that given nothing causes the system to change, then later states are identical with earlier states (except position). But if the system is forced to undergo a change, it is due to some causes, and its state turns gradually from the initial state, the state it has just before the interaction, to its final state, the state it will have just after the interaction has stopped. The alteration of the state happens in virtue of some of its quantities gain new values; however, some of the values remain untouched, these are the conserved quantities, like energy, of the system.

The first five principles constitute the classical tenets of individuation in the sense that they form the conditions for naming, reference and predication and therefore for any objective description of nature. *One would therefore assume that belief revision should take place along lines in which as many as possible of these ontological principles are not violated.* Belief revision must be conservative as long as it does not interfere with getting to the truth. One can make an analogy between hypothesis discovery and Racine's requirements to the classical drama: the unity of space, time, and action. A relevant hypothesis is one where it as the explanans fits most, if not all, of the above ontological principles of individuation and where it has the same ontological index as the explanandum.

By the same ontological index we mean that the explaining fact and the fact being explained belong to the same realm of reality. We do not seek a relevant hypothesis of the growth of some cancer forms in New York in the changes of the Down Jones' Index even if we could find a high correlation between these phenomena. A causal mechanism can only be expected to be discovered in case the assumed cause and effect are describable, not by the same terms, but in the same language. Indeed, the proper ontological index is not something we know a priori. It is experience which eventually has told us what belongs to the same or to different ontological categories. For instance, we know that whatever causes a change in a physical system, it has to be described in terms of position, time, velocity, mass, charge and energy, and perhaps something further. It will not be something in terms of, say, emotions, wishes, motives, intentions, etc.

Thus, whenever we discover a new hypothesis we use ontological principles to guide us. Nevertheless, it also seems as if quantum mechanics and theory of general relativity together violate most of the above principles but not all of them. Quantum mechanics breaks the first four, and the general relativity theory the last but one. For instance, a pair of particles in a singlet state which are spatially separated seems not to have separable properties until a certain measurement has taken place. This is a violation of (i). The Big Bang and radioactive decay in general have no causes. This excludes (ii) from being universally valid. And (iii) is broken in such cases where the ascription of dynamical and kinematical properties to a quantum system are subordinated Heisenberg's indeterminacy as when position and momentum, or two spin components, cannot be applied simultaneously to an atomic object with any infinite accuracy. The transition of an electron from one energy state to another within an atom happens discontinuously in space and time, and thereby such a quantum jump disobeys (i) and (iv). The quantum mechanical encroachment on the classical principles as conditions for an objective description seems to prove that these principles in both their origin and foundation are subjective in nature. Finally, theoretical studies of quantum particles in strong gravitational fields indicate that even conservation of energy cannot always act as a principle of individuation. It is found that an accelerating observer will detect the existence of particles where the static observer will persistently maintain that none are present. An observer participating in a uniform acceleration sees him- or herself immersed into a sea of radiation which has a spectrum as the emission from a hot body. The temperature of the radiation would be proportional to the rate of acceleration. In contrast the observer at rest will all the same see only empty space.

Better it will be to see the principles of individuation as still valid presuppositions for the identification of a certain domain of objects that are subject to classical physics. For we can understand at least some of them as reflecting some basic conditions for all human knowledge including the comprehension of an ontological separation of the known object from the knowing subject, and the grounding of this separation in practical actions and observation. The principles themselves may nevertheless change as our knowledge captures new land, but some of concepts by which it is done are basically the same.

Indeed, the principles of individuation would not have been violated by quantum mechanics if our knowledge of quantum objects had used quite different concepts than classical mechanics. But such a situation would be quite incomprehensible because – as also Bohr has maintained – every understanding of nature must rest on those fundamental categories which were already procured and elaborated in the pre-scientific experience as part of man's general practice of separation and orientation. Those categories are, for instance, space, time, thing, and cause. Following Bohr we can say that concepts in classical physics like time, position, momentum, and energy are more or less explicit formulations of such basic notions.

Since the principles of individuation do not stand up in quantum mechanics, we have to introduce new ways of seeking out an atomic system, and therefore new conditions for attributing such an object those dynamical and kinematical properties which classical mechanics uses in the description of macroscopic objects. In fact, we seem to have two options: either one can say that the violation of the principles is a shortcoming of quantum mechanics and that the orthodox quantum mechanics therefore is incoherent (and replace it with a theory like Bohm's quantum mehanics), or one can argue for some operational principles of individuation. Here Bohr came forward with an interpretation on which atomic objects can only be singled out as individuals in relation to actual performed experiments. It is impossible to ascribe the classical concepts of a state to atomic objects, unless it happens in a certain situation of experimental observation. Outside these experimental situations it does not make sense to attribute them dynamical and kinematical properties as we know from classical physics because no such properties can be said to belong to them intrinsically. This is due to the fact that various observational situations exclude one another so that they leave no possibility for a simultaneous attribution of definite value to an object whose state is not in an eigenstate for that observable. The ascription of dynamical and kinematical properties are therefore incompatible or complementary. It does not make sense to say that atomic objects satisfy the predication of classical properties in a well-defined manner since we cannot individuate such a system because of the quantum of action.

2.2 Cognitive goals

The aim of methods for scientific discovery is to output hypotheses that satisfy some fixed *cognitive goal*. Suppose we take *correctness* to be a fixed relation between the world and the hypothesis the scientist is investigating. Hence, the scientist applies certain theoretical and/or experimental methods to formulate hypotheses correct or incorrect of the aspect of the world under consideration. Correctness could, for instance, be exhausted in theoretical truth, but could in principle also denote other cognitive goals such as empirical adequacy, simplicity, unification, consistency, practical usability, etc.

The immediate question is why anyone would wish to entertain any other cognitive goal but truth. The answer is, that even while the scientist chooses to adopt background assumptions to limit the range of scientific success, the evidence may still globally underdetermine the truth. Say that a hypothesis is globally underdetermined by the evidence iff the following occurs: two possible worlds assign opposite truth-values to the same hypothesis discovered by the scientist in such a way that the data (emitted from the possible worlds and) received by the scientist remains the same forever regardless of which world is the actual world. Hence there may be no way for the scientist to decide whether the hypothesis is true or false. The truth may depend on something other than the evidence, like the underlying (metaphysical) reality. Instead the scientist may settle for a more lenient goal like empirical adequacy, simplicity or coherence to mention but a few of the ones typically advocated. According to van Fraassen it suffices that there exists a world in which all the observational sentences of the hypothesis is true. Empirical adequacy is clearly a more lenient goal than truth and may be conceived as less ambitious. To van Fraassen, science is not capable of delivering the truth whatever the truth might be, so there is no need to pursue it. On the other hand, some ambitious philosophers advocating simplicity seem to forget that simplicity itself is not a truth tracking principle unless one can provide an epistemologically convincing argument to the effect that either simplicity exhausts theoretical truth or is a truth-tracking recommendation. A truth-tracking recommendation, however, concerns methodology rather than epistemology. Methodology is the study of the methods and their recommendations by which science arrives at its posited truths.

2.3 Methodological prescriptions

Another way to proceed once truth is underdetermined and simplicity substitutes truth is to argue that simplicity is interesting in its own right. For an example, (Curd, 1980) argues that simplicity is a methodological appraisal of a hypothesis and at the same time linked to the problem of underdetermination, but before long insists that the methodological appraisal is interesting in its own right under the dichotomy of economy of research:

> Simplicity is not necessarily a sign that we have unlocked the mysteries of nature but a prudent preference for theories that are easier to work with and to test (Curd, 1980, p. 214).

But "economy of research" sounds much more like methodology than epistemology (i.e., fixing the cognitive goal) – two things which should not be equivocated among. The same reasoning governs coherence or any other cognitive goal. Without some sort of epistemological argument or call to reliability to secure, say, the simplicity or coherence of the underlying reality, the approach suggested by Curd, and also by Horwich, implies an initial indifference to the question of whether following the proposed rules or embracing the methodological maxims has anything whatsoever to do with finding the truth, now or ever. For all that has been said, inquiry may proceed forever "rationally" according to the dictates, and yet chase its tail for eternity, headed nowhere. Levi considers this issue and explicitly endorses what he calls the "myopic" perspective on inquiry, i.e., gaining truths and avoiding errors. He calls the alternative "messianic". As an example consider simplicity in more detail.

Simplicity is actually one of the cognitive goals which allows for methodologically operationalizable counterpart. For instance, Peirce originally thought to measure simplicity by logical means only. There have been some subsequent attempts to formulate such a logical scale, and Schlesinger presented an idea fairly close to Peirce's such that ceteris paribus the less the theory says the simpler it hitherto is. Hence a partial ordering induced with respect to entailment is a simplicity ordering and accordingly one theory is simpler than another if it is logically entailed by the other but not vice versa. (Glymour, 1980) has shown that this idea is not highly entertainable especially when paired with two other plausible requirements. First, that only the simplest theory entailing a body of evidence be confirmed by the evidence and that the theory must be finitely axiomatizable while the observational consequences are not required to be so. The problem is that a theorem of Tractenbrot dictates that for a certain interesting class of cases there is no simplest theory entailing a non-finitely axiomatizable set of sentences.

Another attempt stems from Goodman who provided a measure for simplicity for first order axioms based on (i) the number of occurrences combined with the arity of the primitive predicates and (ii) the symmetrical properties of the predicates. The proposal immediately furnishes a simplicity measure for any theory independent of axiomatization. This is due to the fact that the simplicity of a theory is construed to be the smallest value of the simplicity measures of its axiomatizations. The problem with this account is that it is far from evident what conditions should be met such that there is a simplest theory entailing a given body of evidence.

Methodologists concerned with methodological recommendations, i.e., simplicity, consilience, predictiveness etc. for holding a particular belief, have paid little or no attention to matters of reliability. They might respond that the proposed norms are motivated not hypothetically, as means for finding the truth, but categorically, as ends in themselves. This brings us back to the above considerations. A methodologist strongly committed to the proposal that his principles are normative and yet indifferent to correctness may prove to fail insofar it can be demonstrated that the intuitively justified recommendations are impediments to discovering correctness when it could have been reliably found by a method violating them. When the principles stand in the way of finding the correct answer, the principles are said to be restrictive. It seems that when intuitive inductive rationality principles compete with reliability, the principles are going to lose. In other words, if inductive rationality does not aid in finding the correct answer, at least it should stay out of the way. One way to respond to these charges is the traditional somewhat Cartesian and rationalistic answer that the canons are self-evident to the cognitive eye and insofar addresses truth. But when the principles are clearly intended to be myopic, i.e., gaining truth and avoiding error, they require an independent epistemological argument. One question remains: why worry about reliability in any significant way, Carnap's confirmation theory concerned confirmation rather than getting to the truth?

Suppose that the cognitive goal is truth and a the scientist "has a preference for true hypothesis". The problem then seems to be the privileged status of truth. Truth is not operational like some of the other goals (like simplicity, for instance) and hence it is hard to see how one could impose a methodological recommendation of such a nature. There is a different possibility however. As a methodological recommendation, one could advocate reliability of the method in such a way that if the method is applied, then one is guaranteed to arrive at, sooner or later (depending on the success criteria) the correct answer given the proper background assumptions. Methodologists like Gold (1967), Osherson (1986), and Kelly (1996), have provided strong results for (*ID*) which go show how inductive discovery indeed may be carried out reliably. The question remains whether these results may be extended to (*IBE*) and especially (*AD*).

2.4 Semantical definitions

The fourth class of constraints on scientific discovery is set up by some semantical definitions. Here we can isolate six of them: (i) semantical invariance; (ii) semantical variance; (iii) semantical inclusion, (iv) semantical disjointedness, (v) semantical expansion and (vi) semantical correspondence.

- Semantical Invariance

For any discovered hypothesis *h*, *h* is semantically invariant with respect to the evidence language *E*, if and only if, the terms contained in *h* do have both the same intension as well as extension as those terms contained in the evidence language of *E*.

- Semantical Variance

For any discovered hypothesis *h*, *h* is semantically variant with respect to the evidence language *E*, if and only if, some of the terms contained in *h* do not have the same intension but the same extension as those terms contained in the evidence language *E*.

- Semantical Inclusion

For any discovered hypothesis *h*, *h* is semantically disjoint with respect to the evidence language *E*, if and only if, the terms contained in *h* have the same intension as those terms contained in the evidence language *E* but where the extension of *E* is a proper part of the extension of *h*.

- Semantical Disjointedness

For any discovered hypothesis h, h is a semantically disjoint with respect to the evidence language E, if and only if, some of the terms contained in h have a different extension but the same intension as those terms contained in the evidence language of E.

- Semantical Expansion

For any discovered hypothesis h, h is a semantical expansion[1] with respect to the evidence language E, if and only if,

1. h includes the relevant evidence terms of E,
2. h furthermore includes terms such that:
   a) the intension of these terms is not included in the evidence language of E,
   b) the extension of these terms is not included in the evidence language of E.

- Semantical Correspondence

For any discovered hypotheses h_1 and h_2 and for all the evidence of h_1 and h_2 respectively, E_1 and E_2, h_2 corresponds with h_1, if and only if:

a) there is a non-empty intersection $D_C = E_1 \cap E_2$, s. t.
b) every model M_2 of h_2 is isomorphic to a model M_1 of h_1 within D_C but M_2 of h_2 is not isomorphic to M_1 of h_1 outside D_C.

In other words, there exists a correspondence between two hypotheses h_2 and h_1 if and only if all the interpretations of h_1 are isomorphic to the interpretations of h_2 within D_C are such that h_1 and h_2 differ empirically with respect to their domains outside D_C but are empirically (and elementary) equivalent with respect to D_C.

Now, we may illustrate the various principles with some examples. The first one seems quite obvious. Assume that the hypothesis is that "all ravens are black". If the language of E contains terms like "raven" and "black", then both terms keep the same meaning in the language of h.

More interesting, however, is semantical variance. Examples of this principle are numerous in numbers. Each time a new theoretical term is being introduced into a hypothesis, its extension is covered by some observational terms but its intension differs from theirs. A simple illustration is the one Hempel considers in his *The Theoretician's Dilemma* (Hempel, 1965). Wood floats on water, and iron sinks in it. Our h is that a certain object flows on a liquid if its specific gravity is lesser than the one of the liquid, and it sinks in the liquid if its specific gravity is greater. Here "wood", "water", "iron",

[1] It is assumed that the semantical expansion is consistent.

"sinking" and "floating" are terms in the language of *E*, whereas the terms in the language of *h* are "specific gravity", "weight" and "volume". Eventually the language of *h* can become part of the language of *E* with respect to yet another discovery. In fact, the example with specific gravity should perhaps be analyzed as a two-step discovery procedure. First we have the ascription of weight and volume to every physical stuff and object, then these terms are included in the language of E, and based on these terms the one of specific gravity is defined in the language of *h*.

To illustrate semantical inclusion one can think of examples from physics. Initially Maxwell electromagnetism was confined to certain electrical phenomena, but after 1865 also included light, etc. Another example would be the quantum of action which by Planck in 1900 was connected to the black-body radiation, but which first was extended around 1913 to spectroscopy, the structure of atoms and specific heat, and later from 1928 and onwards to molecules (chemical bonds), atomic nuclei, stellar energy, and so on. A third example would be the notion of electron which had two different origins in electrodynamics by Lorentz and others, and the cathode ray by Thomson. Very quickly one realized that electrons are "everywhere"; for instance, in radioactivity, the Zeeman effect, metallic conductivity, etc. A final example is taken from chemistry. Originally organic and unorganic bonds were considered to be different, but around the 1830's it became clear that *C*-bonds were not different from other bonds and that "vis vita" apparently has no reality.

An example of semantical disjointedness can be found in the change of the application of the notion of natural movement between the Aristotelian and the Newtonian physics. Natural motion is the kind of movement that requires no explanation. For Aristotle such a force-free motion was vertical movement, whereas Newton it was uniform motion in a straight line ("horizontal movement"). Another example has been reported by Hacking (1983). In 1936 two groups of experimentalists discovered a new kind of particle in cosmic rays, an energetic particle with a mass between the electron and proton. Another characteristic of the particle was that the Bethe-Heitler energy-loss formula for electrons did not apply to this particle. The particle was dubbed "mesotron", a name which soon was shortened to "meson". However, a year before the discovery, in 1935, H. Yukawa had developed a theory of the strong nuclear force in which a particle having a mass similarly between the electron and proton was acting as the messenger particle. This particle was then identified with the mesotron by Bohr and others, although there were problems from the very beginning with the divergence between the lifetime calculated by Yukawa and the one actually being observed. First after the Second World War other particles were discovered which fitted such better to Yukawa's predictions. These particles were called π-mesons in contrast to the 1936 particle that became a μ-meson. Eventually, it became clear that a π-meson and a μ-meson belonged to two very different kinds of particles. It was therefore quite confusing to continue to group both together as mesons. So what happened was that the name "meson" stuck to the post-war particles, whereas the pre-war particle, the one that was originally named mesotron (meson), was renamed "muon".

An example of semantical correspondence, which also goes to show semantical variance, may be seen in the Newtonian vs. Einsteinian conception of mass m:

Newton (h_N)	Einstein (h_E)
$m = constant$	$m = m_0 / \sqrt{1 - u^2 / c^2}$

where m_0 denotes the mass of the body at rest, u denotes the relative velocity of the body and c denotes the speed of light. Taking the latter definition of semantical variance first, mass constitutes the same property of an object; it is the same set of physical objects which satisfy the extension of the predicate mass, but in classical physics mass is constant while mass is the special theory of relativity varies with the frame of reference. Hence the extension of the term mass in h_N and h_E remains fixed while the intension changes.

For semantical correspondence consider the case where u^2/c^2 goes towards 0. Then Einsteinian mass becomes indistinguishable from Newtonian mass. In other words under these conditions the two "evidences" E_N and E_E have a non-empty intersection such that the models of the theories become isomorphic.

3. THREE METHODS FOR SCIENTIFIC DISCOVERY

Once the principles, goals, prescriptions and definitions within which a discovery process takes place have been laid down we continue to analyze how the three different methods of scientific discovery fare with respect to these dimensions of change. Suppose initially that epistemic hypothesis of correctness[2] is defined relative to the background knowledge of possible worlds in which the actual could turn. Then epistemic correctness entails that:

[2] We assume a rather weak correctness relation given the general problems pertaining to defining truth.

- *the hypothesis* h *is verified by evidence up to some time* n,
- *possibly the hypothesis will be verified by future evidence in the sense that there exists possible worlds in the proposition corresponding to* h *which are consistent with existing evidence.*

If some hypothesis makes claims transcending the background knowledge there may just be an empty intersection between the hypothesis and the background assumptions. This implies that there will be no way of deciding the truth value of the hypothesis and in turn the hypothesis will remain forever globally underdetermined by the evidence and the background knowledge. *It follows immediately that global underdetermination makes reliable inquiry impossible.*

3.1 Inductive discovery (ID)

Inductive generalizations over finite evidence sequences were essentially what the whole tradition after Reichenbach was against. Inductive generalizations were thought unreliable basically due to Sextus Empiricus' classical diagonal argument against inductive inference with a finite time convergence criterion. Later Hume cited Sextus and Popper took Hume's argument to be conclusive but at the same time confused a mathematical issue with a psychological one. Reliable inductive inference is indeed possible as methodologists like Osherson, *et al.* (1986), Glymour (1980), Kelly (1996) and others have proved. Logical reliabilism dictates that an *ID*-method is reliable if the evidence and the background knowledge logically entail the correctness of the hypothesis.

$$\delta : \omega^{<\omega} \rightarrow H_e.$$

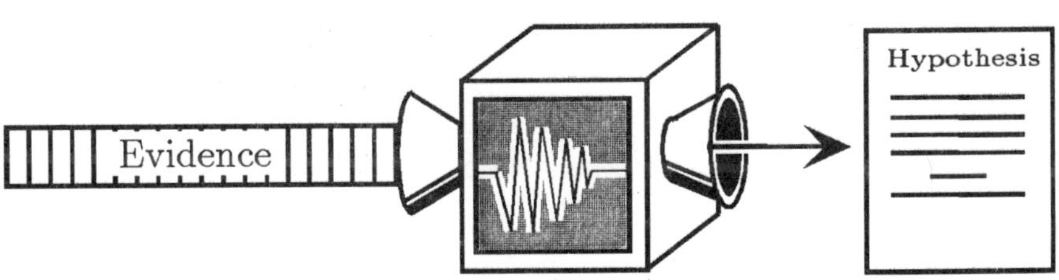

Figure 1. Inductive discovery methods. Inductive discovery methods take finite evidence sequences as inputs and return hypotheses.

Now say that an *ID*-discovery method observes ever growing segments of worlds and outputs propositions. In inductive discovery problems one usually specifies a countable set of empirical propositions[3] that the discovery method is asked to identify for some notion of convergence which may be everything from limiting to finite time convergence. This type of discovery may be done reliably and there is a host of systematic completeness results to support this claim.

But a critical methodologist may respond that the philosophical problem of discovery is that in real discovery problems we do not have a countable set of empirical propositions whose truth values must eventually be settled by the methods conjectures. Furthermore real discovery typically requires leaps out of the evidence language forming new concepts in the hypothesis language to account for the evidence. This is essentially just to say that the critical methodologist is thinking of *AD*-discovery rather than *ID*-discovery when thinking of discovery, because viewing *ID*-discovery from the point of view of the principles, goals, prescriptions and rules the following is noticeable: none of the above ontological principles are violated, or rather those ontological principles which at any given time form the foundation and the world-scope of the background knowledge are unchanged. This means that if our background knowledge already rets on a violation of any of the above principles, the *ID*-discovery will be based on this violation. *ID*-discovery may fit whatever cognitive goal one finds acceptable, it may be defined to be reliable, and finally it satisfies the semantical definition of invariance.

3.2 Inference to the best explanation (*IBE*)

Some philosophers of science have argued, that even though no formal logic of discovery can be given, some guidelines to discovery may still be provided under the heading of inference to the best explanation (Peirce, 1958; Thagard, 1988; Lipton, 1991). Peirce originally labeled it *retroductive* inference.

The most commonly embraced schema for retroductive inference runs accordingly:

1. Some surprising (or currently inexplicable) phenomena *a,b,c* are observed.

[3] Or some complexity class if the discovery problems are posed in a first order setting or recursion theoretic setup.

2. Phenomena a,b,c would (logically follow) be explained from a new empirical hypothesis h_e in conjunction with a set of already accepted premises.
3. Therefore there is some plausibility in h_e, and it is worth while seriously entertaining it.

Note that the content of the hypothesis h_e is already given in the premise (2) and, for instance, Peirce denies the existence of a genuine algorithm for generating theories: the above schema cannot be construed literally as accounting for the sequence of steps leading to the discovery of a new hypothesis.[4] Instead, the above suggests an enumeration of hypotheses from which only one at a time is drawn, examined, and hopefully entertained. So the algorithm has to specify conditions powerful enough to uniquely select one hypothesis in case of more than one abductively eligible – and so pick out the best explanation.

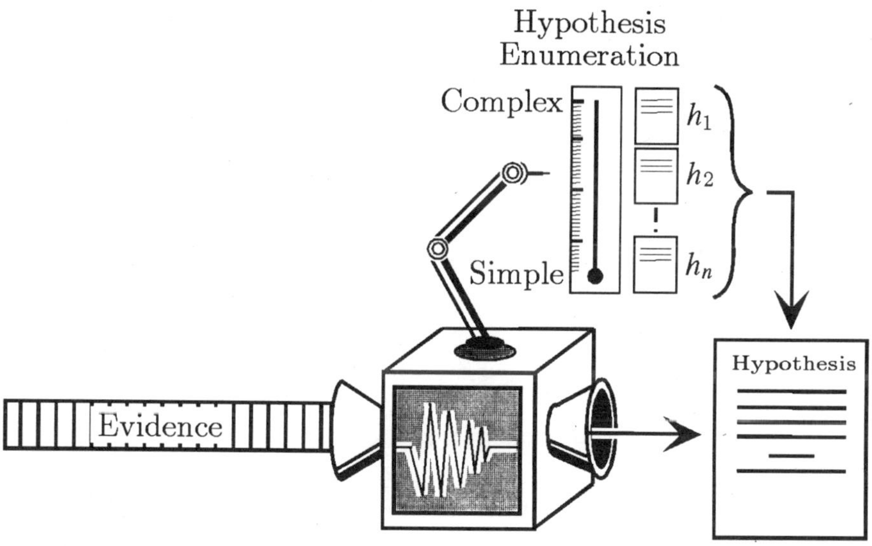

Figure 2. Inference to the best Explanation. IBE-methods enumerate possible hypotheses according to some "rational" standard, take the recalcitrant evidence in and return a new "hypotheses" explaining the evidence in accordance with the standard.

Accordingly inference to the best explanation is more like hypothesis assessment (where finally entertained hypotheses are justified) than genuine discovery. In case the evidence underdetermines the hypothesis and com-

[4] In this respect, Peirce does not seem to diverge much from the views of Popper, Hempel et. al. In addition Peirce claims that the generation of hypotheses involves a distinct psychological component, a view resembling the one of the hypothetico-deductivists.

peting hypotheses suggests themselves, enumerate the possible hypothesis according to some "rational" standard of audacity, simplicity etc. and proceed to investigate whether one of the hypotheses in the enumeration can account for the recalcitrant evidence according to this more lenient cognitive goal. In general call any hypotheses present in the enumeration for an *entrenched* hypothesis.

Adopting the most entrenched hypothesis does, according to Peirce, neither make it true nor even inductively probable but does provide the scientist with a working hypothesis, conjecture or guideline worth testing and elaborating. This lead Curd one step further to the aforementioned view of construing Peirce's proposal as a logic of prior assessment, where by prior assessment is meant something on the order of pursuit-worthiness according to some locally chosen pre-testing virtues like simplicity, epistemic entrenchment or precision. Hence one could hold Curd responsible for advancing the claim that the point of scientific inquiry is to formulate theories (or explanations) possessing desirable internal properties. In case of multiple abducibles, apply the measure defined in terms of the desirable properties or pre-testing virtues and choose the one with the highest score. The one with the highest score is hitherto the best explanation.

The primary problem of *IBE* is to define a "proper" entrenchment enumeration of the hypotheses. "Creative" intuition may be an aid but one has to chose the entrenchment criterion cautiously. The criterion should be truth tracking if we want to remain in myopia and it requires an independent epistemological argument to ensure that the criterion is behaving accordingly.

But this latter consideration calls to a comparison in the four dimensions. Assume a set of entrenched hypotheses which are either inductively or abductively discovered. The "best" of these hypotheses is typically taken to be the one, which accounts for all evidence. If, however, more than one hypothesis accounts for the evidence, i.e. in case they are empirically underdetermined by available data, then various constraints on the selection among the relevant hypotheses are called for insofar one favors a singular explaining hypothesis. As a rule one should select the hypothesis which violates less ontological principles, the one, which involves fewest breaks of the semantical definition of invariance and expansion, and the one, which obeys the definition of semantical correspondence. The latter definition of correspondence ensures at least consistency with available evidence while semantical invariance hopefully guarantees no semantical leaps outside the evidence language and the same goes for not violating at all, if possible, the ontological principles. Second, since truth is defined relative to background knowledge, possible leaps out of this set of accessible possible worlds due to either lack of semantical correspondence or violation of ontological principles, make evaluation of the enumerated hypotheses impossible due to global underdetermination.

Implicitly, there is in this recipe for choice a preference for a hypotheses-enumeration based on inductively generated conjectures rather than abductively generated ones. True, and the reason for this is that even though the rule of semantical correspondence can be satisfiable in abductive discovery, maintaining the ontological principles may prove complicated as will be momentarily clear below.

3.3 Abductive discovery (AD)

Consider Hempel's argument against "a logic of discovery":

> An adequate rule of induction would therefore have to provide, for atomic theory and for every other conceivable case, mechanically applicable criteria determining unambiguously, and without any reliance on the inventiveness or additional scientific knowledge of the user, all those new abstract concepts which need to be created for the formulation of the theory that will account for the given evidence. Clearly, this requirement cannot be satisfied by any set of rules, however ingeniously devised [...] (Hempel, 1965, p. 6).

We have seen that Hempel's argument as solely an argument against reliable inductive inference is fallacious given *ID*-discovery. But when Hempel by the phrase "all those new concepts" means that the hypothesis language includes something not present in the evidence language then indeed he may have a point. The hypothesis language must include some theoretical terms underdetermined by the evidence and hence impossible to formulate in the evidence language.

New theoretical concepts can be introduced transcending the current body of background knowledge while yet others remain within the given understanding of things. A case in point would be the formulation of the hypothesis of electron spin. Bohr considered the spin conjecture as a welcome supplement to the current magnetic core theory. Pauli remained rather skeptical pertaining to the spin hypothesis due to the fact that it actually required the theory of quantum mechanics for its proper justification, which was not part of the background knowledge at the time of the conjecture. In such cases two paradigms are competing and the abduction is then dependent upon whether the conjecture is made within the paradigm or outside it. Hence distinguish between paradigmatic and trans-paradigmatic abduction.

3.3.1 Paradigmatic abduction

In paradigmatic abductive discovery there is no intention of violating the fundamental ontological principles given by the background knowledge at any given time. Reviewing the Hempelian quote, the primary aim of paradigmatic abduction is to provide new theoretical terms. But this is both a semantical and methodological challenge. Under the general assumption that a paradigmatic abduction does not violate the ontological principles it should neither undermine the cognitive goal given the intention of myopic inquiry.

The two primary semantical principles governing paradigmatic abduction are semantical variance and expansion. Recall, first that semantical variance implies a discrepancy in the intension of the terms present in the hypothesis language while the extension of those terms remain the same as the extension of the terms contained in the evidence language. Second, semantical expansion means the addition of new terms to the hypothesis language. Returning to Hempel's Theoretician's Dilemma, this may essentitally be viewed as a case of paradigmatic abduction. Neither "specific gravity", "weight" nor "volume" are parts of the evidence language as such. Yet "specific gravity" consists of merging two concepts intended to explain a *functional* rather than a causal law between the flotation of some bodies and sinking of others. Let us consider the example in closer details.

How do we by abductive reasoning get to the hypothesis that a solid body either floats on or sinks in a liquid depending on its specific gravity? Assume that words like "wood", "iron", "stone", "wine", "water", and "space" belong to the language of E together with predicates like "is sinking", "is floating" and "is heavy". Based on this vocabulary we can make the following observations based on inductive reasoning:

1. Wood floats on water, iron and stone sink in it; likewise wood floats on wine, iron and stone sink in it.

The steps in the abductive reasoning are then:

2. All things have a volume in the sense that they occupy a smaller or greater part of space.
3. All things have weight in the sense that they are more or less heavy.

Those two steps are cases of semantical invariance. For instance, instead of talking about things or matter occupying a part of space or being spatially confined to a certain area, we introduce a term like "volume" to cover the same property. Hence we introduce a new hypothesis in terms of "specific gravity".

4. All things have a specific gravity defined as the quotient of its weight and volume.

This step is formed by applying semantical variance. Finally we get to the hypotheses:

5. A solid body floats on a liquid if its specific gravity is less than that of the liquid; and
6. A solid body sinks in a liquid if its specific gravity is greater than that of the liquid.

Thus, paradigmatic abductive reasoning leading to a functional hypothesis seems to be successful by the application of semantical variance. To facilitate, however, a properly causal hypothesis semantical variation is inadequate since the effect events are described in different terms from the cause events. Let two phenomena *A* and *B* be highly correlated. Quite often such a correlation is brought about by a common cause *C*. Here the description of *A* and *B* is included in the evidence language of *E* whereas the description of *C* transcends the language of *E* both with respect to intension as well as extension. The reason for this intensional and extensional divergence is to be found in the attempt to provide a causal hypothesis. Two contingent events statistically correlated other than conceptually, i.e., causally, implies that those two events can be described intensionally and extentionally independent of one another, and hence there must exist a non-observable common cause which gives rise to the introduction of a new theoretical term. This is a case of semantical expansion.

The important methodological prescription is unification or rather unification falls out of the endeavor of satisfying the cognitive goal for myopic inquiry. Unification of the introduced terms facilitate subsumption under general law-like connections.

3.3.2 Trans-paradigmatic abduction

In contrast to cases of paradigmatic abduction we have cases of trans-paradigmatic abduction where the discovered hypothesis transcends the immediate empirical agreement between the two paradigms. Consider a paradigmatic matrix, call it Γ, for which yet another matrix Δ covering the very same domain of research is emerging. Now this uniform domain given in terms of observational agreement up to a certain empirical threshold is ensured by satisfaction of semantic correspondence; semantic correspondence guarantees linguistic commensurability up to an observational bound. Take for instance the transition from classical mechanics to quantum mechanics. It was Bohr's view that consistent quantum mechanics only could be established insofar there was a correspondence between the classical mechanics and the new paradigm. Starting out Bohr required a "formal analogy between the quantum theory and the classical theory" given in terms of determining the line spectrum of an atomic system in such a way that

> [...] in the limit where the motions in successive stationary states comparatively differ very little from each other, will tend to coincide with the frequencies to be expected on the ordinary theory of radiation from the motion of the systems in the stationary states (Bohr, 1918, p. 8).

This formal analogy led Bohr to the insight that classical concepts like energy, momentum, position, frequency and wave-length must re-figure in quantum mechanics. If not, there is no hope of minimum empirical equivalence facilitating that those two paradigms are speaking of the same thing, i.e., have the same domain of research (Faye, 1991). Whatever trans-paradigmatic abductions one may find from here, like quantum mechanics, they must be based on this correspondence requirement; because otherwise the abductions, may, for all we know, be completely arbitrary. To avoid arbitrariness hinges on empirical equivalence up to a certain point and from here onward the soundness of the trans-paradigmatic abduction of the actual hypothesis is in general indexical on the cognitive goal. Initially, assume the following ordering of some frequently advocated cognitive goals in terms of epistemic lenience:

1. Epistemic correctness,
2. Empirical adequacy,
3. Simplicity,
4. Unity,
5. Coherence.

Clearly, the ordering is not complete and neither indisputable in terms of ranking the goals. Nevertheless, this does not obscure the primary point to be made for some other ordering. Moving from one paradigm to another in such a way that the cognitive goal is only empirical adequacy leave us without virtually any constraints on the ontological principles, since whatever theory that merely satisfies semantical correspondence and empirical adequacy will be capable of becoming the new paradigm.

Pertaining to epistemic correctness, the ontological principles must be kept intact below the empirical threshold, but above this boundary condition of empirical equivalence the new entities in the ontology given by the new theoretical terms may be relative to this particular paradigm only and to no other paradigm either preceding or succeeding it. Realize that the definition of epistemic correctness does not necessarily imply correctness in the future for the empirical hypotheses below the threshold. Hence there is no guarantee that the abductively generated hypothesis with some new theoretical terms is true all the way into the infinite future even if it was true in the past. In other words, succeeding paradigms may include ontological principles falsifying the principles of the current one entertained.

Suppose metaphysical correctness is defined in such a way that a hypothesis is metaphysically correct in the world just in case the hypothesis covers correctly the future course of the world from here to eternity.

Hence, insofar one gets hold of the metaphysical correctness it reaches into the indefinite past and into the indefinite future for the empirical part. Methodologically this implies that one remains rather conservative with respect to the selection of ontological principles. Ontological principles which held in the former paradigm must by definition of metaphysical correctness also hold in the current paradigm and in any paradigm succeeding it.

4. CONCLUSION

We have attempted to capture a number of features characterizing and dictating how scientific discovery is indeed guided. And for inductive discovery at least guided in an logically reliable way. Inductive discovery both stays within boundaries of the evidence language and usually pre-defines a set of empirical hypotheses to which the method is asked to converge given the available evidence and the background knowledge. Here inductive discovery differs from abductive as well as inference to the best explanation by the possible introduction of new theoretical terms and, especially for abductive inference, no pre-defined identification class. Now inference to the best explanation does not seem to cast too much light over the identificational problems because it is more akin an assessment engine than a genuine discovery engine due to the hypotheses enumeration. Whether abductive discovery can be carried out in a logical reliable way remains an open question. We have attempted to sketch some guidelines along which an analysis of abductive discovery should proceed. Particular emphasis should go to methodological conservatism and the semantical principle of correspondence. Both designed to block identificational arbitrariness. A deep philosophical question remains because global underdetermination may at any time emerge when at least trans-paradigmatic abduction is considered. And, again, global underdetermination equals the impossibility of reliable inquiry. The only way known to avoid underdetermination is to allow for background assumptions; but background assumptions are indexical on the paradigm entertained at any given time hence only metaphysical correctness will guide us all the way. So the prospects of reliable trans-paradigmatic abduction requires metaphysical insight!

REFERENCES

Bohr, M., 1918, On the quantum theory of line-spectra, *Det kgl. Danske Vid. Selsk. Skrifter. Naturvidenskab og Matematisk Afdeling*, Række 8, IV, Copenhagen, 1918-22.

Curd, M., 1980, The logic of discovery: an analysis of three approaches, *Scientific Discovery, Logic and Rationality*, T.D. Nickles, ed., Reidel, Dordrecht, pp. 201-221.

Faye, J., 1991, *Niels Bohr: His Heritage and Legacy. An Anti-realist View of Quantum Mechanics*, Kluwer, Dordrecht.

Glymour, C., 1980, *Theory and Evidence*, Princeton University Press, Princeton.

Gold, E.M., 1967, Language identification in the limit, *Information and Control* 10:447-474.

Hacking, I., 1983, *Representing and Intervening*, Cambridge University Press, Cambridge.

Hanson, R., 1958, *Patterns of Discovery*, Harvard University Press, Cambridge.

Hempel, C.G., 1965, *Aspects of Scientific Explanation*, Macmillan, New York.

Hendricks, V.F., 1997, Epistemology, methodology and reliability, Doctoral Thesis. Department of Philosophy, University of Copenhagen.

Hendricks, V.F., 1999, *The Convergence of Knowledge*, in preparation.

Hendricks, V.F., and Pedersen, S.A., 1998, The epistemology of limiting convergence, submitted for publication in *Erkenntnis*.

Hendricks, V.F., and Pedersen, S.A., 1998, Assessment and discovery in the limit scientific inquiry, *LMPS99*-submission.

Kelly, K., 1996, *The Logic of Reliable Inquiry*, Oxford University Press, New York.

Kelly, K., Schulte, O., and Hendricks, V.F., 1996, Reliable belief revision, in: *Logic, Methodology, and Philosophy of Science X*, Kluwer, Dordrecht.

Lipton, P., 1991, *Inference to the Best Explanation*. Routledge and Kegan Paul, London.

Osherson, D., Stob, M., and Weinstein, S., 1986, *Systems that Learn*, MIT Press, Cambridge, MA.

Peirce, C.S., 1958, *Charles S. Peirce: Selected Writings*, P. Wiener, ed., Dover Publications, New York.

Thagard, P., 1988, *Computational Philosophy of Science*, MIT Press, Cambridge, MA.

Fictionalism and the Logic of "As If" Conditionals

Claudio Pizzi
Dipartimento di Filosofia e Scienze Sociali Università di Siena, Siena, Italy,
claudio_pizzi@rcm.inet.it

Abstract: The paper accepts Hans Vaihinger's suggestion that "as if" conditionals contain a counterfactual element and aims to developing a formal analysis of them by using conditionals expressing a consequential connection between the clauses. The key idea is that "as if" counterfactuals are explicit counterfactuals with true consequents, whose rendering is unproblematic at the propositional level. The most interesting problems however arise when the "as if" concerns not a comparison between facts but a comparison between gradable predicates. The solution proposed here relies on Seuren's theory of comparison and make an essential use of existential quantification over extents, while existential quantification over predicates allows to render the notion of similarity which appears to be involved in many ordinary uses of "as if". It is stressed that the proposed analysis gives a solution also to the problem of so-called countercomparatives, which has been treated by David Lewis in the framework of counterpart theory.

1. HANS VAIHINGER'S COUNTERFACTUAL ANALYSIS IS "AS IF"

Hans Vaihinger's *Die Philosophie des "Als Ob"* (1911) belongs to the class of the important books which are more often quoted than read. The book contains the original formulation of the philosophical perspective known as fictionalism, and may still be an interesting source of suggestions not only for epistemologists who work in the field of model construction but also for logicians and linguists who aim to reconstruct the logical form of "as if" statements. Let us simply remark, in the present context, that Chapter XIII has the title "The linguistic form of the fiction. Analyis of 'As If'". Vaihinger insists here on the falsity of fictions. In the "als ob" construction, he claims, "it is clearly expressed that the given real, which is a singular element, is compared to another, whose impossibility or irreality is expressed in the meantime" (p. 96). For instance, he adds, there is a juridical fiction whose formula says "this heir is to be treated as if he died before his father", a proposition which should be intended to mean: "if this heir had

Model-Based Reasoning in Scientific Discovery, edited by L. Magnani, N.J. Nersessian, and P. Thagard, Kluwer Academic/Plenum Publishers, New York, 1999

died before his father he should be treated as such" (hence he has no right to receive any part of the legacy).

The suggested paraphrase contains a counterfactual conditional, so it appears that there is a counterfactual element in "as if" – statements. It is not obvious, however, how such form should be correctly represented, and perhaps it is not by chance that the problem has been widely neglected up to now (for two exceptions see Shuford, 1964, and Adams, 1993).

2. FICTIONAL CONDITIONALS IN A PROPOSITIONAL FRAMEWORK

Let us begin by assuming a propositional formal language which consists of propositional variables *p,q,r...*, symbols $\neg$, $\wedge$, $\vee$, $\rightarrow$, $\leftrightarrow$, for standard truth-functional operators, symbols $\Box$, $\Diamond$ for modal operators of possibility and necessity, and a symbol ">" for non-truth-functional conditional such that

(1) $\Box(A \rightarrow B)$ implies $A > B$ but not *vice versa*

(2) $A \rightarrow B$ does not imply $A > B$

A neglected feature of conditionals is that they may or may not transmit explicit information about the truth or falsity of the clauses.

Let us call *explicit conditional* any conjunction whose form is

(3) $\delta \wedge (A > B)$

where δ stands for members of the set $\{A, \neg A, B, \neg B\}$
and *fully explicit conditional* any conjunction whose form is

(4) $\delta \wedge \delta' \wedge (A > B)$, where both δ and δ' stand for members of the set $\{A, \neg A, B, \neg B\}$.

In simple (non-explicit) conditionals the truth or falsity of the clauses may be suggested by the mood, by the context or by the content of the clauses, but is not explicitly expressed. For instance, $p \wedge (p > q)$ is the form of *since*-conditionals or *factual* conditionals while, if p is simply suggested to be true, some other symbolization as $[p]\ p > q$ should be used, where $[p]$ indicates that the sentence used by the speaker presupposes that he is believing p (see Pizzi, 1981, 1990).

Unfortunately, the verbal mood alone is not enough to determine which is the truth value of the antecedent, as the following table shows:

Factual	Afactual	Counterfactual
Indicative	Indicative or Subjunctive	Subjunctive

When the antecedent, however, contains a past subjunctive verb, we may "normally" be sure that the antecedent is a counterfactual one.[1]

Suppose now that from my window I see my friend John running in the street. Since I have never seen John running in my life I say:

(5) John is running, as if he were chased by the police.

It may turn out, surprisingly, that John is indeed chased by the police. After acquiring this information, however, I cannot use any more the "as if" construction but the "since" (factual) construction and so I will say "John is running *since* he is being chased by the police". So the use of "as if" is appropriate only if the conditional is afactual or counterfactual.

In the afactual case (the case in which the speaker does not know the truth value of the antecedent) the form of the "as if" statement will be simply given by

(6) $(p > q) \wedge q$

But the case in which "as if" conditionals are fictional is the case in which the conditional is *counterfactual*. Here, if we want to distinguish between explicit and non-explicit conditionals, we have two possible formalizations, a stronger and a weaker one, namely

(7) $(p > q) \wedge \neg p \wedge q$

and

(8) $[\neg p\,]\,(p > q) \wedge q$

[1] The example of a factual conditional having the past subjunctive in the antecedent is proposed by Anderson (1951). Suppose a detective sees a dead body showing the symptoms of strychnine and says: if this person had taken strychnine he would show exactly the symptoms he shows now. For what will be said in the present section, the use of "as if" would be inappropriate in such cases, so that such counterexamples could be neglected in the present analysis.

An English ortholinguistic paraphrase of (3) in the strong sense would then be

(9) John is not chased by the police, but if John were chased by the police he would be running, and he is actually running.

Notice that if we could perform a correct abduction from the consequent to the antecedent, the consequence relation expressed in the counterfactual here would have the same strength of an "if and only if": so that the truth of the consequent would entail the truth of the antecedent. As a result of the abduction, the antecedent turns out to be true, and cannot be considered afactual or counterfactual. Thus, a conclusion we may draw is the following: the truth of fictional conditionals excludes the possibility of an abduction from the consequent to the antecedent.

What we have presupposed up to now is that the form of "as if" conditionals is given by "*x* is A as if *y* were B" where "*x* is A" expresses a fact, or a true proposition. The proposed analysis would be surely inadequate if "*x* is A" were replaced by a non-assertive statement, for instance a normative statement such as the one occurring in the juridical example quoted at the beginning.

The introduction of a deontic operator "O" in the language here, however, yields an alternative between two possible variants of (6):

(10) $(p > \mathrm{O}q) \wedge \mathrm{O}q$

and

(11) $(p > q) \wedge \mathrm{O}q$

The alternative mirrors an ambiguity in the language itself. When for instance we say

(12) We ought to behave as if God existed

are we saying that if God existed we would behave in some morally praiseworthy way or that we *ought* to behave in some morally praiseworthy way? As a matter of fact the problem may be deeper here, since the intended meaning of (12) is likely to be that we are to behave as if *we knew* that God exists, so that some epistemic operator should also come into the picture.

Normative "as if" are important in Vaihinger's philosophy since his methodology rests essentially on what he calls "fictional judgments" , namely on prescriptions having the form

(13) *a* is to be considered by scientists as if it were *b*.

The examples (12) and (13) suggest that a logical analysis of both should employ not only deontic operators but epistemic operators for knowledge and belief. A paraphrase of (13) for example could be:

(14) For every *x*, if *x* is a scientist and *x* knew that *a* is *b*, the beliefs of *x* about *a* would be the beliefs which *x* ought to have about *a*.

What is clear, anyway, is that propositional language is inadequate to render the logical form of (12) and (13) and that their formalization requires first or second order languages. In what follows we will neglect the question of deontic-epistemic operators and we will concentrate on the use of quantifiers in the analysis of "as if" statements.

3. "AS IF" CONDITIONALS AS CONSEQUENTIAL CONDITIONALS

As a premise to the further investigations on "as if", something should be said about the logical properties of the corner operator. To begin with, in order to give sense to factual conditionals and to give support to the proposed taxonomy, we cannot accept that among the theorems involving the corner there is

(15) $(p \wedge q) \rightarrow (p > q)$

which equals the equivalence

(16) $(p \wedge q) \leftrightarrow (p \wedge q \wedge (p > q))$

According to (15) then every fully explicit conditional $p \wedge q \wedge (p > q)$ is trivialized since it is equivalent to the conjunction $p \wedge q$.

As is well known, (15) is one of the theorems validated by David Lewis' semantics for counterfactuals, since it corresponds to what Lewis calls the Centering Condition (see Lewis, 1973, pp. 26-28). On the contrary, what we want to grant is that some kind of connection exists between the clauses, as it was common to assume in the Chisholm-Goodman-Reichenbach (CGR) tradition. According to the CGR view $p > q$ is true if and only if q follows from p coinjoined with laws of nature and background information (see Goodman, 1947).

The properties of ">" should be such as to avoid all the paradoxical features of the Stalnaker-Lewis conditionals. Furthermore, the logic of ">" should validate at least the following wffs:

(17) $(p > q) \rightarrow \neg(p > \neg q)$

(18) $\Diamond(\neg p \wedge \neg q) \rightarrow ((p > q) \rightarrow \neg(\neg p > q)$

(19) $\neg(p > \neg p)$

None of such wffs is a theorem for strict conditionals, relevant conditionals or Stalnaker-Lewis conditionals. However, it may proved (see Pizzi 1991) that such laws are validated in simple extensions of standard modal logic by defining the corner operator as follows:

(20) $A > B =_{Df} \Box (\mathbf{w}(A) \wedge A) \rightarrow B) \wedge (\Diamond B \rightarrow \Diamond A)$

where the intuitive meaning of **w**(A),**w**(B), **w**(C)... is that they stand for the most informative part of the background knowledge which is compatible with the suppositions A,B,C....

4. COUTERCOMPARATIVES AND SEUREN'S THEORY OF COMPARISON

The definition proposed in §2 provides a simple analysis of fictional conditionals at the propositional level. But propositional language is inadequate when the "as if" expresses a comparison concerning the predicates and not the truth-value of the propositions. Sometimes only a nuance distinguishes conditionals of the first kind and conditionals of the second kind. Suppose I say for instance

(21) John is running as if he were chased by the police

(21) is superficially similar to (5) (only the comma being omitted), but what is compared in (21) is not *the fact* that John is running but *the way* in which he is running.

In front of such examples we have to take into account two facts:

1. Some predicates are gradable (tall, short, heavy...) and some are absolute (square, living, being a stone...)
2. The same word in natural languages may be used both in gradable and in absolute sense

("is running" is an example, but so are common predicates such as "red", since it is appropriate both to say that a body is red and that a body is red to a certain degree).

What is clear is that "as if" – conditionals are not generally translatable into propositional form when they contain gradable predicates in essential way. For instance:

(22) This body behaves as if it were under the influence of a magnetic field.

Here we meet a predicate - "to behave"- which is not equivalent to any absolute predicate, and no propositional paraphrase will do.

In the discussion of this problem it may be useful to recall that in the literature about counterfactuals we meet a problem which may be usefully treated in this connection, even if it is apparently unrelated to it. The problem is treated by Nelson Goodman in relation to what he calls *countercomparative conditionals.*

In his seminal 1947 paper Goodman identifies different kinds of counterfactuals. Beyond the standard ones, he classifies the counteridenticals (if *x* were identical to *y*...) the *counterlegals*, the *semifactuals* (even if A, B) and, last but not least, the *countercomparatives*. Countercomparative suppositions amounts to supposing that the subject has less or more of some property than it actually has. The celebrated counterfactual supposition

(23) If Cleopatra's nose had been longer (than it was)...
is indeed a countercomparative supposition.

Countercomparatives give rise to an intriguing problem of logical parsing. There is a wide agreement about the fact that it is better to paraphrase any conditional whose form is:

(24) If *a* had P, *b* would have Q

into the ortholinguistic:

(25) If it were the case that a has P it would be the case that *b* has Q
since "*a* had P" and "*b* would have Q" are not statements at all.

Goodman remarked that such countercomparatives as

(26) If I had come one minute later I would have lost the train
translated into the ortholinguistic form provided by (25) become

(27) *If it were the case that* I came one minute later than I did, *it would be the case that* I lost the train.

But obviously (27) cannot be the correct analysis of (26) since the antecedent is self-contradictory.

This puzzle recalls an old story reported told by Bertand Russell. Rossi says to Bianchi:

(28) I thought that your yacht was longer than it is

And Bianchi replies

(29) No, my yacht is as long as it is!

The problem focused here is similar to the one of countercomparatives, with the only difference that the comparison is in the scope not of a conditional but of a belief operator.

What we realize in this connection is that there are two points to be added to Goodman's analysis.The first is that the supposed comparison may be an *equality comparison*:

(30) If my yacht were as long as yours...

The second one is that the comparison may take place in the consequent and not in the antecedent :

(31) If Rossi were wise, he would have more money.

Consequently it may happen that we have an equality comparison in the consequent. For instance

(32) If John were chased by the police, he would run at the same speed he is running.

According to the a rush paraphrase in the ortholanguage, (32) should be translated into

(33) *If it were the case that* John is being chased by the police then *it would be the case that* John runs at the same speed as he runs.

Here the underlined consequent is not contradictory but tautologically true. So we find that Goodman's analysis yields not only unwanted contradictions but unwanted tautologies, which are also oddities to be eliminated. Such unwanted tautologies, as (33) shows, crop up in the paraphrase of fictional conditionals, so we find that a solution to the problem of countercomparatives involves a solution to the problem of fictional conditionals.

Goodman's paper (1947) contained a proposal to solve the problem of countercomparatives. In a note of his paper he remarks that the correct paraphrase of (26) is

(34) $\exists t$ (t is an instant and I arrived at t and if I had arrived later than t I would have lost the train).

The drawback of Goodman's paraphrase, however, according to David Lewis (1973) is that it presupposes that we have at our disposal a metric for every property. Lewis is ironic about any attempt to generalize this metric approach to every property. Let us consider

(35) If my yacht were longer I would be happier.

Lewis remarks that in this case I should introduce a unity measure for pleasure – call it *hedon* – and paraphrase (35) into

(36) $\exists x,y$ (my yacht is x feet long and I enjoy y hedons and (my yacht is longer than y feet > I enjoy more than y hedons).

Lewis' answer to the problem of countercomparatives is given by employing on one side counterpart theory and on the other side a bidimensional semantics in which any statement is relativized to two possible worlds. In other words, the truth conditions for an arbitrary A are defined for propositions whose form is "A is true at j with respect to i "(for instance "My yacht is longer at j than it is at i"). According to counterpart theory (35) means that the counterpart of my yacht at j is longer than my yacht at the actual world i.

Even neglecting the complications due to double indexing, however, it is not a mystery that counterpart theory is itself a source of puzzles.

Alvin Plantinga (1974) noticed that counterpart theory leads to justify such absurdities as:

(37) Socrates could have been taller than Socrates

which appears to be an explicit contradiction in ordinary language. On the same line of thought, we could give legitimacy to such odd conditionals as

> (38) If Plato was as tall as Plato, Socrates could have been taller than Socrates.

where it turns out that the words "Plato" and "Socrates" refer ambiguously to counterparts and to real entities.

However, what is sound in Lewis' analysis is the refusal to accept metrization as an obligatory step in paraphrase. It is clear that even in scientific language we may have comparisons without numerically expressed metric. We may compare for instance the growth of an organisms without a quantitative metric of growth and we may say that something is more probable of something else without an exact metric based over frequencies or possible worlds. To sum up, what we need for the analysis of countercomparatives is not counterpart theory but an appropriate theory of comparison are that will take care of both metric and non metric comparison. Luckily, a theory of this kind has been developed, even if not in rigorous terms, by such linguists as Peter Seuren (1973) and James McCawley (1981).

Seuren's theory (which develops some of Leibniz' intuitions and, more directly, Ross, 1968) states that every comparison amounts to having a joint consideration of something which has a property to a certain extent and something which does not have the same property to the same extent. To say, for instance,

> (39) John is taller than Bill.

equals

> (40) John is tall to an extent to which Bill is not.

The latter statement is suitably represented by a quantification over extents in the following way:

> (41) For some extent *e*, John is tall to extent *e* and Bill is not tall to extent *e*.

The general definition of comparative predicates is then given by

> (42) x is P-er than $y =_{Df} \exists e$ (x is P to e and y is not P to e).

We have to notice, however, that there is an ambiguity between a weak and a strong sense of the term "extent". In the first sense, which here we endorse, having a property to a certain extent *e* means having it to any lower extent but not necessarily to a maximum extent. It appears, then , that if we should introduce meaning postulates for the ordering relation "<" among extents , one of them would be as follows:

(43) $\forall e\, \forall e'$ (If $e' < e$ and x is P to extent e) $\rightarrow$ (x is P to extent e')).

The advantage of taking the notion of extent in a weak sense as a primitive are is that we can define the strong notion in terms of the weak one. The strong notion is the notion of a maximum extent (M-extent) to which x is P, which is coincident with the extent beyond which *x* is not P:

(44) x is P to M-ext $e =_{\mathrm{Df}} x$ is P to extent e and $\forall e'$(if $e < e'$ then x is not P to extent e').

According to (42) if *x* is P-er than *y* it may be that there is some extent to which they are both P, but not a maximum extent to which they are both P.

Intuitively, then, the equality comparison of *x* and *y* with respect to the same property P amounts to saying that *x* and *y* are P to the same M-ext:

(45) x is as P as y $=_{\mathrm{Df}} \exists e$ (x is P to M-ext $e \wedge y$ is P to M-ext e)

Notice, however, that a different approach to the definition of equality comparison would take place by reducing

(46) *x* is as tall as *y*

to the equivalent

(47) *x* is not taller than *y* and *y* is not taller than *x*

Now, since not (*x* is P-er than *y*) is equivalent to something whose form is

$\neg \exists e(A \wedge \neg B)$, which equals $\forall e(A \rightarrow B)$, by simple transformations the form of (47) reduces to

(48) $\forall e$ (x is P to $e \Diamond y$ is P to e)

and, taking care of the existential presuppositions, to

(49) $\exists e$ (x is P to $e \wedge y$ is P to e) $\wedge$ $\forall e$ (x is P to e §y is P to e)

By standard transformations it may be proved that (49) implies and is implied by (45), so that they, as a matter of fact, are equivalent.[2]

Of course we may be willing to say such things as

(50) Bill is as tall as Evelyn is nervous.

which in the proposed analysis amounts to

(51) $\exists e$ (Bill is tall to M-extent e and Evelyn is nervous to M-extent e).

Let us simply add that the extent comparison may be applied to relational predicates, for instance

(52) Bill depends on Evelyn to extent e.

The notion of similarity itself may undergo an extent comparison:

(53) Bill is similar to Evelyn to extent e.

where the extent e expresses the amount of properties which are common between a and b. When two objects are *completely* similar they share every property, so they are identical for Leibniz's law.

A problem of course should be raised concerning metric comparison, found for instance in

(54) Jack is 20 cm taller than David.

which from Seuren's viewpoint should be paraphrased into

(55) For 20 cm Jack is taller than David.

If we want to work with a metric notion of extent we may take arithmetical subtraction as primitive along with suitable axioms for it and define "<" as

[2] It is trivial to prove that (49) implies (45) by instantiation of the universal quantifier. In the reverse direction: let us use the simbol e^x to indicate the M-extent to which x is P and let us suppose that (39) is true. Then we distinguish three cases: (a) $e=e^x$ (b)$e < e^x$, (c)$e^x < e$. In case (a) it is obvious that (39) implies (45). In case (b) we have that "x is P to e" has the same truth value as "y is P to e" since they are both true in view of (43). In case (c) "x is P to e" and "y is P to e" are also equivalent since they are both false in view of the definition of M-extent. So y is P to e § y is P to e is true for every e.

(56) $e' < e =_{\mathrm{Df}} e - e' \neq 0$

Thus (53) could be rendered as

(57) $\exists e \exists e'$ (M-ext *e*-M-ext *e'*=20 cm $\wedge$ John is tall to M-ext *e* $\wedge$ David is tall to M-ext *e'*).

which implies

(58) $\exists e$ (Jack is tall to *e* and Bill is not tall to *e*).

so the metric comparison embodied in (57) implies the non-metric comparison embodied in (58), as it appears to be intuitively sound.

We have simply to remark that, as it happens, the formal translation of comparatives is more analytical than is needed by ordinary language comparatives. There is, for instance, an ambiguity in such simple statements as for instance

(59) Planes are safer now than thirty years ago.

while we have two possible paraphrases at our disposal:

(60) For every plane *x*, there is an extent to which *x* is safe now but was not safe thirty years ago.

or

(61) There is an extent such that every plane is safe to that extent now but thirty years ago every plane was not safe to that extent.

Armed with such tools, let us now go back to the equality comparison embodied in fictional conditionals. According to our analysis a counterfactual such as

(62) John runs more than if he were chased by the police.

becomes (in the explicit version)

(63) John is not chased by the police $\wedge$ $\exists e$ (John runs with speed to extent *e* and if John were chased by the police he would not run with speed to extent *e*).

Notice that Russell's puzzle is also solved along similar lines since

(64) I believed that your yacht is longer than it is.

becomes in the proposed analysis

(65) $\exists e$ (I believed that your yacht is long to M-extent *e* and your yacht is not long to M-extent *e*).

So, for what has been said, the fictional conditional

(66) John runs as if he were chased by the police.

becomes

(67) John is not chased by the police $\wedge$ $\exists e$ (John runs with speed to M-extent *e* and if John were chased by the police he would run with speed to M-extent *e*).

5. A FURTHER STEP: QUANTIFYING OVER PROPERTIES

One may have the impression that the examples introduced in the preceding sections are biased since fictional conditionals are normally more ambiguous than our examples suggest. It is contextually clear, in the example of running John, that the equality comparison concerns speed, and not other qualities of running as, for instance, elegance of movement. Unfortunately, however, language is very often more fuzzy.

Consider for instance

(68) Joe behaves as if he were ill.

It appears that, due to the vagueness of the word "behaves", we cannot paraphrase the given statement as

(69) If Joe were ill he would behave to the extent such that...

The speaker who speaks of "way of behavior" is referring to some properties of behavior. Let us suppose that such properties are gradable. So we might have recourse to existential quantification over properties and obtain

(70) Joe is not ill and $\exists e \exists e' \exists e''$∃P∃Q∃R(Joe's behavior has property P to M-extent *e*, Q to M-extent *e'*, R to M-extent *e* "and if Joe were ill his behavior would have property P to M-extent *e*, Q to M-extent *e'*, R to M-extent *e*").[3]

In case the properties referred to are not gradable, the quantification over extents is vacuous. But in view of our definition of similarity in terms of common predicates the preceding statement (70) turns out to have the same meaning as

(71) Joe is not ill and ∃*e* (if Joe were ill, his behavior would be similar to extent *e* to the one he actually has).

so that we are back again to a quantification over extents.

6. FICTIONAL CONDITIONALS AND METAPHORS

We may agree with Vaihinger that fictional conditionals not only belong to scientific language but are indispensable to it. The following statement for instance is scientifically acceptable:

(72) The mass centre of a system moves as if the mass were located in the mass centre and all the forces were applied to it.

According to our analysis the logical paraphrase of (72) is (omitting for sake of simplicity the supplementary information that the supposition is false)

(73) ∃*e* ∃*e'*∃*e''* ∃P∃Q∃R (the movement of any system S has property P to M-extent *e*, property Q to M-extent *e*, property R to M-extent *e*... and if all the forces were applied to the mass center S, the movement of S would have property P to M-extent *e*, property Q to M-extent *e*, property R to M-extent *e*...).

To use scientific language means not only to use "regimented" language but also to remove vagueness. To begin with, we have to take as values of P,Q,R... predicates which are determinates under determinables (speed, direction ecc) in order to allow for metric or non-metric comparisons. In the second

[3] Technically speaking, the distinction between gradable and not gradable properties may be neglected since, if P is not gradable the existential quantification ∃eP*x* equals P*x* .

place, the value "true" may be assigned to the fictional conditional only after checking that there is at least one value of the quantified variables which grants that the consequent is inferred from the antecedent via well-established laws of nature and the minimally revised background information.

It is clear that value "true" or "false" cannot be assigned to every "as if" statement entertained in ordinary language. Suffice it to say that very often the "as if" construction in ordinary language occurs in the vague and over-simplified form

(74) *x* is as if it were B.

"As if" statements of such kind do not even have the standard form which we have defined for "as if statements", namely "*x* is A as if *y* were B". Unless we paraphrase them in some contextually determined way, many such statements are not very different from metaphorical statements. As is well known, most metaphors are embodied in the language itself in the form of "dead metaphors" (*the hand of the clocks/the legs of the table...* in Nietzsche's words, language is an "army of metaphors"). But it may be claimed that "live" metaphors are fictional conditionals in disguise. For instance one could claim that

(75) Marriage is a zero-sum game.

might be paraphrased into some "as if statement" as

(76) Married people behave inside the marriage as if it were a zero-sum game.

and this means, according to our analysis,

(77) If marriage had the properties of a zero-sum game, married people would behave inside marriage as they actually behave.

(77) should be then "regimented" into the schema of (70), but we realize that it is difficult, or even impossible, to find here some values of the variables P,Q,R.. which allow us to give the value "true" or "false" to (75).

It might be, however, that things are indeed more complex. Vaihinger's reply would perhaps be that (75) has no truth value because it is a fictional judgment, namely a proposition whose form is prescriptive: marriage ought to be considered as if it were a zero-sum game. Another possibility is that metaphors have some more hidden form, for instance

(78) If marriage were a game, it would have the properties of a zero-sum game.

The proposal of Alan Tormey (1983) is in fact that such metaphors as the Shakespearean

(79) Juliet is the sun.

should be recast as

(80) If Juliet were a celestial body she would be the sun.

Here again we see that it is not easy, and perhaps not even useful, to assign the truth-value "true" to such statements, which do not belong to the realm of science. What the preceding remarks suggest, however, is that counterfactuals cannot be avoided in the analysis not only of "as if" statements but also of metaphors or perhaps, more generally, of analogical reasoning. The relations among such different uses of counterfactuals however are not easy to be grasped, and their study may proposed as a progressive task for further inquiries in this area.

REFERENCES

Adams, E., 1993, On the rightness of certain counterfactuals, *Pacific Philosophical Quarterly* 74:1-10.

Anderson, A.R., 1951, A note on subjunctive and counterfactual conditionals, *Analysis* 12:35-38.

Goodman, N., 1947, The problem of counterfactual conditionals, *Journal of Philosophy* 44:113-128.

Grund, C., 1982, Metaphors, counterfactuals and music, in: *Intensional Logics: Theory and Applications*, I. Niiniluoto and E. Saarinen, eds., North Holland, Amsterdam, pp. 28-53.

Lewis, D., 1973, *Counterfactuals*, Blackwell, Oxford.

McCawley, J.D., 1981, *Everything that Linguists always Wanted to Know about Logic but were Ashamed to Ask*, Blackwell, Oxford.

Pizzi, C., 1981, "Since", "even if", "as if", in: *Italian Studies in the Philosophy of Science*, M. Dalla Chiara, ed., Reidel, Dordrecht, pp. 73-87.

Pizzi, C., 1990, Stalnaker-Lewis conditionals: three grades of holistic involvement, *Logique et Analyse* 131/132:311-329.

Pizzi, C., 1991, Decision procedures for logics of consequential implication, *Notre Dame Journal of Formal Logic* 32:618-636.

Plantinga, A.,1974, *The Nature of Necessity*, Oxford University Press, Oxford.

Ross, J.R., 1968, A proposed rule of tree pruning, in: *Modern Studies in English*, D.A. Reibel and S.A. Schane, eds., Englewood Cliffs, N.J., pp. 288-299.

Seuren, P., 1973, The comparative, in: *Generative Grammar in Europe*, F. Kiefer and N. Ruwet, eds, Reidel, Dordrecht, pp. 528-564.

Shuford, H.R., 1964, The logical connector "as if", *Journal of Philosophy* 61:697-698 .

Tormey, A.,1983, Metaphors and counterfactuals, in: *Essays on Aesthetics: Perspectives on the Work of Monroe C. Beardsley*, J. Fisher, ed., Temple University Press, Philadelphia, pp. 528-564.

Vaihinger, H., 1911, *Die Philosophie des "Als Ob"*, Reuther und Reichard, Berlin (translated by R. Ogden, Paul Trench and Trubner, London, 1924).

Scientific Modeling: A Multilevel Feedback Process

Jan M. Zytkow
Department of Computer Science, University of North Carolina, Charlotte, NC; also Polish Academy of Sciences, Warsaw, Poland, zytkow@uncc.edu

Abstract: Model construction is one of the key scientific activities. In distinction to the majority of the previous machine discovery systems, model formation applies in theory-rich context. Our long-term goal is automation of model construction. This paper reports on exploratory work towards that goal. We start from the distinction between models and theories, which is critical to the presented approach. We also distinguish between modeling and two scientific activities, which are different but which support modeling: construction of operational definitions and experimentation. Then we present the basic steps of scientific model construction, outlining data structures and an algorithm which, using a number of feedback loops, incrementally develops a model of a natural phenomenon. A walk through example is used to present the algorithm: motion of a cylinder that rolls downwards on an inclined plane.

1. MODELING: A WALK-THROUGH EXAMPLE

Scientific modeling aims at detailed understanding of concrete classes of physical structures or processes. On one end of the spectrum, modeling is construction of empirical equations from data about the modeled situation. Automated discovery systems such as BACON (Langley, et al., 1987) and FAHRENHEIT (Zytkow, 1996) generate such models by making experiments and by induction of equations from data. But models can be constructed in theory-rich situations, using the appropriate pieces of theoretical knowledge and conducting experiments in order to verify the theoretical explanation. In this paper we analyze the main steps of theory-based model construction on a walk-through example. It is based on a famous experiment of Galileo. We describe work in progress, far from being complete.

1.1 Galileo's experiment

Galileo investigated the motion of objects rolling down an inclined plane. A ball, released at height h on the inclined plane, depicted in Figure 1, rolled down and finally reached the bottom of the inclined plane above the point

Model-Based Reasoning in Scientific Discovery, edited by L. Magnani, N.J. Nersessian, and P. Thagard, Kluwer Academic/Plenum Publishers, New York, 1999

marked as Q. Galileo was not able to measure the accurate time at which the ball reached the bottom, but he could indirectly measure the final velocity. He attached a "jumping board" to the bottom of the inclined plane. Reaching the bottom, the ball assumed horizontal velocity at the jumping board and then fell through the air to hit the floor at point P. Galileo measured the distance PQ and used this distance to calculate the velocity of the ball at the bottom of the slope. After a sequence of experiments in which he varied the height h from which the ball started to roll, he derived an empirical equation for velocity v at the bottom of the inclined plane:

$$v = \sqrt{ah}, \qquad a = constant \qquad (1)$$

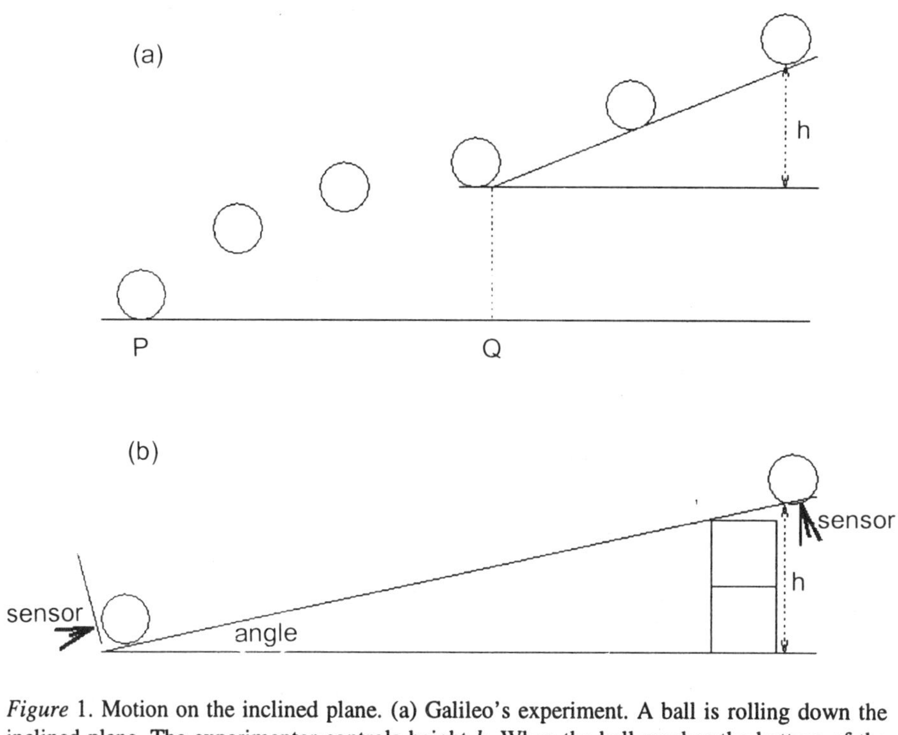

Figure 1. Motion on the inclined plane. (a) Galileo's experiment. A ball is rolling down the inclined plane. The experimenter controls height h. When the ball reaches the bottom of the ramp directly above point Q, it assumes horizontal direction and falls to the floor at point P. The distance PQ is proportional to velocity. (b) Our experiment. A cylinder is rolling down. It covers a fixed distance between a sensor at the top of the ramp and a sensor installed at the bottom. Different cylinders and different angles have been used for data collection.

1.2 Modeling the motion on inclined plane

In this paper we use the knowledge of mechanics as well as data that come from a robotic version of Galileo experiment to develop a model of motion on the inclined plane. We do not pay attention to cognitive modeling and historical accuracy. We focus on the main steps of modeling and on empirical accuracy of a model.

To make it easier to trace the modeling process, let us briefly describe some of the main steps that we will later examine in detail. As a body of mass m rolls down, its kinetic energy grows from zero to $mv^2/2$, where v is the final velocity. At the same time, its potential energy decreases from gmh to zero, where g is Earth acceleration. Both energy terms can be used jointly in the law of energy conservation to produce the equation $mv^2/2 - mgh = 0$. Solved for v, this equation is transformed into

$$v = \sqrt{2gh} \qquad (2)$$

Velocity can be predicted from equation (2). The model can be verified by comparing the predicted values against the measured velocity values.

In this paper we will expand this brief description, analyzing details of many steps with an eye on automation of modeling.

2. MODELING: A MAJOR SCIENTIFIC ACTIVITY

We can divide scientific activities into discovery of basic theories, discovery of structure, construction of operational definitions, construction of experiments, and model construction. In machine discovery, they have been treated in isolation, but in real science each requires the support of others. In this paper, while we focus on model construction, we must take those other activities and their results into account.

2.1 Models and theories are developed in response to opposite tasks

Theory is a product of analysis. It describes simple elements of nature, their states and their interaction viewed from a particular singular perspective of gravity, electromagnetism, and the like. In our example, theory includes (1) terms that describe two types of energy: potential and kinetic, (2) the principle of energy summation: energy of a state is the total of individual energy terms (the summation applies separately to the initial state and the final state), and (3) the principle of energy conservation from state to state. Each form of energy describes a simple aspect of physical states, characterized by a single property of individual bodies, such as height, velocity or temperature.

Model is a product of synthesis and describes a complex structure and interactions between elements of that structure. In the majority of situations, several theories and many interactions must be used to build an adequate model. While physicists are interested in elementary objects and interactions, and look for particularly simple situations, researchers in other domains are occupied with complex objects. Analytical chemists work on chemical samples, geologists investigate the structure of Earth, biologists study organisms. They investigate complex structures which can be understood through their models. In the field of automated scientific discovery, early attempts at computational theory of model construction include Sleeman, et al. (1989) and Gordon, et al. (1995).

2.2 Empirical data

Model predictions must be verified against empirical data, collected by observation or experiment. It is rare that model construction is successful at the first attempt. Mismatch between model predictions and empirical data can be detected and analyzed in search for model improvements that would reduced the mismatch.

For the sake of model verification we used empirical data generated in a robotic experiment (Huang and Zytkow, 1996). In our experiment we used the inclined plane, but precise time measurement turned out easier for us than measurements of velocity. Two touch sensors, that signal the beginning and the end of the process have been placed at fixed points at the top and bottom of the inclined plane at distance *D* (cf. Figure 1b) and attached to the parallel port of a PC. The equivalent of Galileo's equation (2) expressed in terms of time *t*, angle α and Earth acceleration *g* is

$$t = \sqrt{\frac{2D}{gsin(\alpha)}} \tag{3}$$

Instead of a ball, we used nine cylinders of equal external radius, with holes of different radius drilled symmetrically through the axis of each cylinder. Since the fixed locations of touch sensors determined the scope of motion, instead of various locations on the fixed board that Galileo used as his initial states, we varied the angle of the inclined plane. Altogether we used five angles.

Experiment design and experimentation strategies have been the focus of earlier work on automation of scientific discovery, such as Langley, et al. (1987), Kulkarni and Simon (1987), Nordhausen and Langley (1993), Rajamoney (1993), and Zytkow (1996).

2.3 Galileo's operational definition

Consider the way in which Galileo measured the velocity reached by the ball at the bottom of the inclined plane. A jumping board mounted at the bottom of the plane changed the direction of motion to horizontal, without affecting the velocity. Galileo decomposed the parabolic motion in the air into free vertical fall and horizontal motion with constant speed. Since he knew that all bodies fall the same vertical distance in the same time, he could infer that the horizontal distance covered before reaching the floor at point P is proportional to the velocity at the bottom of the inclined plane.

This is an example of an operational definition of velocity and at the same time an example of model application. Galileo used an earlier confirmed model of motion between the end of the inclined plane above Q and P. The model combined free vertical fall with uniform horizontal motion, for which he already knew the theories sufficient for computing the velocity needed in his inclined plane experiment. The model yields the computation of velocity, as proportional to the distance PQ.

Operational definitions have been studied by physicists and philosophers of science (Bridgman, 1927; Carnap, 1936-1937). Research on automation of scientific discovery includes attempts at operational definitions such as Kulkarni and Simon (1987), Zytkow, et al. (1992).

2.4 Our operational definition for time measurement

In our experiment, two touch sensors have been attached to the inclined plane and their wires connected to different pins at the PC's printer port. Sensor signals triggered the time measurement process, providing it with the initial and the final event. Time measurement proceeds by counting fixed time units between two events.

When the cylinder starts to roll, the top sensor is released, creating an interrupt at the printer port, and causing the timer to start up. When the cylinder engages the bottom sensor, the difference in time is measured and recorded:

```
If interrupt-1 then read time-1
if interrupt-2 then read time-2
time := time-2 - time-1
```

After our system made a sequence of experiments, we realized that the PC system clock can only be read from a running program very inaccurately, at 18.2 readings per second. This means that one readable clock tick is about 0.055 seconds. Since the cylinders take between 1.6 and 0.5 seconds to roll from the top to the bottom, the inaccuracy of time reading leads to large error at the order of 10%.

Could we find a more precise time measurement process on the computer? Our immediate choice has been a loop which counts the number of repetitions of a simple constant computation. This works fine on a PC, because it runs a single process. At a time-sharing workstation it would introduce a considerable error.

```
If interrupt-1 then start loop
if interrupt-2 then terminate loop
return the number of cycles
```

The number of cycles *c* gives the time measurement. We could use time *c* to derive empirical equations, but since we wanted to compare empirical and theoretical equations, we calibrated in seconds time measured by *c*. Calibration is a discovery process, too, simple only when it returns a desired constant.

To calibrate the loop, we run it for many seconds, triggering it by releasing the top touch sensors and then depressing the bottom sensor at the inclined plane. The second process has been system clock between the same two events. Sample data obtained in five experiments out of thirty are shown in Table 1:

Time in seconds	Number of Cycles	Average for one cycle
32.42	14,426,343	2.247e-06
41.04	18,244,828	2.250e-06
69.23	30,784,135	2.249e-06
99.45	44,237,041	2.248e-06
138.90	61,806,238	2.247e-06
Mean from 30 experiments:		2.2487e-06
Standard deviation:		0.0018e-06

Table 1. Sample data taken for the timer calibration and calibration results

The calibration procedure reads the PC clock, but now the error of clock reading averages over the period of many seconds, so that calibration can be very accurate. The results show that error has been reduced 40 times. Standard deviation (0.5 error) for the loop has been 0.0007 second, while for direct reading of the PC clock it was 0.028 second.

3. MODEL CONSTRUCTION: A MULTI-LAYER TASK

The construction of each model is oriented on a concrete goal, such as the explanation of an interesting phenomenon. A successful model uses enough of the modeled structure to provide an adequate explanation of the phenomenon, but is simple enough to ensure mathematical representation and solvability, so that empirical predictions can be reached. Given an object (phenomenon, process, structure) O, we distinguish the following modeling steps:

a) make a listing of objects, properties and processes present in O; decide which empirical parameters P of O we want to explain and which we can measure in O; a modeling task depends on how much of the modeled objects we want to represent in the model;
b) create a model-diagram that captures interaction and structure of objects, properties, and processes recognized in O;
c) construct a model-formalism that consists of equations corresponding to the model-diagram;
d) simplify and augment the equations until solvable; use the right number of equations needed to eliminate non-measurable terms. Solve the equations for parameters P;
e) verify the solution against empirical data for P measured for O. Empirical data can take on the form of an empirical equation generated from raw data. In that case, compare empirical equation with model equation. For instance, Kepler laws have been derived as empirical equations. They can be used to verify Newtonian model of planet motion around the Sun.

4. A WALK-THROUGH EXAMPLE IN DETAILS

We will analyze the details of modeling the motion on the inclined plane and draw a number of general conclusions about modeling.

4.1 Modeling task

Only one object changes its properties: the body that rolls down the inclined plane; it undergoes several processes in which body location and velocity change. We want to explain the velocity and time at which the body reaches the bottom of the plane. As two alternatives, we can measure either velocity or time.

4.2 Intro-1: brief description of the situation.

Inclined plane: the object under study is a ball. The behavior is motion from the top to the bottom of the plane. This is a change of state: change of h and change of v.

4.3 Create model-diagram

Model-diagram consists of the initial state (state-1: body located on the top of the inclined plane, zero velocity), and the final state (state-2: body location at the bottom of the inclined plane, positive velocity). Both states are connected by process links for each property whose value is being changed. Each process consists in change of one parameter.

state-1		state-2
$h1=h$	→	$h2=0$
$v1=0$	→	$v2=v$

In general, model-diagram includes:

- objects
- properties of each objects
- processes (one per each object and per each varied property)

For each object included in the model, model-diagram represents all the processes this object is involved in.

Principle 1: consider one process per each object involved and each parameter that changes its value.

Work on qualitative process theory provides a number of alternatives to our model-diagram (Forbus, 1984; Rajamoney, 1993).

4.4 Add model-formalism

Principle 2: for each process and the parameter varied in that process, consider a conserved quantity (quantities) that include that parameter.

Examples of such quantities for velocity are momentum and kinetic energy. Example of such quantity for height is potential energy. Each conserved quantity q is represented by a schema:

$$q(state-1) = q(state-2)$$

The only schema that applies to both the change of h and v is the principle of energy conservation.

The known terms available in a model determine the schemas to be considered. For each schema under consideration, we try to add to the model all other variables that occur in that schema. Before we do that, we must consider whether they can be measured. If sufficient variables are available, other variables may be removed from consideration.

In the case of the inclined plane:

- Energy term that describes the change of h is mgh.
- Energy term that describes the change of v is $mv^2/2$.
- Both terms are substituted into the energy conservation schema.
  The result is: $mv^2/2 - mgh = 0$.

In general, model-formalism must include a sufficient number of equations to eliminate all variables that are be measured.

4.5 Infer observable predictions

The equation of energy conservation can be solved for velocity, producing the solution in the form

$$v = \sqrt{2\,gh} \qquad (2)$$

The equation solving mechanism can solve that equation for time when model-formalism is augmented by an equation which links the velocity and distance with time.

In general, it is practical to require that the equation solving mechanism solves the equation(s) for variables which are measured in the final state. Equation solving can be treated as one of the subtasks in the category of equation transformation (Zytkow, 1990).

4.6 Confront predictions with measurements

The verification of the solved equation against empirical data demonstrates inadequacy of the Galileo model represented by equation (2) and by predictions of time computed from (2). Those predictions are shown in column 6 of Table 2. According to the model, the time should be constant for different cylinders represented as different masses in column 2 of Table 2, whereas the measurements (column 3) show different time for different cylinders.

Angle 1	Mass 2	Robot Stand. 3	Theoret Galileo 4	data error 5	predict theory 6
3.37	56.46	1.657	0.024	1.600	1.164
3.37	100.08	1.685	0.014	1.563	1.164
3.37	141.10	1.586	0.008	1.528	1.164
3.37	176.63	1.551	0.009	1.497	1.164
3.37	203.77	1.528	0.011	1.472	1.164
3.37	224.93	1.520	0.007	1.453	1.164
3.37	239.26	1.507	0.008	1.440	1.164
3.37	251.28	1.508	0.014	1.429	1.164
3.37	254.35	1.483	0.009	1.426	1.164
8.25	56.46	1.004	0.008	1.024	0.745
8.25	100.08	1.002	0.008	1.001	0.745
8.25	141.10	0.981	0.007	0.978	0.745
8.25	176.63	0.948	0.003	0.958	0.745
8.25	203.77	0.955	0.009	0.942	0.745
8.25	224.93	0.950	0.007	0.930	0.745
8.25	239.26	0.935	0.008	0.922	0.745
8.25	251.28	0.927	0.011	0.914	0.745
8.25	254.35	0.910	0.008	0.913	0.745
12.95	56.46	0.809	0.004	0.820	0.596
12.95	100.08	0.815	0.007	0.801	0.596
12.95	141.10	0.780	0.011	0.783	0.596
12.95	176.63	0.771	0.008	0.767	0.596
12.95	203.77	0.737	0.006	0.754	0.596
12.95	224.93	0.723	0.011	0.744	0.596
12.95	239.26	0.708	0.009	0.737	0.596
12.95	251.28	0.703	0.010	0.732	0.596
12.95	254.35	0.702	0.008	0.730	0.596
17.53	56.46	0.709	0.003	0.707	0.514
17.53	100.08	0.685	0.010	0.691	0.514
17.53	141.10	0.662	0.008	0.675	0.514
17.53	176.63	0.641	0.005	0.661	0.514
17.53	203.77	0.619	0.007	0.650	0.514
17.53	224.93	0.613	0.007	0.642	0.514
17.53	239.26	0.608	0.006	0.636	0.514
17.53	251.28	0.606	0.009	0.631	0.514
17.53	254.35	0.604	0.010	0.630	0.514
21.82	56.46	0.615	0.007	0.636	0.463
21.82	100.08	0.604	0.007	0.622	0.463
21.82	141.10	0.587	0.006	0.608	0.463
21.82	176.63	0.567	0.007	0.595	0.463
21.82	203.77	0.552	0.006	0.586	0.463
21.82	224.93	0.547	0.006	0.578	0.463
21.82	239.26	0.546	0.009	0.573	0.463
21.82	251.28	0.551	0.008	0.568	0.463
21.82	254.35	0.536	0.007	0.567	0.463

Table 2. Comparison of data and predictions of theoretical models.

Since the predictions are clearly wrong, the model must be improved. What went wrong? A more incisive observation of the cylinders reveals that they rotate as they roll. Thus, we add the change of angular velocity as an extra process.

A new model-diagram is augmented by angular velocity included both in the initial state (state-1: body located on the top of the inclined plane, zero velocity, zero angular velocity), and in the final state (state-2: body location at the bottom of the inclined plane, positive velocity, positive angular velocity). Both states are connected by process links for each property whose value is being changed:

state-1		state-2
$h1=h$	⟶	$h2=0$
$v1=0$	⟶	$V2=v$
$w1=0$	⟶	$w2=w$

Now, model-formalism yields an equation

$$t = \sqrt{\frac{D(4M - m)}{\sin(\alpha) gM}} \qquad (4)$$

where D is the distance rolled by the cylinder, M is mass of a solid cylinder of the same diameter as the given cylinder (mass m), and g is Earth acceleration.

Actually, to match the experiments, we should use equation 5 that takes into account distance d that the cylinder covered before the top sensor was released

$$time = \sqrt{\frac{D(4M - m)}{\sin(\alpha) gM}} - \sqrt{\frac{d(4M - m)}{\sin(\alpha) gM}} \qquad (5)$$

In our experiment $D=50.2\ cm$ and $d=0.7\ cm$. We used $g=980.03\ cm/s^2$ provided by physicists in our building.

Equation (5) is a far better fit that equation (2). It can be seen in column 5 of Table 2, which represents predictions derived from (5).

Even though the resultant model is more adequate, but a fine comparison against empirical data demonstrates two areas of misfit that call for further model refinements: for the smallest and the largest angle.

According to our background knowledge, those small inadequacies seem to be caused by static friction (for small angles) and by rotation that may be not complete at large angles. The directions of mismatch confirm these conjectures.

Further model improvements must take into account properties such as static and dynamic friction.

In our example, we merely add a known variable (angular velocity) to the model. Other work on automation of scientific discovery proposed various reasons for postulating new variables (Langley, et al., 1987; Kocabas, 1991; Valdés-Pérez, 1993).

Modeler's knowledge of hidden structure of the phenomenon *O* grows until the model satisfies the verification criteria at all levels, or an impasse is reached and modeling must stop until more background knowledge are gained or new observations are made of the modeled phenomenon.

5. FEEDBACK LOOPS: A SUMMARY

The process of model construction meanders through many feedback loops, as solutions to problems specific at a given step may require changes made to the constructions at the earlier steps.

5.1 Objects, properties, processes

The objects observed in *O* become the initial components of the model. New elements are added according to the knowledge-base). For instance, consider modeling a star by a steady-state model. A steady state model means that the values of different parameters that describe a single state have values that satisfy the basic laws.

Since the star surface radiates energy, background knowledge implies that radiation must be accompanied by energy transfer to the surface. The known mechanisms for energy transfer include internal radiation and convection. Modeling a system assumes that they occur, but it also tries to verify them by their observed symptoms in *O*. Since energy is generated and released in different places, we must introduce an energy transfer process (from higher to lower temperature). In that process, energy released is equal to energy generated (not conservation of energy but steady state requirement) when empirical data reveal the existence of absorption lines in the star's spectrum, the star atmosphere can be added to the list of objects. *The ultimate task is to select all objects, properties and processes which are relevant to the modeled phenomenon* O.

5.2 Model-diagram

But many details are reckoned not important, so that model-diagram is an idealization of O. The observed behavior of a physical system can be explained by a combination of many interactions.

5.3 Model-formalism

Basic laws and interactions lead to model-formalism. Each elementary interaction in a model-diagram corresponds to a specific mathematical expression that can be retrieved from the knowledge-base. Those expressions are put together into equation schemes such as conservation of mass, momentum or energy, and term-generating schemas such as additivity for scalar properties, vector additivity for force. When filled with terms specific to a given model, those equations become a model-formalism, that is, a system of equations that link different processes in model-diagram.

5.4 Solving the equations

Equations are refined to yield quantitative predictions. Model-formalism is simplified if the equations are not solvable, by recognizing negligible influence of certain components. Solved equations lead to quantitative predictions. In step e), the predicted and the measured values of parameters P are compared.

The process oscillates between solvability of equations and adequacy of description. An acceptable model is simple enough so that the equations of the model can be solved and complex enough to provide an adequate description of the investigated phenomenon.

5.5 Verification

Verification accompanies each cycle in model construction, providing feedback long before the final solution is reached. The most efficient evaluation occurs at the levels prior to the final verification, according to an AI principle: "evaluate partial solutions as early as possible".

If the equations are empirically disconfirmed, the difference between facts and predictions is analyzed. This can lead to changes in O, in model-diagram, as well as in model-formalism.

Prior to empirical confirmation, the consistency of the model is checked, because equations can be internally inconsistent or inconsistent with the previous knowledge. A solution is physically absurd when, for instance, the value of a measured parameter grows indefinitely. Each component used in the equation has a limited physical range of application. For instance, a particular mechanism of energy transfer applies within certain temperature limits.

Verification is not complete without demonstration that objects not included in the model do not bear observable influence. One of the ways is to construct a more complex model which includes additional objects and their interactions represented in model-formalism by additional expressions. Showing that those additions lead to the observationally indistinguishable results demonstrates their irrelevance. This is not possible when the original model has reached the limit of equations solvability. Still the system can be satisfied that it obtained a viable model if the model fits very well the observed data.

A model may reveal new phenomena and open new questions. Modeling may lead to new hypotheses about hidden structure suggested by a successful model. We also demonstrated, on Galileo's example, how models can be used to design new methods of measurement, that is, new operational definitions, leading to finer measurements and new instruments. In the advanced sciences, the majority of important new phenomena which lead to further scientific revolutions are noticed through differences between predictions of models and the observed phenomena. Those differences can be small, such as the part of precession of Mercury's perihelion, unexplained by the models based on classical mechanics. Similar discrepancies occur in our examples, but they may be explained by reference to the known phenomena.

REFERENCES

Bridgman, P.W., 1927, *The Logic of Modern Physics*, The Macmillan Company.

Carnap, R., 1936-1937, Testability and meaning, *Philosophy of Science* 3:420-471 and 4:2-40.

Forbus, K.D., 1984, Qualitative process theory, *Artificial Intelligence* 24:85-168.

Gordon, A., Sleeman, D.H., and Edwards, P., 1995, Informal qualitative models: a systematic approach to their generation, in: *Working Notes of AAAI Spring Symposium: Systematic Methods of Scientific Discovery*, p. 18-22.

Huang, K-M, and Zytkow, J.M., 1996, Robotic discovery: the dilemmas of empirical equations, in: *Proceedings of the Fourth International Workshop on Rough Sets, Fuzzy Sets, and Machine Discovery*, The University of Tokyo, Tokyo, pp. 217-224.

Kocabas, S., 1991, Conflict resolution as discovery in particle physics, *Machine Learning* 6:277-309.

Kulkarni, D., and Simon, H.A., 1987, The processes of scientific discovery: the strategy of experimentation, *Cognitive Science* 12:139-175.

Langley, P., Simon, H. A., Bradshaw, G. L., and Zytkow, J. M., 1987, *Scientific Discovery: Computational Explorations of the Creative Processes*, MIT Press, Cambridge, MA.

Nordhausen, B., and Langley, P., 1993, An integrated framework for empirical discovery, *Machine Learning* 12:17-47.

Rajamoney, S.A., 1993, The design of discrimination experiments, *Machine Learning* 12:185-203.

Sleeman, D.H., Stacey, M.K., Edwards, P., and Gray, N.A.B., 1989, An architecture for theory-driven scientific discovery, in: *Proceedings of the 4th European Working Session on Learning (EWSL-89)*, K. Morik, ed., Pitman, London, 11-23.

Valdés-Pérez, R.E., 1993, Conjecturing hidden entities via simplicity and conservation laws: machine discovery in chemistry, *Artificial Intelligence* 65:247-280.

Zytkow, J.M., 1996, Automated discovery of empirical laws, *Fundamenta Informatica* 27:299-318.

Zytkow, J.M., 1990, Deriving laws by analysis of processes and equations, in: *Computational Models of Scientific Discovery and Theory Formation*, P. Langley and J. Shrager, eds., Morgan Kaufmann, San Mateo (CA), pp. 129-156.

Zytkow, J.M., Zhu, J., and Zembowicz, R., 1992, Operational definition refinement: a discovery process, in: *Proceedings of the Tenth National Conference on Artificial Intelligence*, AAAI Press, pp. 76-81.

Author Index

Subject Index